Laura Greggio and Alvio Renzini
Stellar Populations

Related Titles

Salaris, M., Cassisi, S.

Evolution of Stars and Stellar Populations
386 pages

2005
ISBN: 978-0-470-09220-0

Stahler, S.W., Palla, F.

The Formation of Stars
865 pages with 511 figures and 21 tables

2004
ISBN: 978-3-527-40559-6

Foukal, P.V.

Solar Astrophysics
480 pages with 199 figures and 13 tables

2004
ISBN: 978-3-527-40374-5

Liddle, A.

An Introduction to Modern Cosmology
200 pages

2003
ISBN: 978-0-470-84835-7

Laura Greggio and Alvio Renzini

Stellar Populations

A User Guide from Low to High Redshift

WILEY-VCH Verlag GmbH & Co. KGaA

The Authors

Dr. Laura Greggio
INAF
Osservatorio Astron. di Padova
Vicolo dell'Osservatorio 5
35122 Padova
Italy

Prof. Alvio Renzini
INAF
Osservatorio Astron. di Padova
Vicolo dell'Osservatorio 5
35122 Padova
Italy

Series Editors

Massimo Stiavelli
Space telecope science institute
3700 San Martin dr
Baltimore MD 21218
USA

Ed Cheng
Senior Partner
Conceptual Analytics, LLC
8209 Woburn Abbey Road
Glenn Dale, MD 20769
USA

All books published by **Wiley-VCH** are carefully produced. Nevertheless, authors, editors, and publisher do not warrant the information contained in these books, including this book, to be free of errors. Readers are advised to keep in mind that statements, data, illustrations, procedural details or other items may inadvertently be inaccurate.

Library of Congress Card No.: applied for

British Library Cataloguing-in-Publication Data:
A catalogue record for this book is available from the British Library.

Bibliographic information published by the Deutsche Nationalbibliothek
The Deutsche Nationalbibliothek lists this publication in the Deutsche Nationalbibliografie; detailed bibliographic data are available on the Internet at http://dnb.d-nb.de.

© 2011 WILEY-VCH Verlag GmbH & Co. KGaA, Boschstr. 12, 69469 Weinheim, Germany

All rights reserved (including those of translation into other languages). No part of this book may be reproduced in any form – by photoprinting, microfilm, or any other means – nor transmitted or translated into a machine language without written permission from the publishers. Registered names, trademarks, etc. used in this book, even when not specifically marked as such, are not to be considered unprotected by law.

Typesetting   le-tex publishing services GmbH, Leipzig
Cover Design   Grafik-Design Schulz, Fußgönheim
Printing and Binding   Fabulous Printers Pte Ltd

Printed in Singapore
Printed on acid-free paper

ISBN Print 978-3-527-40918-1

ISBN ePDF 978-3-527-63663-1
ISBN oBook 978-3-527-63661-7
ISBN ePub 978-3-527-63662-4
ISBN Mobi 978-3-527-63664-8

Contents

Preface *IX*

Abbreviations and Acronyms *XIII*

Color Plates *XV*

1 Firm and Less Firm Outcomes of Stellar Evolution Theory *1*
1.1 A Brief Journey through Stellar Evolution *1*
1.1.1 A 9 $M_\odot$ Star *1*
1.1.2 The Evolution of Stars with Solar Composition *8*
1.1.3 Dependence on Initial Chemical Composition *12*
1.1.4 The Asymptotic Giant Branch Phase *15*
1.2 Strengths and Weaknesses of Stellar Evolutionary Models *18*
1.2.1 Microphysics *19*
1.2.2 Macrophysics *20*
1.3 The Initial Mass-Final Mass Relation *31*

2 The Fundamentals of Evolutionary Population Synthesis *35*
2.1 The Stellar Evolution Clock *35*
2.2 The Evolutionary Flux *38*
2.3 The Fuel Consumption Theorem *39*
2.4 Fuel Consumptions *42*
2.5 Population Synthesis Using Isochrones *46*
2.6 The Luminosity Evolution of Stellar Populations *47*
2.7 The Specific Evolutionary Flux *49*
2.8 The IMF Scale Factor *51*
2.9 Total and Specific Rates of Mass Return *52*
2.10 Mass and Mass-to-Light Ratio *56*
2.11 IMF-Dependent and IMF-Independent Quantities *57*
2.12 The Age-Metallicity Degeneracy *58*

3 Resolving Stellar Populations *61*
3.1 The Stellar Populations of Pixels and Frames *61*
3.1.1 The Stellar Population of a Frame *61*
3.1.2 The Stellar Population of a Pixel *64*

3.2	Simulated Observations and Their Reduction 68

4 Age Dating Resolved Stellar Populations 77
4.1 Globular Cluster Ages 77
4.1.1 Absolute and Relative Globular Cluster Ages 78
4.1.2 Globular Clusters with Multiple Populations 80
4.2 The Age of the Galactic Bulge 83
4.3 Globular Clusters in the Magellanic Clouds 86
4.4 Stellar Ages of the M31 Spheroid 88
4.4.1 The Bulge of M31 88
4.4.2 The M31 Halo and Giant Stream 90
4.5 The Star Formation Histories of Resolved Galaxies 92
4.5.1 The Mass-Specific Production 93
4.5.2 Decoding the CMD 98
4.5.3 The Specific Production Method 102
4.5.4 The Synthetic CMD Method 104
4.5.5 An Example: the Stellar Population in the Halo of the Centaurus A Galaxy 106

5 The Evolutionary Synthesis of Stellar Populations 113
5.1 Simple Stellar Populations 113
5.2 Spectral Libraries 115
5.2.1 Empirical Spectral Libraries 115
5.2.2 Model Atmosphere Libraries 116
5.3 Composite Stellar Populations 116
5.4 Evolving Spectra 118
5.4.1 The Spectral Evolution of a SSP 118
5.4.2 The Spectral Evolution of Composite Stellar Populations 121
5.4.3 There Are Also Binaries 128

6 Stellar Population Diagnostics of Galaxies 133
6.1 Measuring Star Formation Rates 133
6.1.1 The SFR from the Ultraviolet Continuum 134
6.1.2 The SFR from the Far-Infrared Luminosity 136
6.1.3 The SFR from Optical Emission Lines 137
6.1.4 The SFR from the Soft X-ray Luminosity 138
6.1.5 The SFR from the Radio Luminosity 139
6.2 Measuring the Stellar Mass of Galaxies 140
6.3 Age and Metallicity Diagnostics 143
6.3.1 Star-Forming Galaxies 143
6.3.2 Quenched Galaxies 145
6.4 Star-Forming and Quenched Galaxies through Cosmic Times 153
6.4.1 The Main Sequence of Star-Forming Galaxies 155
6.4.2 The Mass and Environment of Quenched Galaxies 163
6.4.3 Mass Functions 164

7	**Supernovae** *171*
7.1	Observed SN Rates *173*
7.2	Core Collapse SNe *175*
7.2.1	Theoretical Rates *176*
7.2.2	Nucleosynthetic Yields *179*
7.3	Thermonuclear Supernovae *184*
7.3.1	Evolutionary Scenarios for SNIa Progenitors *185*
7.3.2	The Distribution of Delay Times *187*
7.3.3	The SD Channel *188*
7.3.4	The DD Channel *191*
7.3.5	Constraining the DTD and the SNIa Productivity *197*
7.3.6	SNIa Yields *201*
7.4	The Relative Role of Core Collapse and Thermonuclear Supernovae *202*
8	**The IMF from Low to High Redshift** *207*
8.1	How the IMF Affects Stellar Demography *208*
8.2	The M/L Ratio of Elliptical Galaxies and the IMF Slope below 1 $M_\odot$ *211*
8.3	The Redshift Evolution of the M/L Ratio of Cluster Ellipticals and the IMF Slope between ~ 1 and $\sim 1.4 M_\odot$ *213*
8.4	The Metal Content of Galaxy Clusters and the IMF Slope between ~ 1 and $\sim 40 M_\odot$, and Above *214*
9	**Evolutionary Links Across Cosmic Time: an Empirical History of Galaxies** *219*
9.1	The Growth and Overgrowth of Galaxies *221*
9.2	A Phenomenological Model of Galaxy Evolution *224*
9.2.1	How Mass Quenching Operates *225*
9.2.2	How Environmental Quenching Operates *227*
9.2.3	The Evolving Demography of Galaxies *229*
9.2.4	Caveats *232*
9.2.5	The Physics of Quenching *234*
10	**The Chemical Evolution of Galaxies, Clusters, and the Whole Universe** *237*
10.1	Clusters of Galaxies *237*
10.1.1	Iron in the Intracluster Medium and the Iron Mass-to-Light Ratio *238*
10.1.2	The Iron Share between ICM and Cluster Galaxies *244*
10.1.3	Elemental Ratios *245*
10.1.4	Metal Production: the Parent Stellar Populations *247*
10.1.5	Iron from SNIa *248*
10.1.6	Iron and Metals from Core Collapse SNe *249*
10.2	Metals from Galaxies to the ICM: Ejection versus Extraction *250*
10.3	Clusters versus Field and the Overall Metallicity of the Universe *252*
10.4	Clusters versus the Chemical Evolution of the Milky Way *254*
	Index *261*

Preface

Galaxies are very complex systems. They are comprised of stars, gas, something else we call dark matter, and are shaped by a large variety of physical processes. This book refrains from addressing all such complexity, and is restricted instead to just one aspect: their stellar populations. Most of the radiation we receive from galaxies is star light, and star light has been detected up to redshift ~ 8, having been emitted when the Universe was just ~ 300 million years old. As most of the information we can gather on galaxies comes from star light, the ultimate aim of this book is to show how much can be learned on galaxy formation and evolution from the study of their stellar populations. Thus, this book discusses only a partial view of galaxies, but a view that provides insights that no other approach can offer. Certainly, complementary information is needed for a full description of galaxies, including how they form and evolve. Other fundamental aspects such as galaxy structure and dynamics, interactions with other galaxies and the environment, the interstellar medium and nuclear activity are however not discussed in this book.

Besides describing how stellar populations work, this book aims to present a few aspects concerning the method. In this kind of study one sometimes encounters important, even fundamental questions which nevertheless involve complicated physics. The struggle involved with understanding the physics can make one agonize over the problems for decades without much progress. Thus, it is better to ask questions that one can answer now, or within a reasonably short amount of time, and postpone the more difficult topics to the future. The complexity of Nature tends to project itself into the tools we develop in our attempts to understand it. There are situations in which an *economy of means* can be more effective than a cumbersome machinery with too many *bells and whistles*, that inevitably becomes too laborious to maneuver. Stellar population tools can offer numerous examples in this sense.

We also attempt to keep the tools we do employ as simple as possible, and include their limits and internal workings. Tools are never perfect, and inevitably imprint their defects into the results. The use of stellar population tools developed by others and downloaded from the web is now widespread, but a *push button* practice should be discouraged, whereby one feeds the data into a *package* and bothers only about the quality of the fit. Instead, *looking inside* the tool is highly recommended, hence becoming familiar with the assumptions and approximations on which it is built.

Indeed, many of our astrophysical measurements are dominated by systematic, as opposed to statistical errors. The accuracy of a measurement, the quality of a fit, are often quantified by a χ^2, which is felt to be an *objective* assessment of the level of confidence of the result. Yet, the *subjective* eye of an experienced astronomer may often grasp the material better than the χ^2 automatically delivered by the package.

For all these reasons it is always advisable to keep the use of models to a minimum, as results can only gain in robustness. This is especially true in the case of stellar population models. It is in this spirit that we have in this *user guide* kept to a minimum the mathematical burden for the reader, relying instead as much as possible on illustrations meant to capture the core of specific astrophysical issues.

This book is meant to illustrate the specific role played by stellar populations in our attempt to understand galaxy formation and evolution. This is an extremely active field of research, where large observational and theoretical efforts are being dedicated. Thus, the traditional approach has been one in which theorists construct simulated populations of galaxies at the various redshifts, and observers deliver data sets such as catalogs of galaxies, luminosity functions, etc. Such data are then translated into mass functions, star formation rates, ages, metallicities, etc., using stellar population tools. Simulated galaxies are finally compared to real ones, similarities are appreciated, and discrepancies are addressed to adjust theoretical models in an attempt to increase the similarities and reduce the discrepancies. In practice, at each iteration adjustments are applied to the parameters used to describe the many physical processes that cannot be modeled from first principles.

This way of proceeding has resulted in great progress over the last two decades, but a different approach is now being explored thanks to the multiwavelength observational mapping of galaxies from the local Universe to almost the reionization epoch. Thus, a phenomenological approach is becoming possible, in which schemes are developed that describe how the population of galaxies at one cosmic epoch maps into the population of galaxies at a later epoch. One of the chapters of this book is dedicated to illustrate this approach, and how far one can reach by using only the stellar population diagnostics of galaxies, providing for each galaxy its stellar mass, star formation rate, stellar ages and metallicities. Establishing in a fully empirical way what are the main evolutionary links across cosmic time can indeed greatly help to identify the key physical processes that drive such transformations.

This is a textbook, not a review, and to make it self-sufficient and more readable no references are given in the text except for the original papers from which figures and equations have been taken. However, at the end of each chapter references to *further readings* are given, mentioning classical and/or recent relevant papers on the subject. The literature on stellar populations has grown immensely, beyond the capacity of individuals to absorb it all. Therefore, the choice of references listed under *further readings* is certainly biased by our own familiarity with specific papers, in many of which we were directly involved. On the other hand, even if extensive, list of references rapidly become obsolete nowadays, whereas online databases are continuously updates and easily browsed.

Galaxy evolution is still a field in extremely rapid development, with results being continuously superseded by wider and more accurate studies. Thus, there are issues which are still unsettled, with conflicting results having been reported, and in some cases no general convergence has been reached. In such cases, we tend to illustrate one particular option that we now feel to be more promising, but we also list references under *further readings* that give alternative views, providing the full range of current options. We do not venture much beyond redshift $\sim 2.5-3$, as in such a domain progress is now taking a sharp upturn, with galaxies being discovered at such high redshifts that they may have contributed to the cosmic re-ionization.

The book starts with a rather unconventional summary of the results of stellar evolution theory (Chapter 1), as they provide the basis for the construction of synthetic stellar populations. Current limitations of stellar evolutionary models are highlighted, which arise from the necessity to parametrize all those physical processes that involve bulk mass motions, such as convection, mixing, mass loss, etc. Chapter 2 deals with the foundations of the theory of synthetic stellar populations, and illustrates their energetics and metabolic functions, providing basic tools that will be used in subsequent chapters. Chapters 3 and 4 deal with resolved stellar populations, first addressing some general problems encountered in photometric studies of stellar fields. Then some highlights are presented illustrating our current capacity of measuring stellar ages in objects such as Galactic globular clusters, the Galactic bulge, and nearby galaxies. Chapter 5 is dedicated to the construction and exemplification of synthetic spectra of simple as well as composite stellar populations, drawing attention to those specific aspects of synthetic spectra that may depend on less secure results of stellar evolution models. Chapter 6 illustrates how synthetic stellar populations are used to derive basic galaxy properties, such as star formation rates, stellar masses, ages and metallicities, and does so for galaxies at low as well as at high redshifts. Chapter 7 is dedicated to supernovae, distinguishing them in core collapse and thermonuclear events, describing the evolution of their rates following an episode of star formation, and estimating the supernova productivity of stellar populations and their chemical yields. In Chapter 8 the stellar initial mass function (IMF) is discussed, first showing how even apparently small IMF variations may have large effects on the demography of stellar populations, and then using galaxies at low and high redshifts and clusters of galaxies to set tight constraints on possible IMF variations in space or time. In Chapter 9 the phenomenological model of galaxy evolution is presented and discussed, and, finally, in Chapter 10 we discuss chemical evolution on the scale of galaxies, clusters of galaxies and the whole Universe.

We are especially indebted to the many colleagues and friends with whom we have collaborated through the years, and from whom we have learned much of what we know on the evolution of stars and/or galaxies. We would like to mention here Nobuo Arimoto, Ralf Bender, Thomas Brown, Enrico Cappellaro, Marcella Carollo, Andrea Cimatti, Emanuele Daddi, Mark Dickinson, Natascha Föster

Schreiber, Mauro Giavalisco, Bill Harris, Icko Iben Jr., Simon Lilly, Claudia Maraston, Sergio Ortolani, Giampaolo Piotto, Marina Rejkuba, Mike Rich, Bob Rood, Allen Sweigart, Daniel Thomas, Monica Tosi, Gianni Zamorani, and Manuela Zoccali.

We especially thank Claudia Maraston for having calculated for us the spectral energy distributions of composite populations that are used to illustrate Chapter 5 of this book. We would also like to thank those authors that have allowed us to reproduce some of the figures that appear in this book, including Andrea Bellini, Thomas Brown, Michele Cappellari, Santino Cassisi, Will Clarkson, Emanuele Daddi, Simona De Grandi, Renato Dupke, Valentino Gonzalez, Henry McCracken, Claudia Maraston, Antonio Marin-Franch, Alessio Mucciarelli, Sergio Ortolani, Maurilio Pannella, Yingjie Peng, Giampaolo Piotto, Lucia Pozzetti, Marina Rejkuba, Giulia Rodighiero, Andrew Stephens, Daniel Thomas, Paolo Ventura, and Norbert Werner.

Padova, June 2011 *Laura Greggio and Alvio Renzini*

Abbreviations and Acronyms

ACS	Advanced Camera for Surveys
AGN	Active Galactic Nuclei
AGB	Asymptotic Giant Branch
AIC	Accretion Induced Collapse
BPS	Binary Population Synthesis
BS	Blue Straggler
BSG	Blue Supergiant
CC Supernovae	Core Collapse Supernovae
CC Clusters	Cool Core Clusters
CDM	Cold Dark Matter
CE	Common Envelope
CMD	Color–Magnitude Diagram
CO	Carbon-Oxygen
DD	Double Degenerate
DTD	Distribution of the Delay Times
E-AGB	Early Asymptotic Giant Branch
E-ELT	European Extremely Large Telescope
ELT	Extremely Large Telescope
FIR	Far Infrared
FWHM	Full Width Half Maximum
GOODS	The Great Observatories Origins Deep Survey
GRB	Gamma Ray Burst
GWR	Gravitational Wave Radiation
HB	Horizontal Branch
HBB	Hot Bottom Burning
HMXB	High Mass X-ray Binary
HRD	Hertzsprung–Russel Diagram
HST	Hubble Space Telescope
ICM	Intracluster Medium
IGM	Intergalactic Medium
ISM	Interstellar Medium
IMF	Initial Mass Function
IMFM	Initial Mas – Final Mass

IR	Infrared
LBV	Luminous Blue Variable
LMC	Large Magellanic Cloud
LMXB	Low-Mass X-ray Binary
LPV	Long Period Variable
MICADO	Multiconjugate adaptive optics Imaging Camera for Deep Observations
MLT	Mixing Length Theory
MS	Main Sequence
NICMOS	Near Infrared Camera and Multi-Object Spectrometer
ONeMg	Oxygen-Neon-Magnesium
P-AGB	Post Asymptotic Giant Branch
PAH	Polycyclic Aromatic Hydrocarbon
PN	Planetary Nebula
PSF	Point Spread Function
RGB	Red Giant Branch
RL	Roche Lobe
RSG	Red Supergiant
SD	Single Degenerate
SDSS	Sloan Digital Sky Survey
SED	Spectral Energy Distribution
SFH	Star Formation History
SFR	Star Formation Rate
SGB	Subgiant Branch
SMG	Sub-millimeter Galaxy
SN	Supernova
SNIa	Supernova of type Ia
SNII	Supernova of type II
sSFR	Specific Star Formation Rate
SSP	Simple Stellar Population
SW	Super Wind
TE	Thermal Equilibrium
TO	Turnoff
TP-AGB	Thermally Pulsing Asymptotic Giant Branch
TRGB	Red Giant Branch Tip
UV	Ultraviolet
WD	White Dwarf
WFPC2	Wide Field Planetary Camera 2
WHIM	Warm/Hot Intergalactic Medium
WR	Wolf–Rayet
ZAMS	Zero Age Main Sequence

Color Plates

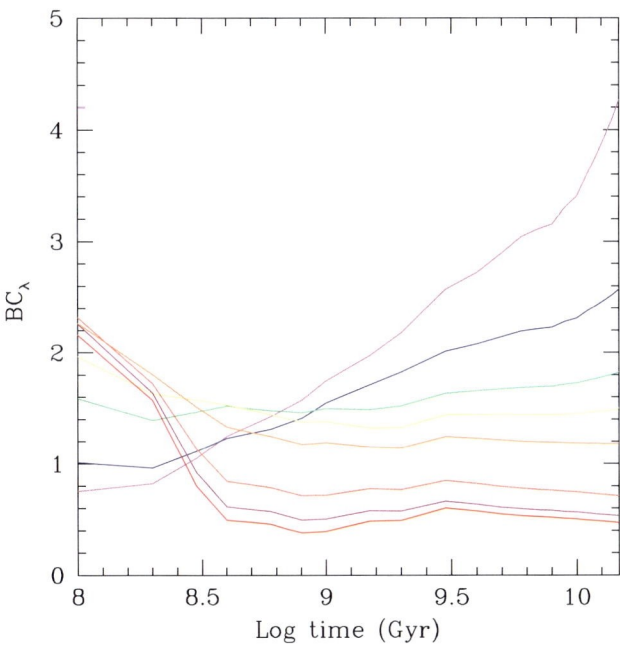

Figure C 3.1 The bolometric correction for a SSP of solar metallicity as a function of age, for various broad bands in the Johnson–Cousin photometric system. On the intercept with the right, vertical axis one can identify the bolometric correction for the bands U, B, V, R, I, J, H, and K, from top to bottom, respectively (figure is based on the synthetic populations by Maraston, C. (2005, *Mon. Not. R. Astron. Soc.*, 362, 799)).

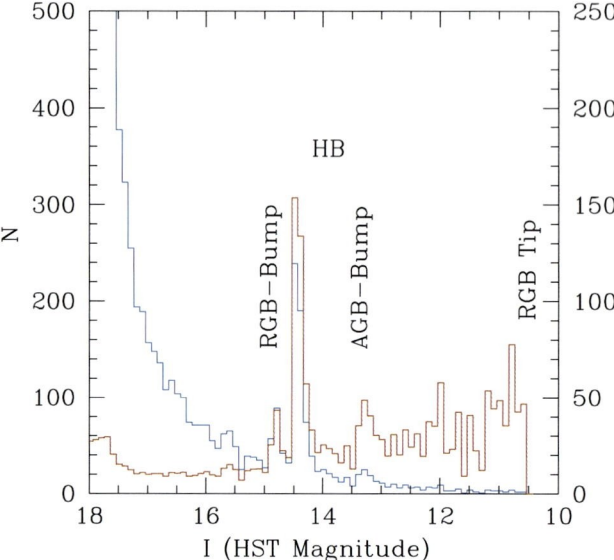

Figure C 3.2 The *I*-band luminosity function of the globular cluster 47 Tuc (blue line) with HB, RGB-bump, AGB-bump and RGB tip labeling the corresponding features. The red line shows the contribution of each magnitude bin to the total *I*-band luminosity of the cluster ($\propto N \times 10^{-0.4I+\mathrm{const.}}$, where N is the number of stars in the bin) (data from Brown, T. et al. (2005, Astron. J., 130, 1693)).

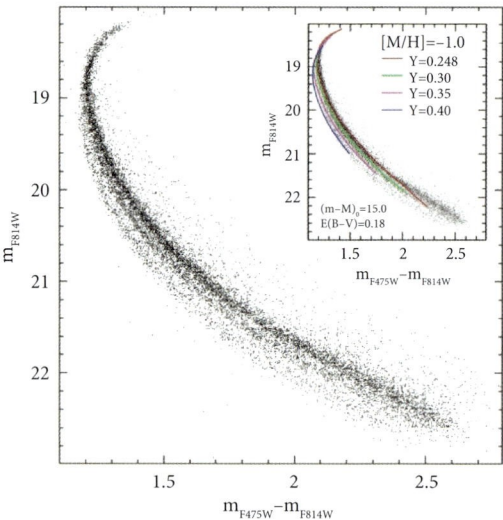

Figure C 4.3 The main sequence and turnoff region of the HST/ACS color-magnitude diagram of the globular cluster NGC 2808. The insert shows overplotted isochrones with different helium abundances, from near primordial up to $Y = 0.40$. The latter value appears to be appropriate for stars on the bluest main sequence (source: Piotto, G. et al. (2007, Astrophys. J., 661, L53)) (Reproduced by permission of AAS.).

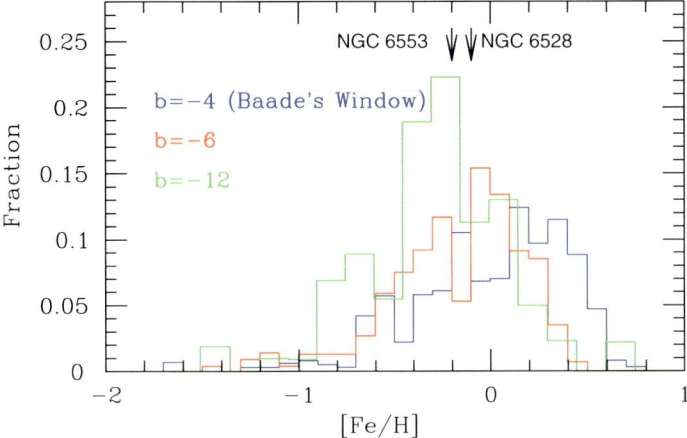

Figure C 4.4 The metallicity distribution function of RGB stars in three distinct fields of the Galactic bulge, along its minor axis. The galactic latitude of the fields is indicated. The metallicity [Fe/H] of the two bulge clusters NGC 6528 and NGC 6553 is marked with arrows (data from: Zoccali, M. et al. (2008, Astron. Astrophys., 486, 177)).

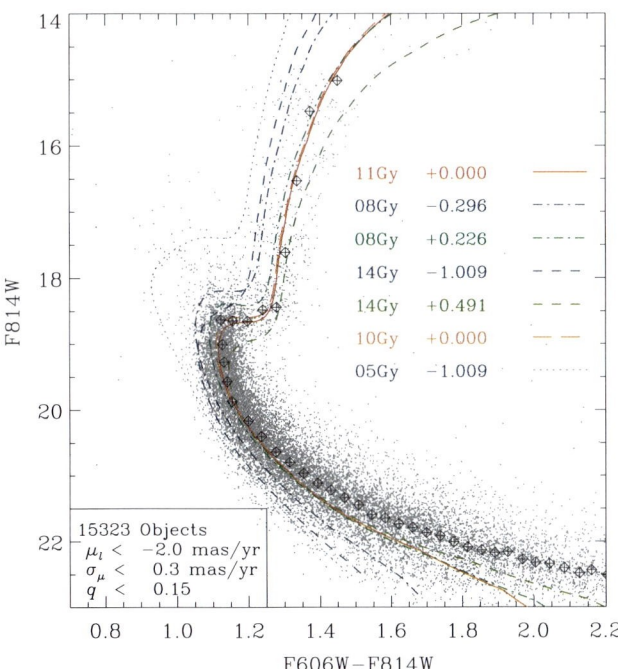

Figure C 4.6 The HST/ACS color-magnitude diagram of a Galactic bulge field with proper motion selected stars to ensure an almost pure bulge membership. Isochrones with different ages and [Fe/H] metallicities as indicated in the insert are overplotted (source: Clarkson, W. et al. (2008, Astrophys. J., 684, 1110)) (Reproduced by permission of AAS.).

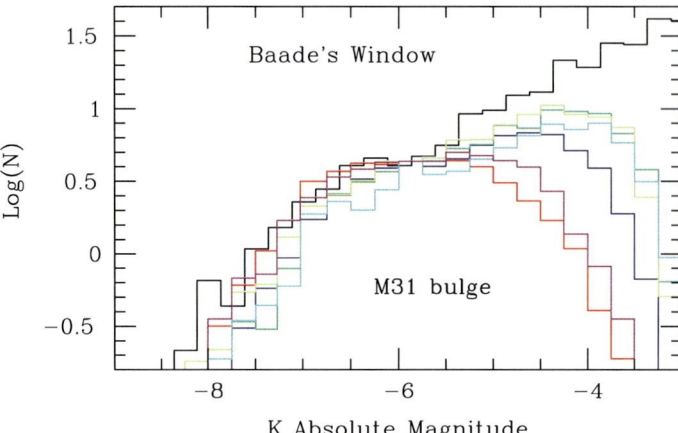

Figure C 4.10 The *K*-band HST/NICMOS2 luminosity function in the bulge of M31 from several fields at different levels of surface brightness. For comparison the luminosity function of the Galactic bulge (in black) from 2MASS data is overplotted (adapted from Stephens, A. *et al.* (2003, *Astron. J.*, 125, 2473)).

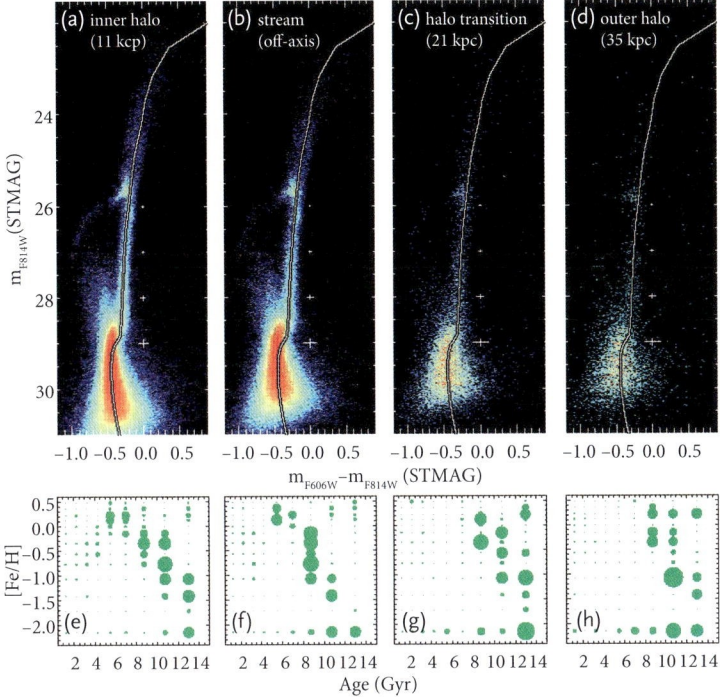

Figure C 4.11 (a–d) The HST/ACS color-magnitude diagrams for four different fields in the M31 halo, namely: three fields on the M31 minor axis, at a projected distance of ∼ 11, 21 and 35 kpc from the center of the galaxy, plus a field in the M31 giant stream. (e–h) For each field the best fit star formation history obtained using the StarFish code by Harris, J. and Zaritski, D. (2001, *Astrophys. J. Suppl.*, 136, 25); the size of the circles being proportional to the weight of each SSP (age, [Fe/H]) component in the synthetic composite population. The curves in (a–d) show the ridge line of the CMD of 47 Tuc (source: Brown, T. (2009, *ASP Conf. Proc.*, 419, 110).

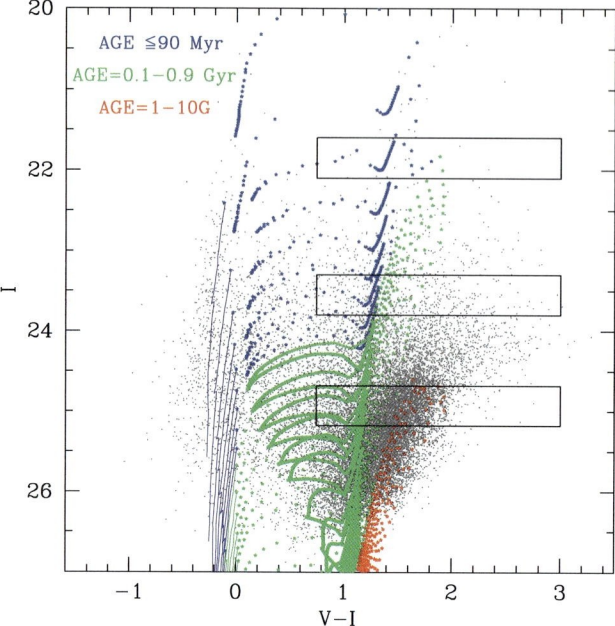

Figure C 4.16 Observed CMD of the dwarf galaxy NGC 1705 (gray dots) with superimposed stellar evolutionary tracks (with $Z = 0.004$) with various turnoff ages as labelled in the upper left corner, shifted with $(m - M)_0 = 28.54$ and $E(B - V) = 0.045$. The MS portion of the tracks is drawn as a solid line, while in the post-MS portion the density of points is higher where the evolution is slower. The three boxes are shown to illustrate how the CMD traces the SFH: while the brightest box samples only 30 Myr old stars, the others collect a wide range of stellar ages. Although the different age components are in principle separated by color, in practice it is very difficult to tell them apart on the observational CMD. Data for the CMD in the figure are from Annibali, F. et al. (2003, *Astron. J.*, 126, 275) and the tracks are derived from the Fagotto, F. et al. (1994, *Astron. Astrophys. Suppl.*, 105, 29) database).

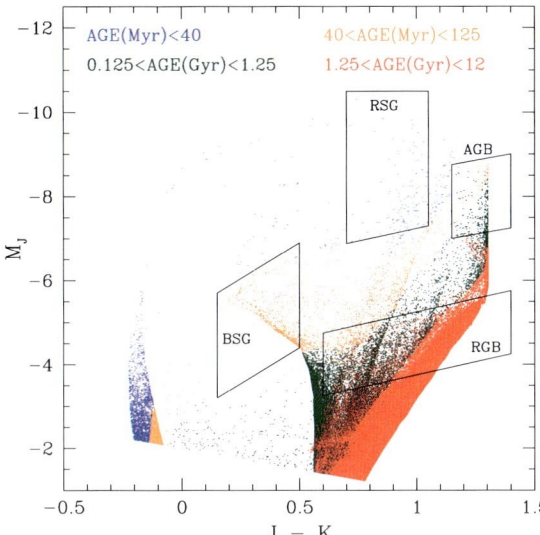

Figure C 4.18 Synthetic CMD in the infrared bands produced with a constant SFR over 12 Gyr and an age-metallicity relation in which $Z = 0$ at $t = 0$, $Z = 0.02$ at $t = 7.5$ Gyr and $Z = 0.025$ at $t = 12$ Gyr. The simulation contains 200 000 stars brighter than $M_K = -2$ and corresponds to a total star formation of 2.9×10^8 $M_\odot$, having adopted a straight Salpeter IMF. The four diagnostic boxes target the labeled age ranges, and contain 122 (RSG), 598 (BSG), 340 (AGB) and 22 198 (RGB) stars, respectively. A theoretical modeling of the TP AGB phase is included in the simulation (simulation computed by G.P. Bertelli).

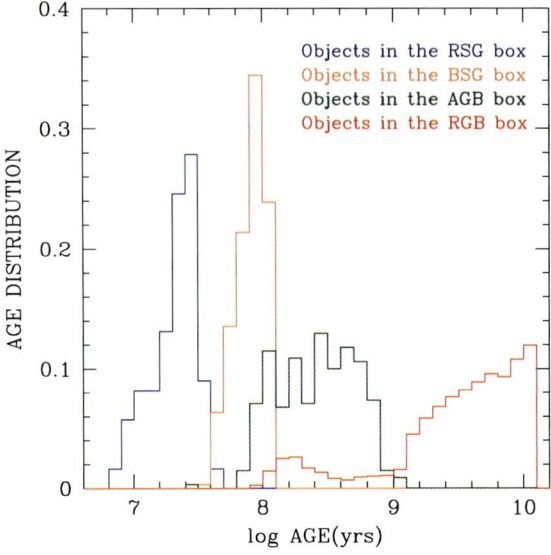

Figure C 4.19 Age distributions of the synthetic stars in the four diagnostic boxes in Figure 4.18, normalized to the total number of stars in each box. The color of the lines encodes the box membership.

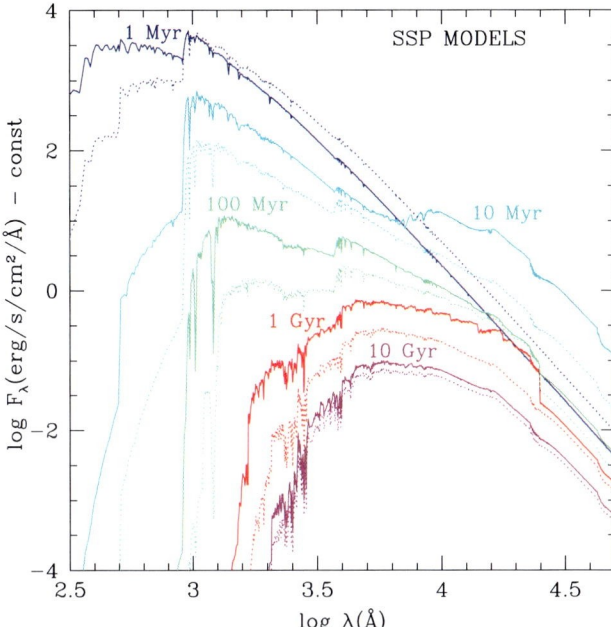

Figure C 5.1 The spectral evolution of a solar metallicity SSP from 1 Myr to 13 Gyr. The dotted lines correspond to ages 3 times older than the spectrum preceding them, except for the last one which refers to an age of 13 Gyr. Purely red HBs are assumed at late epochs (figure constructed using M05 database models: Maraston, C. (2005, Mon. Not. R. Astron. Soc., 362, 799)).

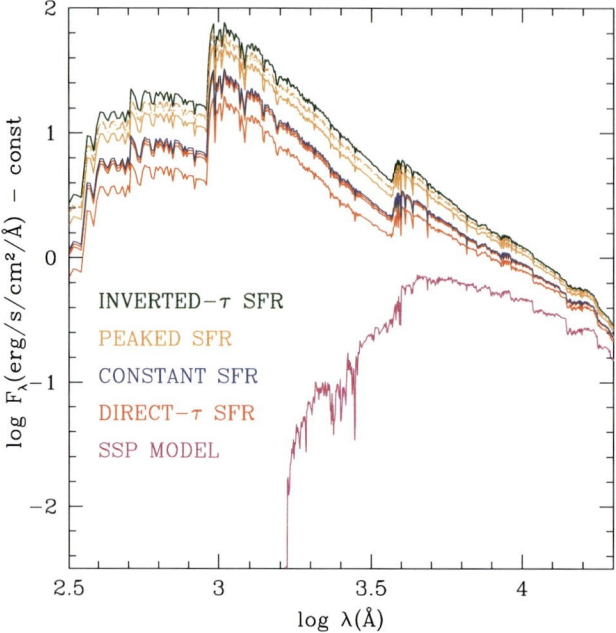

Figure C 5.4 Spectra of 1 Gyr old stellar populations with different past star-formation histories (SFH), but normalized to have formed the same mass of stars. Top to bottom the spectra refer to exponentially increasing SFRs with $\tau = 0.5$ Gyr, peaked SFH with $\tau = 1, 4, 7$ Gyr (the last two ones perfectly overlap), SFR = const., and exponentially decreasing SFR, with $\tau = 1, 4, 7$ Gyr (the latter two also overlap). For comparison the spectrum of a 1 Gyr old SSP is also shown (synthetic spectra kindly provided by C. Maraston).

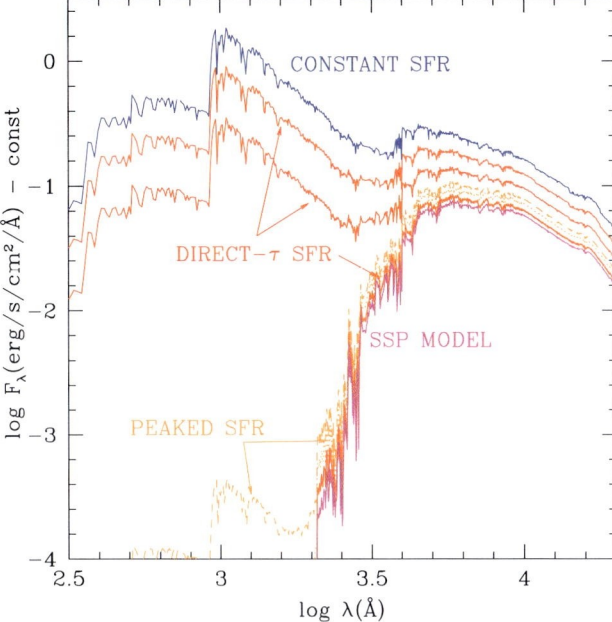

Figure C 5.5 The same as Figure 5.4 (without the inverted τ models) but for $t = 13$ Gyr. Displayed top to bottom are the spectra for a population with SFR = const., three spectra for direct-τ models with $\tau = 7, 4$, and 1 Gyr, then spectra for peaked SFHs with the same τ (largely overlapping). The spectrum of an SSP model for a 13 Gyr old population is also shown. Notice that the scale of the flux coordinate is consistent with that of Figure 5.4 (synthetic spectra kindly provided by C. Maraston).

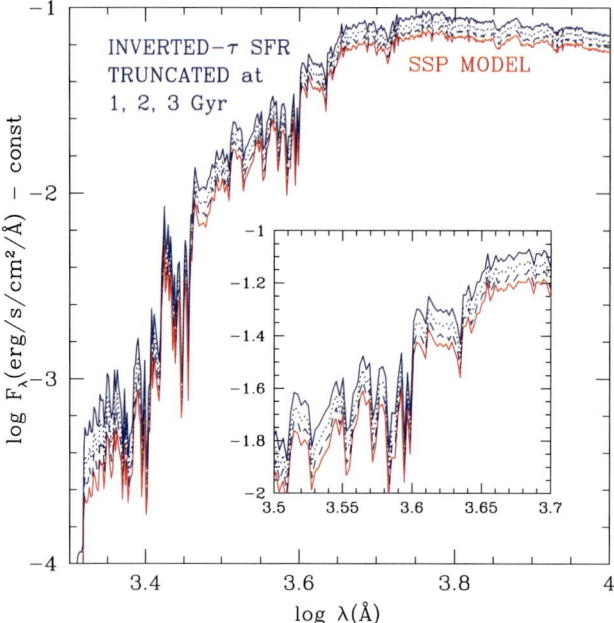

Figure C 5.6 The SSP spectrum at an age of 13 Gyr shown in Figure 5.5, now in an expanded scale, is compared to the spectra of $\tau = 0.5$ Gyr inverted-τ models with star formation truncated after 1, 2 and 3 Gyr. The insert shows a blow-up of the spectral region around the Balmer/4000 Å break, to further illustrate the close similarity of all four spectra (synthetic spectra kindly provided by C. Maraston).

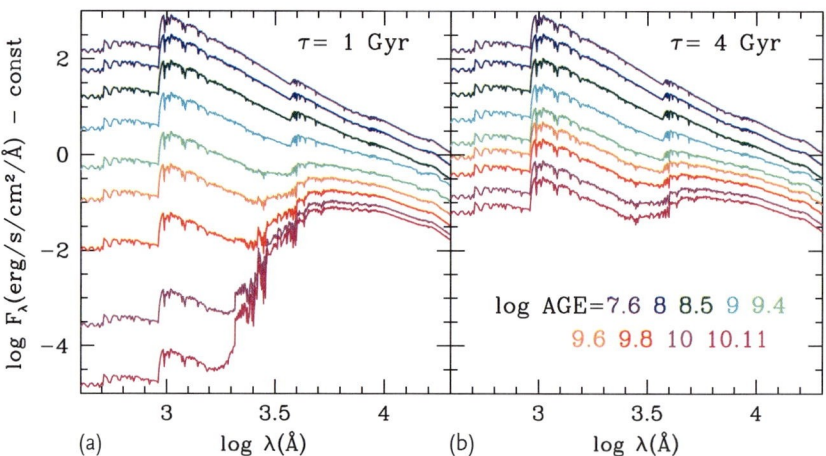

Figure C 5.7 The spectral evolution of direct-τ models for the $\tau = 1$ Gyr (a) and $\tau = 4$ Gyr (b) for ages between 40 Myr and 13 Gyr (top to bottom) as indicated (synthetic spectra kindly provided by C. Maraston).

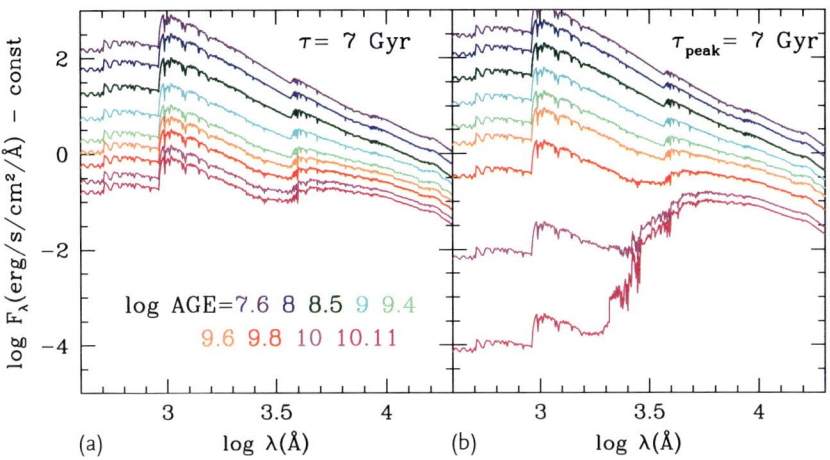

Figure C 5.8 The spectral evolution of direct-τ (a) and peaked-SFR (b) models, both with $\tau = 7$ Gyr, is illustrated for ages from 40 Myr to 13 Gyr (synthetic spectra kindly provided by C. Maraston).

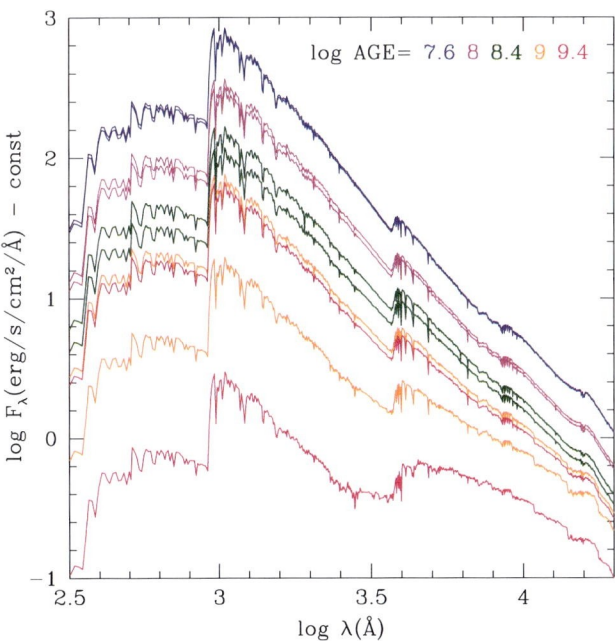

Figure C 5.9 A comparison between the spectral evolution of direct- and inverted-τ models, respectively with $\tau = 1$ and 0.5 Gyr, and for ages between 40 Myr and 2.5 Gyr. Ages are coded by color as indicated, with the upper spectrum for a given age always referring to the inverted-τ model (synthetic spectra kindly provided by C. Maraston).

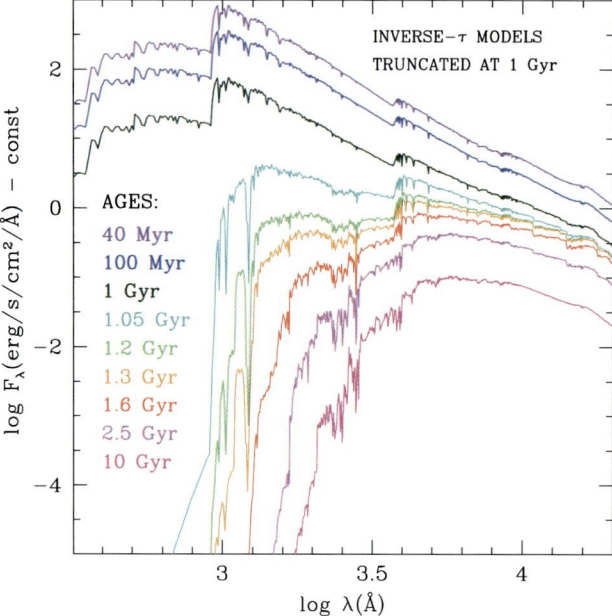

Figure C 5.10 The spectral evolution of models with exponentially increasing SFR up to 1 Gyr with $\tau = 0.5$ Gyr, and SFR $= 0$ thereafter. Ages are indicated, and apply top to bottom with increasing age (synthetic spectra kindly provided by C. Maraston).

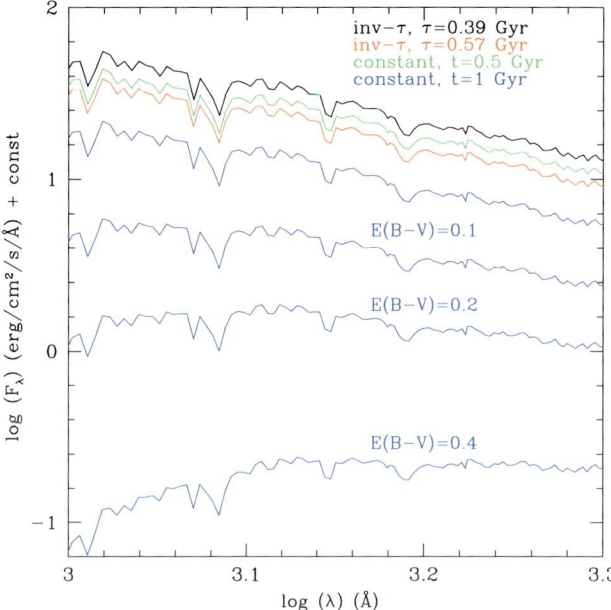

Figure C 6.1 The UV spectrum between 1000 and 2000 Å of complex populations with various star formation histories, namely SFR = const. for a duration of 0.5 and 1 Gyr, and for exponentially increasing SFR ($\propto e^{-t/\tau}$) also for a duration of 1 Gyr and two values of τ as indicated. For all four SFHs the same amount of gas is turned into stars, and the spectra refer to the time $t = 1$ Gyr after the beginning of star formation ($t = 0.5$ Gyr for the spectrum in green). The upper four spectra are unreddened, whereas the lower three ones correspond to the SFR = const., $t = 1$ Gyr case, with various amounts of reddening, as indicated (synthetic spectra kindly provided by C. Maraston).

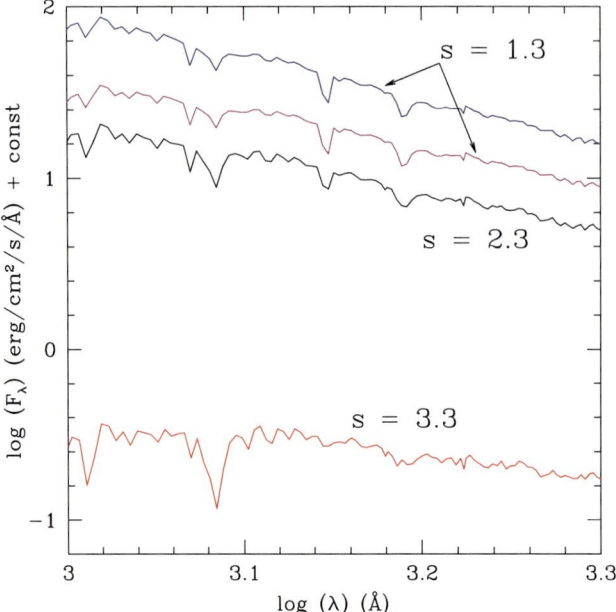

Figure C 6.2 UV spectrum of complex stellar populations obtained with a constant SFR operating for 1 Gyr, for three slopes of the IMF: $s = 2.3$ (as in Figure 6.1), a *flat* IMF with $s = 1.3$ and a *steep* IMF with $s = 3.3$. The case of the flat IMF reddened with $E(B - V) = 0.07$ is also shown in purple. Note that the SFR is the same in all cases shown here, the relative differences in the UV flux are those implied by the use of different IMFs (synthetic spectra kindly provided by C. Maraston).

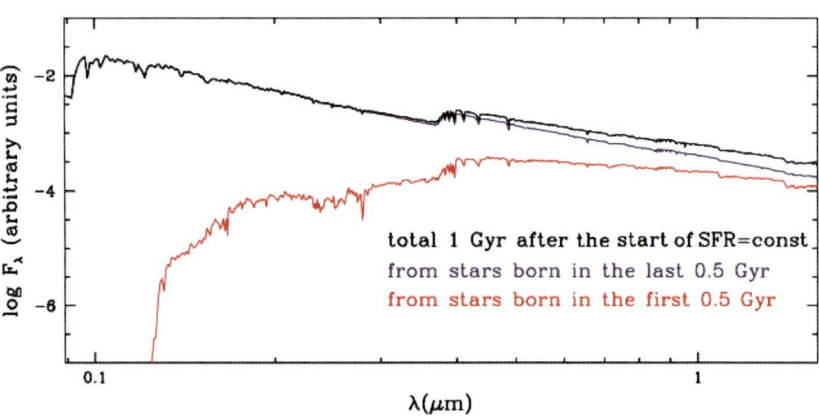

Figure C 6.4 The effect of outshining by the youngest fraction of a composite population: the synthetic spectrum is shown for a constant SFR over 1 Gyr. The contribution of the stars formed during the first and the second half of this period are shown separately as indicated by the color code, together with the spectrum of the full population (source: Maraston, C. et al. (2010, Mon. Not. R. Astron. Soc., 407, 830)) (Reproduced by permission of the Royal Astronomical Society.).

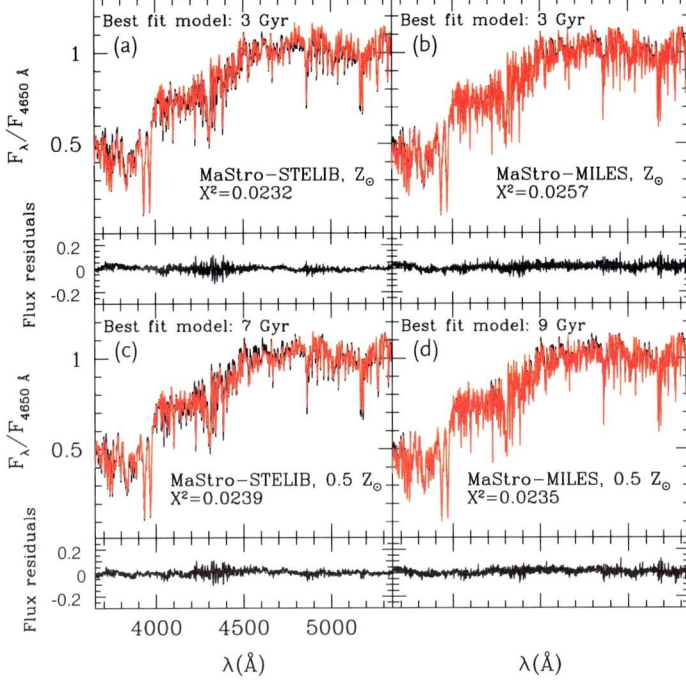

Figure C 6.5 (a–d) The integrated spectrum of the open cluster M67 obtained by coadding the spectra of its individual stars weighted by the luminosity function (black lines). Superimposed are the best fit synthetic spectra (red lines) for the ages and metallicities indicated in each panel, and for two different choices of the stellar spectral library. The corresponding residuals are plotted below each spectrum, and the reduced χ^2 of each fit is indicated (courtesy of C. Maraston).

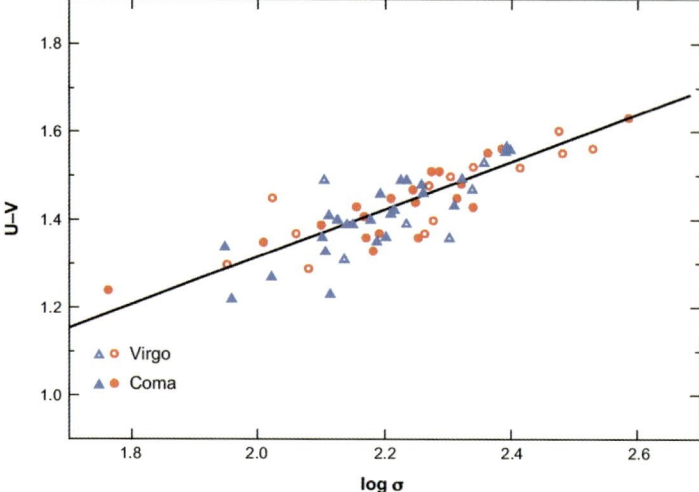

Figure C 6.6 The relation between the $(U-V)$ color and the central velocity dispersion (σ) for early-type galaxies in the Virgo (open symbols) and Coma (filled symbols) clusters. Red circles represent ellipticals, blue triangles represent S0 (adapted after Bower, R.G. et al. (1992, Mon. Not. R. Astron. Soc., 254, 613)).

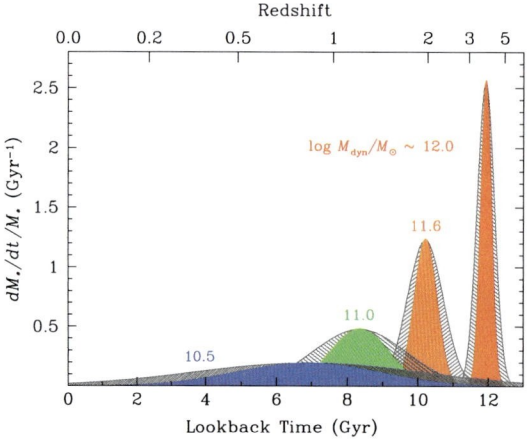

Figure C 6.8 The *average* specific SFR as a function of lookback time (and corresponding redshift) for quenched galaxies in the local Universe, as inferred from a large sample of such galaxies and the analysis of the distributions of their Lick/IDS indices. The various stellar masses are indicated by the labels. The gray hatched curves indicate the range of possible variation in the formation timescales that are allowed within the intrinsic scatter of the derived [α/Fe] ratios. Note that these star formation histories are meant to sketch the star formation history averaged over the entire galaxy population (within a given mass bin), in particular reflecting the distribution of their quenching epoch. The star formation histories of individual galaxies are expected to be very different from these averages, for example increasing with time until star formation is suddenly quenched (source: Thomas, D., et al. (2010, Mon. Not. R. Astron. Soc., 404, 1775)) (Reproduced by permission of the Royal Astronomical Society.).

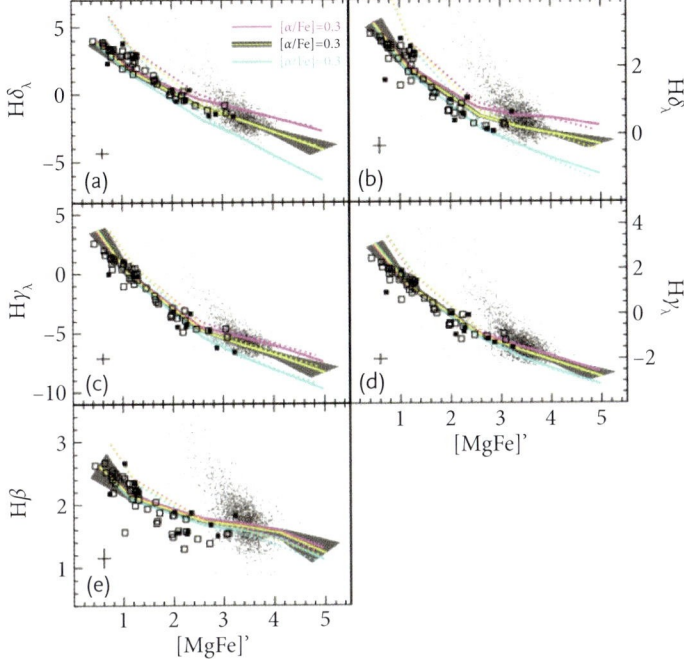

Figure C 6.9 (a–e) The Lick/IDS Balmer-line indices for 13 Gyr old synthetic populations with three different values of the α-element enhancement, and total metallicity [Z/H] in the range from −2.25 to 0.67. Note that the [MgFe]' index is fairly insensitive to [α/Fe], cf. Figure 6.7. Data for galactic globular clusters are shown as filled and open squares. The small black points refer to ellipticals galaxies from the SDSS database (source: Thomas, D., et al. (2011, Mon. Not. R. Astron. Soc., 412, 2183)) (Reproduced by permission of the Royal Astronomical Society.).

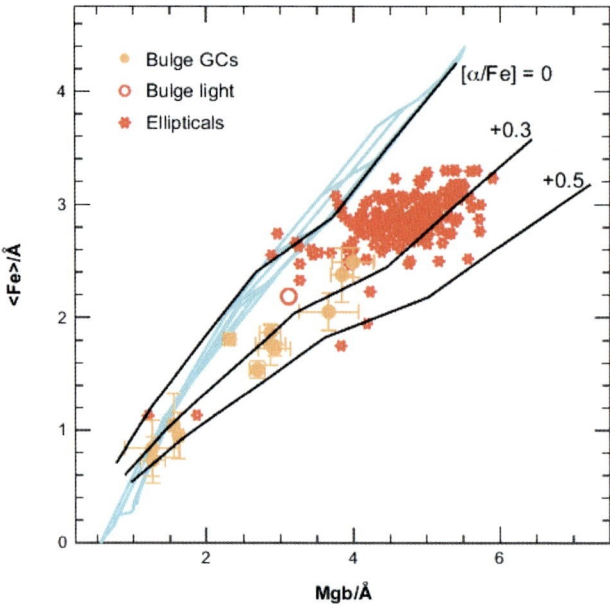

Figure C 6.10 The ⟨Fe⟩ index versus the Mgb index for a sample of galactic halo and bulge globular clusters spanning the full range of globular cluster metallicities, the Galactic bulge integrated light in Baade's Window, and for a sample of local elliptical galaxies. Superimposed are synthetic models (black lines) for an age of 12 Gyr, [Fe/H] increasing from -2.25 to $+0.67$, and various α-element enhancements as indicated. The cyan grid shows the effect of varying the age from 3 to 15 Gyr (adapted from Maraston, C. *et al.* (2003, *Astron. Astrophys.*, 400, 823).

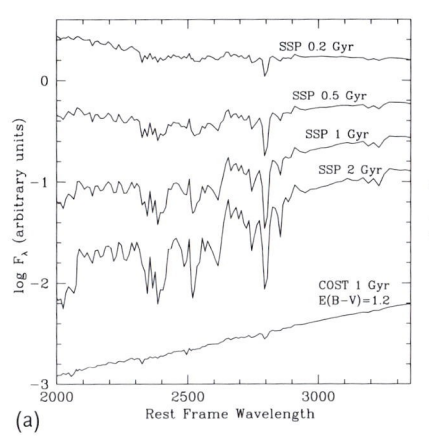

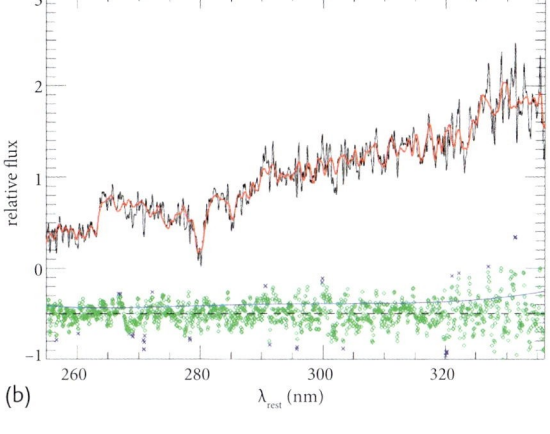

Figure C 6.11 (a) The near-UV spectrum of a solar metallicity SSP as it ages from 0.2 to 2 Gyr. Notice the ~ 2 orders of magnitude dimming of the UV flux over this time interval, and that the Mg+Fe feature at $\lambda = 2650-2850$ Å appears only for ages older than ~ 500 Myr. Also shown is the spectrum of a population with SFR $=$ const. and reddened with $E(B-V) = 1.2$. (b) The stacked near-UV spectrum of seven passively evolving galaxies at $1.6 < z < 2$. The black line is the observed spectrum and the red one is the best fitting SSP template with an age of 1 Gyr and solar metallicity. The green points are the residuals (arbitrarily shifted for convenience) and the blue line is the estimated 1σ noise. Blue crosses indicated bad pixels that were excluded in the fit ((a) used models from Maraston, C. (2005, *Mon. Not. R. Astron. Soc.*, 362, 799), (b) is from Cappellari, M. et al. (2009, *Astrophys. J.*, 704, L34)) (Reproduced by permission of the AAS.).

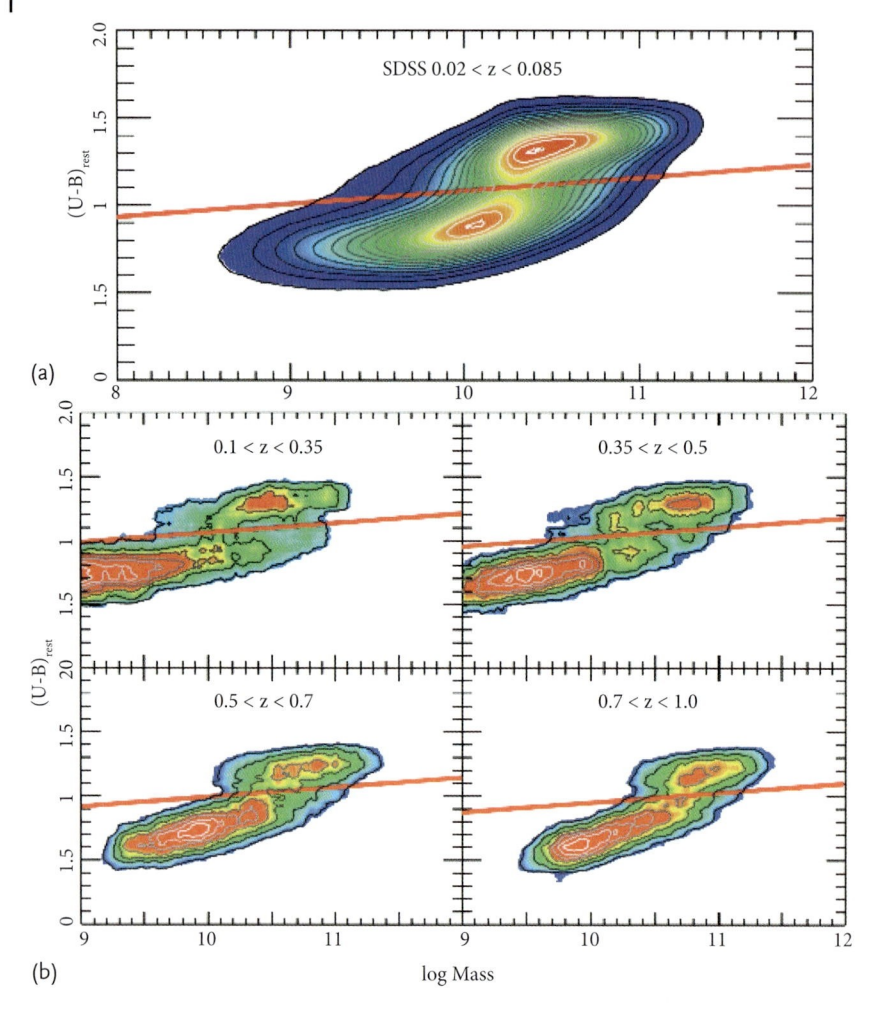

Figure C 6.12 The distribution of galaxies in the rest-frame $U - B$ color versus stellar mass plots, for large samples of galaxies at various redshifts. Part (a) refers to the nearby Universe, whereas (b) refer to various redshift slices as indicated. The redshift-dependent red line divides the red sequence galaxies from the blue cloud ones (source: Peng, Y. et al. (2010, Astrophys. J., 721, 193)) (Reproduced by permission of the AAS.).

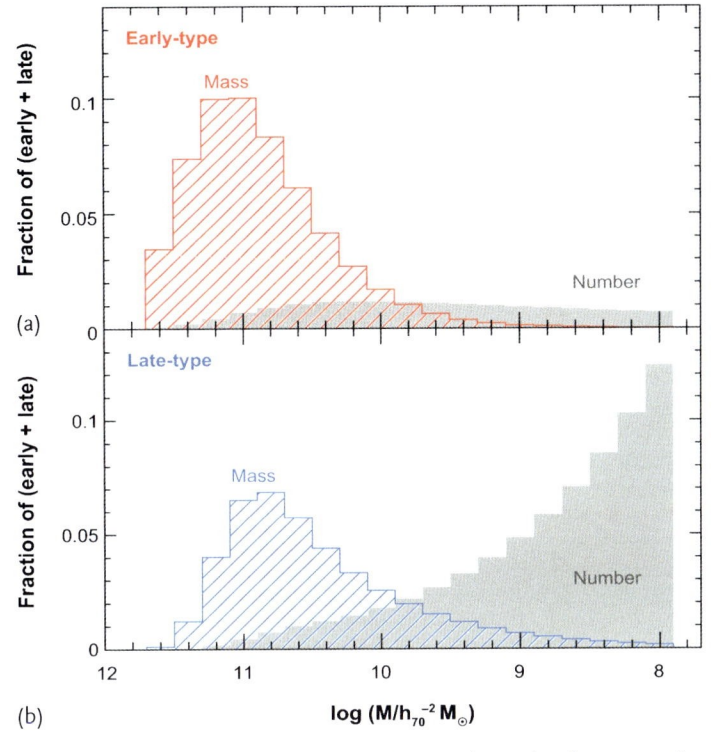

Figure C 6.13 The contributions to the total stellar mass and to the number of galaxies by (a) early-type (red sequence) and (b) late-type (blue cloud) galaxies in the local Universe. The relative areas are proportional to the contributions of the early- and late-type galaxies to the total stellar mass and to the number of galaxies. The galaxy database is the same as in Figure 6.12a (figure based on the SDSS data, Baldry, I.K. et al. (2004, Astrophys. J., 600, 681)).

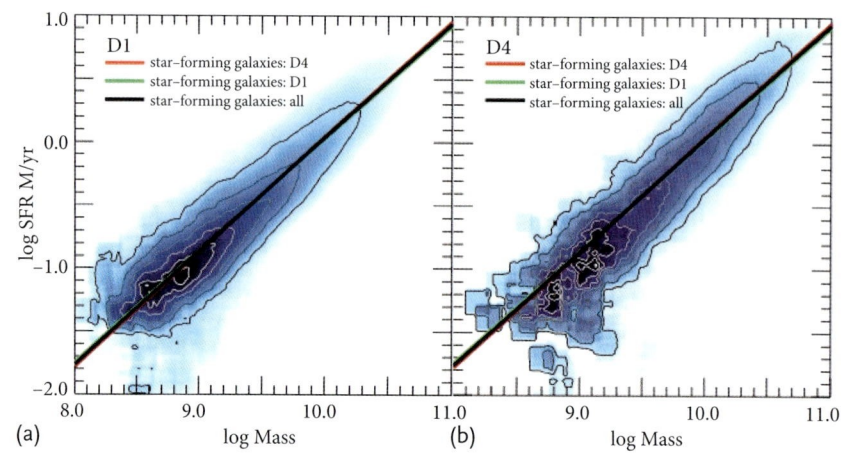

Figure C 6.14 The SFR versus stellar mass for a large sample of star-forming (blue cloud) galaxies in the nearby Universe, as derived from the Hα luminosity corrected for extinction. Part (a) refers to galaxies in the low environmental density quartile and (b) to the high-density quartile. Best fit linear relations for the whole galaxy sample as well as for the low- and high-density quartiles are shown in (a) and (b), though they can be barely distinguished, meaning that the SFR–$M_\star$ relation is virtually independent of environment. The galaxy database is the same as Figure 6.12a,b (source: Peng, Y. et al. (2010, Astrophys. J., 721, 193)) (Reproduced by permission of the AAS.).

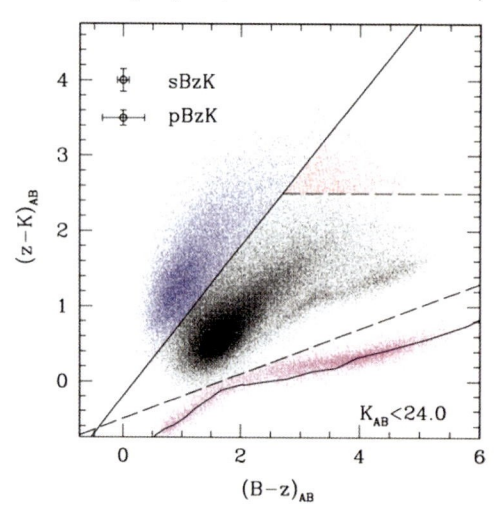

Figure C 6.15 The $z - K$ versus $B - z$ plot for a sample of objects down to $K_{AB} \simeq 24$. Candidate star-forming galaxies at $1.4 \lesssim z \lesssim 2.5$ (called sBzKs) lie to the left of the diagonal line, whereas candidate passively evolving (quenched) galaxies in the same redshift interval (called pBzKs) lie within the wedge in the upper-right. Most galaxies in the central part of the plot are at $z \lesssim 1.4$. Stars occupy the sequence in the lower part. Typical error bars for sBzKs and pBzKs are indicated (source: McCracken, H. et al. (2010, Astrophys. J., 708, 202)) (Reproduced by permission of the AAS.).

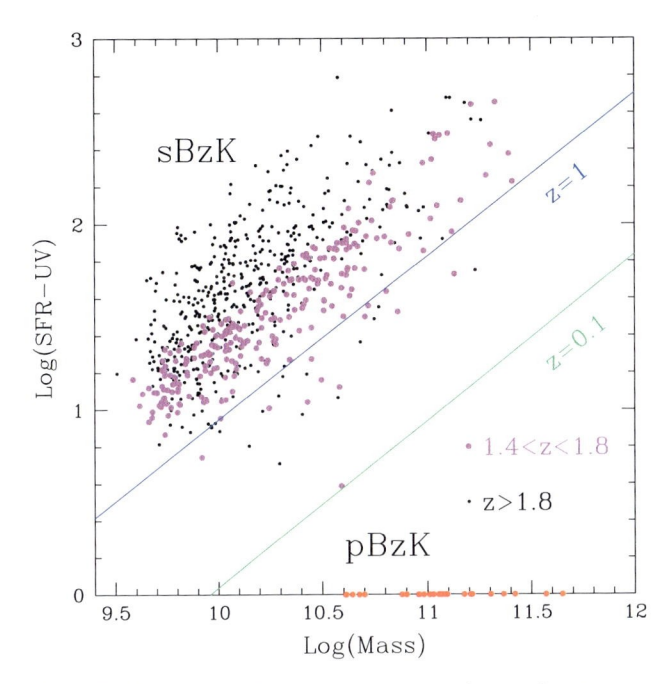

Figure C 6.17 The SFR–$M_\star$ relation for a sample of sBzK-selected star-forming galaxies, color-coded by their redshift as indicated. Best fit linear SFR–$M_\star$ relations for star-forming galaxies at $z \sim 1$ and $z \sim 0.1$ are also shown. The pBzK-selected passively evolving (quenched) galaxies at $z > 1.4$ are shown at log SFR $= 0$ (red symbols) though their SFRs are likely to be substantially lower. The sBzK sample is almost complete above $M_\star \sim 4 \times 10^9\ M_\odot$ and the pBzK sample above $M_\star \simeq 4 \times 10^{10}\ M_\odot$. Units are $M_\odot \mathrm{yr}^{-1}$ and $M_\odot$ (figure used GOODS data from Daddi, E. et al. (2007, Astrophys. J., 670, 156)).

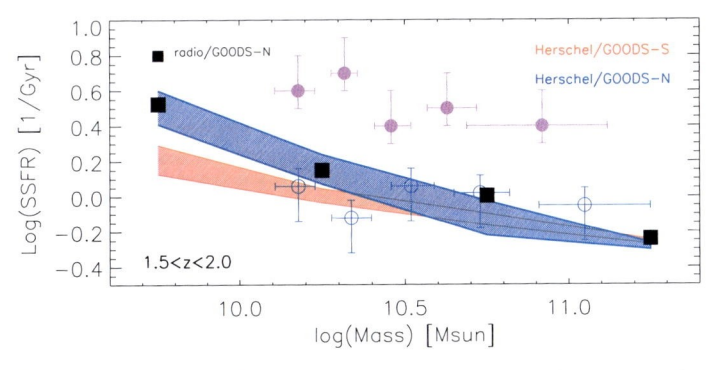

Figure C 6.19 The sSFR–$M_\star$ relation for a 4.5 μm selected sample down to AB magnitude 23 of $1.5 < z < 2.0$ galaxies in the GOODS fields (shaded areas) with SFRs measured from stacked Herschel data at 100 μm plus models for the far-IR SED. Filled and open circles represent the same data at $<z> = 2.1$ and 1.6 shown in Figure 6.18a,b. Filled squares represent SFRs from 1.4 GHz stacked radio data (adapted from Rodighiero, G. et al. (2010, Astron. Astrophys., 518, L25, and private communication)).

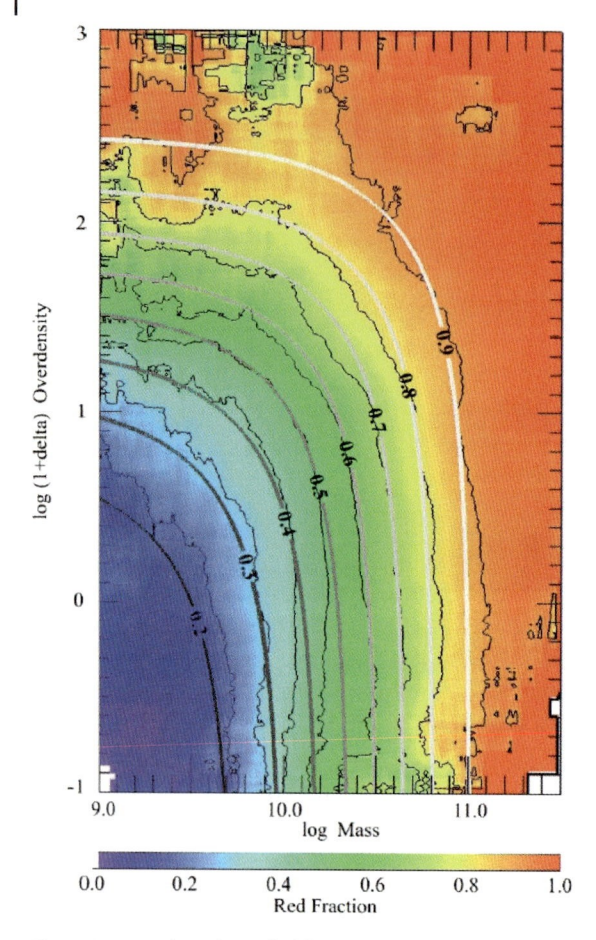

Figure C 6.21 The color-coded fraction of red/quenched galaxies in the local Universe as a function of galaxy stellar mass and environment as measured by the local overdensity (source: Peng, Y. *et al.* (2010, *Astrophys. J.*, 721, 193)) (Reproduced by permission of the AAS.).

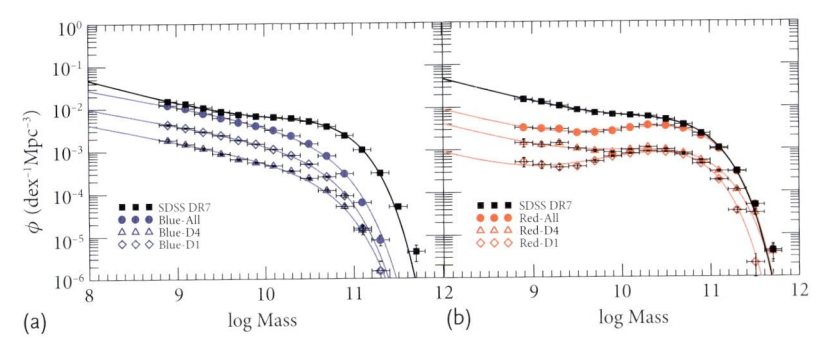

Figure C 6.22 The mass functions of galaxies in the local Universe. (a) The global mass function is shown as a black line with solid squares (replicated also in (b)); the other lines show the mass function for the star-forming (blue) galaxies, for the whole such population as well as for the low- and high-density quartiles, respectively D1 and D4. (b) The same for the quenched (red) galaxies. A Salpeter-diet IMF was adopted (source: Peng, Y. et al. (2010, Astrophys. J., 721, 193)) (Reproduced by permission of the AAS.).

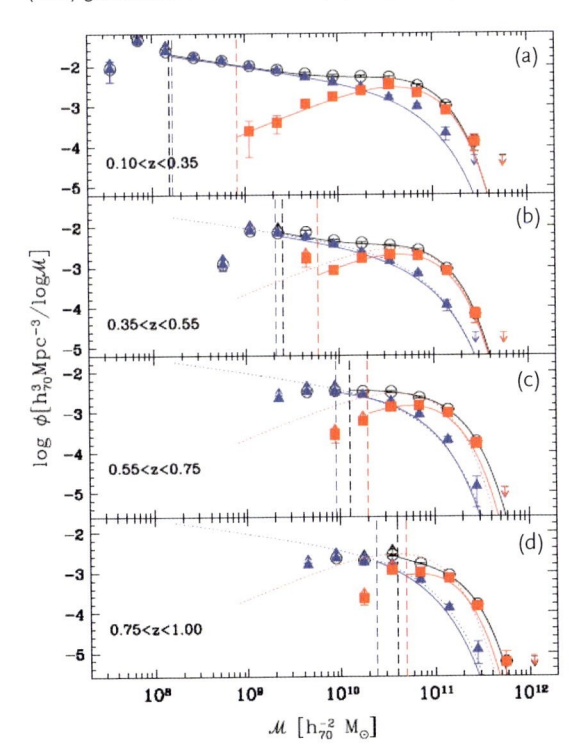

Figure C 6.24 (a–d) Galaxy stellar mass function by galaxy types and in various redshift bins. Quenched (red) galaxies and star-forming (blue) galaxies are shown as red squares and blue circles, respectively. Continuous lines are Schechter fits. Dotted lines in each panel show the Schechter fits to the first redshift bin. Dashed vertical lines in each redshift bin indicate the limits of the mass completeness (source: Pozzetti, L. et al. (2010, Astron. Astrophys., 523, A13)) (Reproduced by permission of © ESO.).

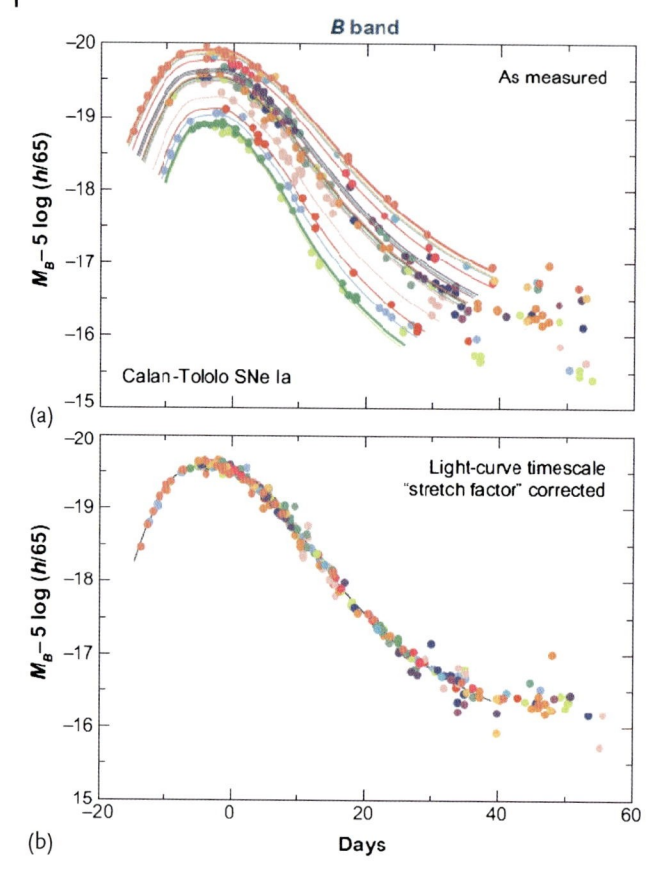

Figure C 7.1 Light curves of SNIa events from the Calan–Tololo survey (a). (b) Shows the light curves of the same events as in (a), after applying the *stretch* correction (source: Frieman, J.A. et al. (2008, *Annu. Rev. Astron. Astrophys.*, 46, 385)) (Reproduced by permission of Annual Reviews Inc.).

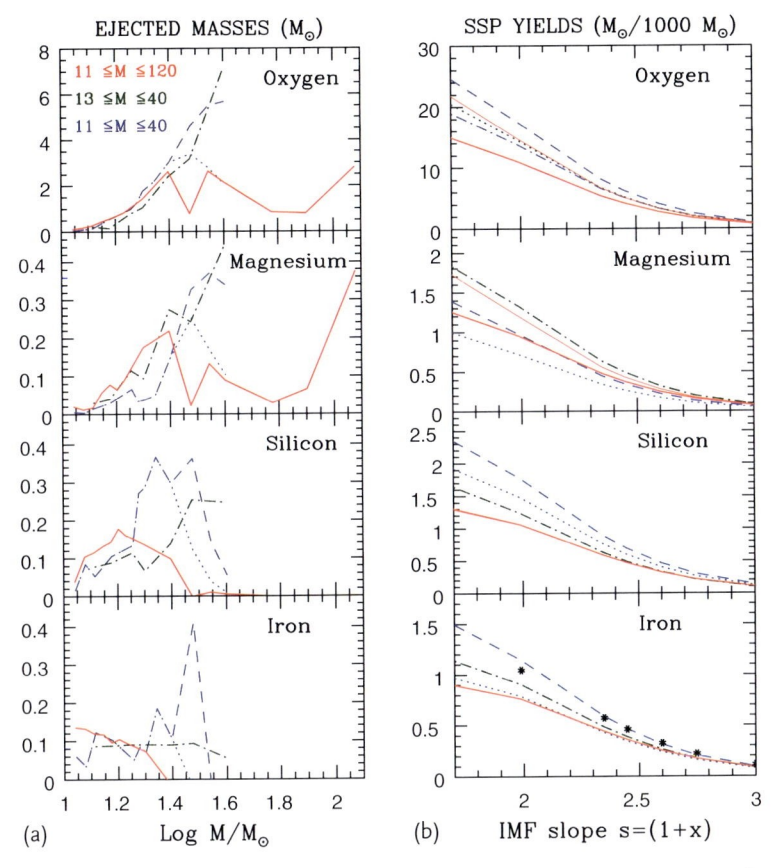

Figure C 7.5 Stellar (a) and SSP (b) yields from CC SNe from four sets of solar composition literature models: LC in red, KUNTO in green, and WW in blue, the dashed lines showing the effect of enhancing the explosion energy for stars with $M \geq 30\,M_\odot$ (the acronyms refer to the references given at the end of this caption). The ranges of initial stellar mass covered by the four sets are labeled in the legend in (a). The ejected masses and the SSP yields refer to the newly synthesized material, that is, the original amount of the element present in the ejecta has been subtracted. The SSP yields, computed for a family of *diet* IMFs adopt an upper limit of 40 $M_\odot$ (thick lines), or 120 $M_\odot$ (thin line, relative to LC models). Black dots in (b) for iron show a semiempirical estimate of the Fe yield, assuming that CC progenitors range from 8 to 120 $M_\odot$, and that each event provides 0.057 $M_\odot$ of Fe, which is the average yield of the events plotted in Figure 7.6 (stellar yields are from Limongi, M., and Chieffi, A. (2006, *Astrophys. J.*, 647, 483, LC); Kobayashi, C., *et al.* (2006, *Astrophys. J.*, 653, 1145), KUNTO; Woosley, S.E., and Weaver, T.A. (1995, *Astrophys. J. Suppl.*, 101, 181, WW)).

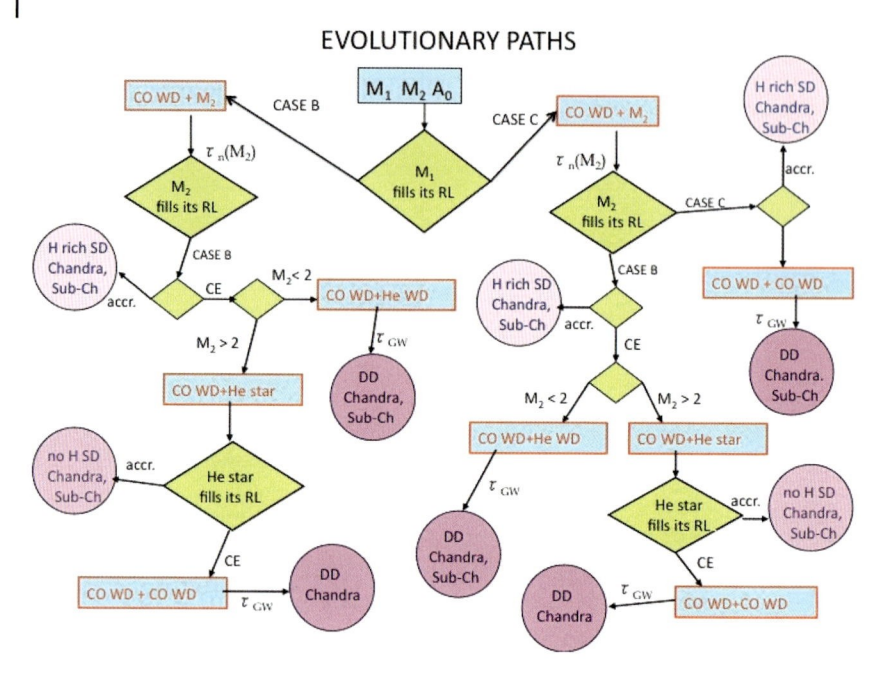

Figure C 7.7 Evolutionary paths for close binaries with intermediate-mass components, leading to SNIa explosions. The RL overflow can occur when the expanding star is in the shell hydrogen burning phase (Case B), or in the double shell burning phase (Case C). See text for more details.

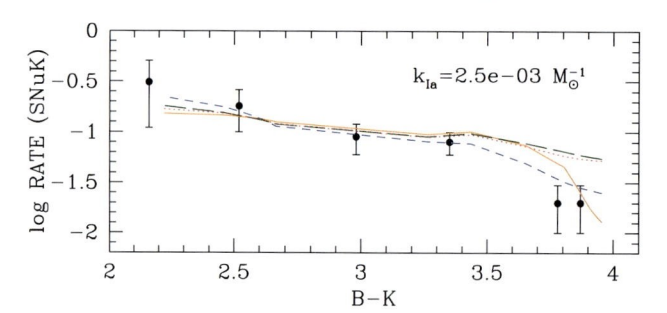

Figure C 7.12 SN Ia rate per unit K-band luminosity as function of the $B - K$ color of the parent galaxy. Black dots show the rates determined separately for E, S0, Sa, Sb, Sc and Irregular galaxies; colored lines connect models calculated with the SFR given by Eq. (5.6) and τ_{SF} ranging from 0.1 to 20 Gyr, plus three models with $\psi(t) \propto e^{t/\tau}$ and $\tau = 5, 3$ and 1 Gyr (right to left). The colors of the model lines encode the DTD: SD Chandra (orange), DD CLOSE (blue and green), and DD WIDE (red) models. The DD CLOSE models adopt a minimum secondary mass of 2.5 $M_\odot$ and a distribution of the GWR delays proportional to $\tau_{GW}^{-0.975}$ (blue) and $\tau_{GW}^{-0.75}$ (green). The DD WIDE model adopts a minimum secondary mass of 2 $M_\odot$ and a distribution of the separations proportional to $A^{-0.9}$. The value of the productivity assumed is indicated (updated from Greggio, L. and Cappellaro E. (2009, in *Probing Stellar Populations out to the Distant Universe, AIPC, 1111, 477*) with M05 stellar population models).

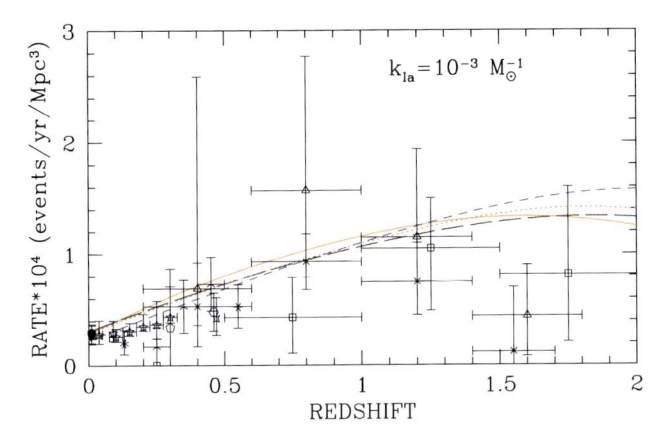

Figure C 7.13 SN Ia rate per unit volume as a function of redshift. Points are literature data. Colored lines are theoretical rates computed with Eq. (7.30) and for the same DTDs as in Figure 7.12 with the same color encoding. A productivity of 1 event every 10^3 $M_\odot$ of parent stellar population matches the level of the observed rates (updated from Blanc, G. and Greggio, L. (2008, New Astron., 13, 606)).

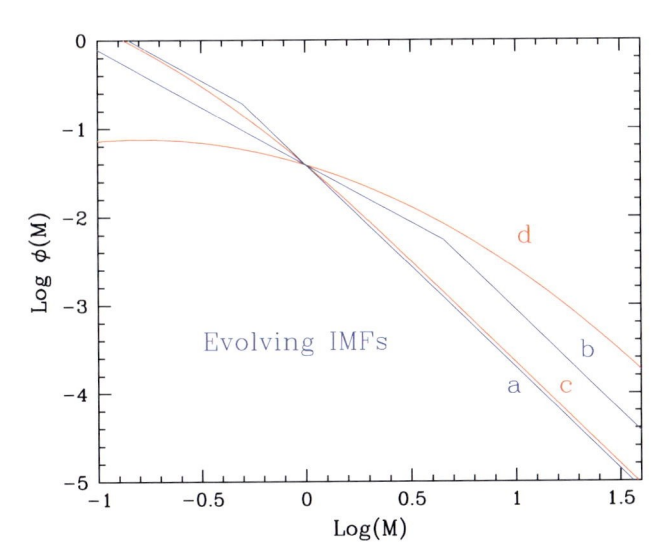

Figure C 8.3 Examples of evolving IMFs, for a two-slope IMF and a Chabrier-like IMF. Lines (a) and (c) represent the local IMF. The other lines show modified IMFs to explore a hypothetical evolution with redshift, with the break mass and the characteristic mass M_c having increased to $\sim 4\,M_\odot$, lines (b) and (d) respectively for the two-slope and the Chabrier-like IMF. All IMFs have been normalized to have the same value for $M = M_\odot$.

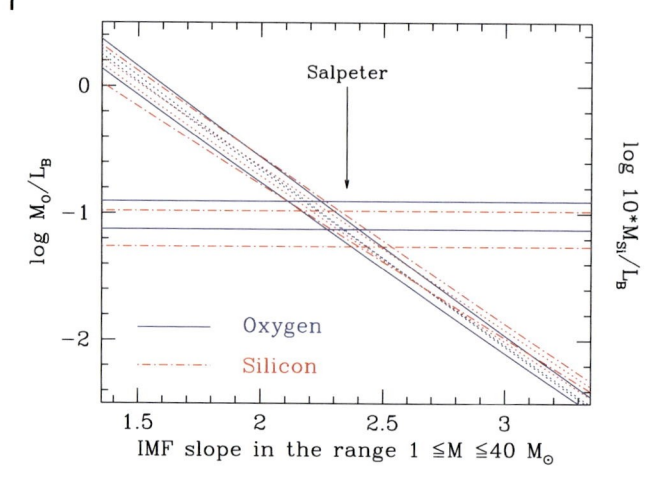

Figure C 8.7 The oxygen and silicon mass-to-light ratios as a function of the IMF slope for a ~ 12 Gyr old, near-solar metallicity SSP with oxygen and silicon yields from standard nucleosynthesis calculations. Different lines refer to the different theoretical yields shown in Figure 7.5a,b. The horizontal lines show the uncertainty range of the observed values of these ratios in clusters of galaxies, with central values as reported in Chapter 10, and allowing for a ~ ±25% uncertainty.

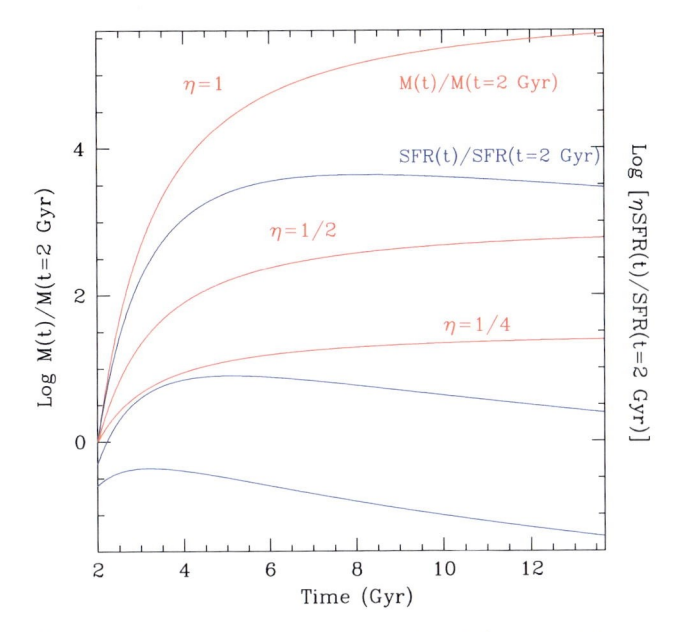

Figure C 9.1 The growth with cosmic time of the stellar mass normalized to its initial value at $t = 2\,\mathrm{Gyr}$ ($z \sim 3$), following Eq. (9.2), and for three values of η as indicated. Also shown is the corresponding evolution with time of the SFR, following Eq. (9.3), for the same values of η. The three curves are initially offset by a factor η to show the initial difference in SFR (i.e., at $t = 2\,\mathrm{Gyr}$). One can appreciate that SFRs for a given mass and time that differ by only a factor of 4 lead to vastly different evolutionary paths (source: Renzini, A. (2009, *Mon. Not. R. Astron. Soc.*, 398, L58)) (Reproduced by permission of the Royal Astronomical Society.).

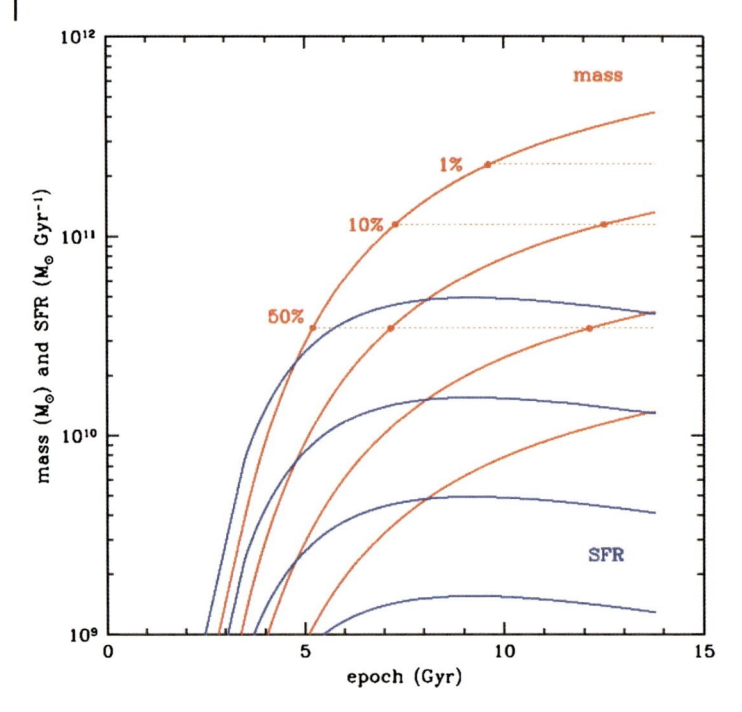

Figure C 9.2 The growth with cosmic time of the stellar mass and SFR of galaxies following Eq. (9.4) for cosmic time $t \geq 3.5$ Gyr, and constant specific SFR for cosmic time $t \leq 3.5$ Gyr. The specific SFR rapidly declines between $z = 2$ and $z = 0$ but the actual SFR predicted by Eq. (9.4) remains nearly the same because the decrease in specific SFR is compensated by the secular increase in mass. The dots along the mass lines indicate at which mass the survival probability (avoiding quenching of star formation) has to drop below 50%, 10% and 1% in order for the mass function of star-forming galaxies to remain nearly stationary (see text). As the mass of a galaxy grows via star formation, the chance to be "mass quenched" rapidly increases (source: Peng, Y. et al. (2010, Astrophys. J., 721, 1931)) (Reproduced by permission of the AAS.).

Figure C 9.4 (a) The relative environmental quenching efficiency ε_ρ as a function of overdensity $\log(1 + \delta)$, and for several redshift bins. At low density virtually all galaxies are star forming, whereas in the highest density regions, with overdensity exceeding ~ 1000, virtually all galaxies are quenched. (b) The growth of structure through cosmic times from $z = 1$ to $z = 0.1$, as described by the evolution of the median overdensity of all galaxy inhabited regions (middle line), as well as for the highest and lowest density quartiles (upper and lower line, respectively). Dashed lines show the corresponding best fit relations. Large circles refer to the local ($z \sim 0.1$) Universe. Small points represent the galaxies used to map the density field (source: Peng, Y. et al. (2010, *Astrophys. J.*, 721, 1931)) (Reproduced by permission of the AAS.).

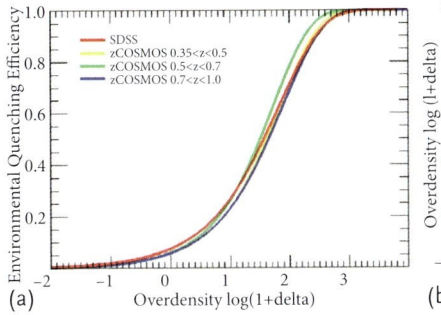

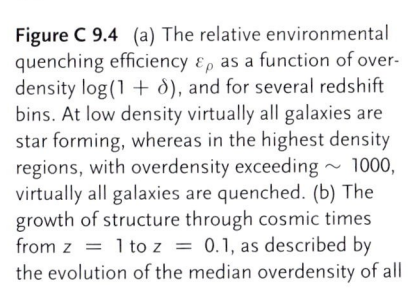

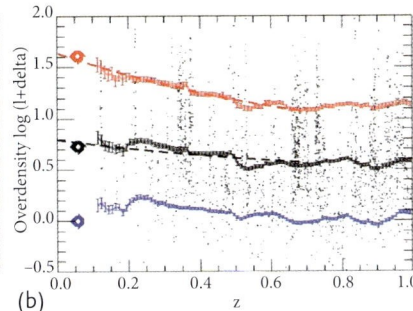

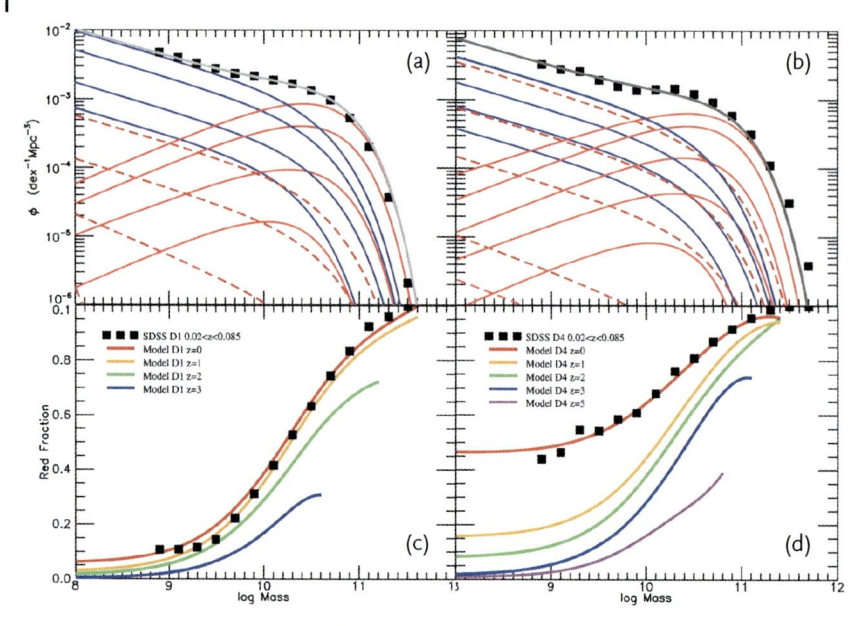

Figure C 9.5 The phenomenological model evolution of the mass functions (a,b) and the quenched (red) fraction (c,d), respectively, in the low-density quartile (a,c) and high-density quartile (b,d). Red solid lines show the evolution of the mass function of quenched galaxies that are mass-quenched, while red dashed lines refer to galaxies that are environment-quenched. The blue lines show the evolution of the mass function of star-forming galaxies. Mass functions are shown bottom-up for $z = 3, 2, 1, 0$, and also for $z = 5$ in (b,d). The top gray line is the final, cumulative mass function (star-forming plus quenched galaxies) of the model, with overplotted the empirical mass function from SDSS data (black squares) (source: Peng, Y. et al. (2010, Astrophys. J., 721, 1931)) (Reproduced by permission of the AAS.).

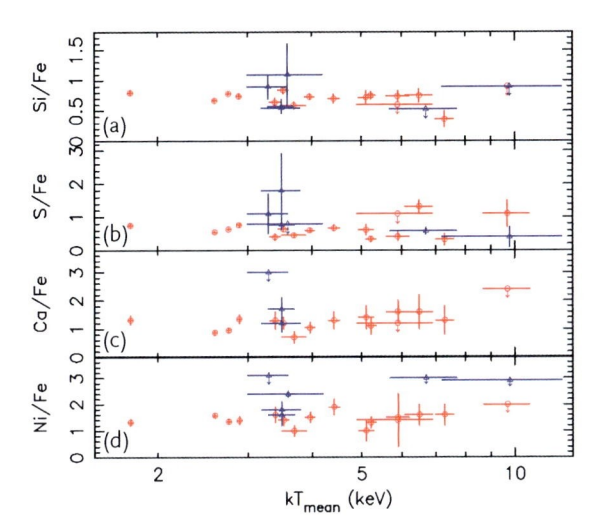

Figure C 10.6 The silicon (a), sulfur (b), calcium (c) and nickel-to-iron ratios (d) in the ICM (relative to the solar ratios) as a function of ICM temperature. No appreciable trend with ICM temperature (i. e., cluster richness) is apparent (source: Werner, N. *et al.* (2008, *Space Sci. Rev.*, 134, 337)).

1
Firm and Less Firm Outcomes of Stellar Evolution Theory

Stellar evolution theory is a mature science which provides the backbone to the theory of stellar populations and the construction of their synthetic luminosities, colors and spectra. In turn, synthetic stellar populations offer the virtually unique possibility to estimate the ages of stellar systems from star clusters to whole galaxies, their star formation rates and star formation histories, and their mass in stars. This chapter is no substitute to a whole textbook on stellar evolution, of which there are many. It rather offers an introduction to stellar evolution, meant to familiarize the reader with the pertinent nomenclature and a few key concepts. Particular attention is devoted to highlight the strengths and weaknesses of current methods, assumptions and results of stellar evolution theory, in particular concerning their impact on synthetic stellar populations. Further readings on the treated issues are listed at the end of this chapter.

1.1
A Brief Journey through Stellar Evolution

1.1.1
A 9 $M_\odot$ Star

Figure 1.1 shows the calculated evolutionary path on the Hertzsprung–Russell diagram (HRD) of a 9 $M_\odot$ star of solar composition. Letters mark critical points in the course of evolution and Table 1.1 gives the corresponding times to reach them. Specifically, the various points mark the following events:

- A: Beginning of steady hydrogen burning, zero age main sequence (ZAMS).
- C-C′: Exhaustion of hydrogen in the core, and ignition of hydrogen burning in a shell surrounding the hydrogen-exhausted core.
- E: Arrival on the Hayashi line, that is, the envelope is (almost) fully convective.
- F: Ignition of helium burning in the center.
- J: Exhaustion of helium in the core.
- JK: Ignition of helium burning in a shell surrounding the helium-exhausted core.

Stellar Populations, First Edition. Laura Greggio and Alvio Renzini
© 2011 WILEY-VCH Verlag GmbH & Co. KGaA. Published 2011 by WILEY-VCH Verlag GmbH & Co. KGaA.

1 Firm and Less Firm Outcomes of Stellar Evolution Theory

Table 1.1 Evolutionary times for the 9 $M_\odot$ sequence.

Point	Time (10^6 yr)	Point	Time (10^6 yr)
A	0.0000	G	24.3046
B	21.8130	H	24.8211
C	22.5220	I	26.1765
C'	22.5378	J	26.7813
D	22.6435	K	26.8740
E	22.7752	L	26.9380
F	22.8126	M	26.9488

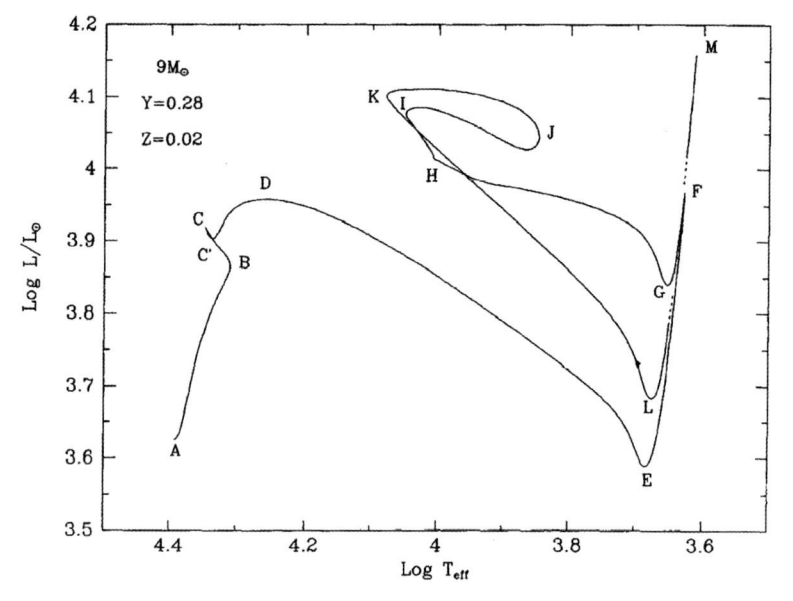

Figure 1.1 The evolutionary track in the Hertzsprung–Russel diagram of a 9 $M_\odot$ star with solar composition from the ZAMS all the way to the asymptotic giant branch phase (hydrogen and helium burning in two separate shells). What happens at the labeled points along the sequence (A–M) is discussed in the text, while times to reach such points are reported in Table 1.1 (source: Renzini et al. (1992, Astrophys. J., 400, 280)) (Reproduced by permission of the AAS.).

- L: Back to the Hayashi line (fully convective envelope).
- LM: Early asymptotic giant branch phase (E-AGB).

Before disclosing what happens at points BDGHI and K (not mentioned in the previous list) it is necessary to introduce a few concepts in the language of stellar model makers. The *core* is the innermost part of the star, where nuclear reactions have greatly altered the original composition. By *envelope* one means the outer part of the star, over (most of) which nuclear burning is negligible and whose composition may still be close to the original one.

Next is the concept of *thermal equilibrium*. A star is said to be in thermal equilibrium (TE) when the total rate of nuclear energy generation (L_N) is almost perfectly equal to its surface luminosity (L_S). TE means that the envelope is able to transfer outside and radiate into space quite precisely the same amount of energy which per unit time is produced by nuclear reactions in the core. If $L_S \simeq L_N$ the star is in TE and its evolution proceeds on a nuclear timescale. Conversely, if $L_S \neq L_N$ the star is out of TE, and its evolution proceeds on a thermal timescale, which usually is much shorter than the nuclear timescale. TE is often broken when the core runs out of fuel or when a new fuel is ignited. A dramatic example of breaking TE is the helium ignition in a degenerate core, called a helium flash. At flash peak $L_N \simeq 10^{10} L_\odot$ while $L_S \simeq 2000 L_\odot$ (and decreases!). The energy that does not escape the star is used to expand the helium core, hence relieving its degeneracy.

Evolutionary phases in TE and those out of TE can be easily identified by plotting L_S versus L_N, or, equivalently, versus the luminosity impinging at the base of the envelope (L_B), as shown in Figure 1.2 relative to the evolution of the same 9 $M_\odot$ star model whose HRD is shown in Figure 1.1. Letter labels in this figure mark the

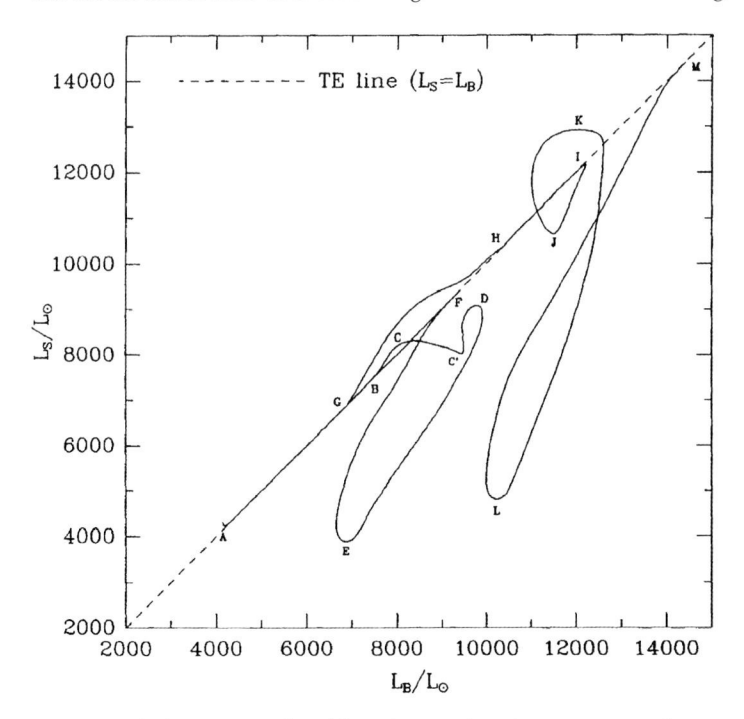

Figure 1.2 The luminosity radiated from the star's surface as a function of the luminosity released by the stellar core and entering the envelope from its base, for the same track shown in Figure 1.1. Labels of relevant points along the sequence are the same as in Figure 1.1 (source: Renzini et al. (1992, Astrophys. J., 400, 280)) (Reproduced by permission of the AAS.).

same events as in Figure 1.1, so that one can identify those phases that are in TE and those which are out of it.

Thus, phase AB (which is most of the main sequence phase) proceeds in strict TE, but as hydrogen approaches exhaustion in the core, nuclear burning starts to fall short of keeping in pace with the rate at which energy is being radiated away, the star starts departing from TE and begins to contract (point B). The virial theorem may help in understanding why the star contracts. If $\Delta L = L_B - L_S \neq 0$ it means that the envelope is either gaining or losing energy, that is:

$$\Delta L = L_B - L_S = \frac{d E^{env}}{dt} = \frac{d(E_{th}^{env} + E_{grav}^{env})}{dt}, \quad (1.1)$$

where the energy of the envelope is split in its thermal and gravitational parts, with

$$E_{grav}^{env} = -G \int_{env} \frac{M_r}{r} d M_r . \quad (1.2)$$

From the virial theorem:

$$2 E_{th}^{env} + E_{grav}^{env} \simeq 0 , \quad (1.3)$$

hence, using this virial relation to eliminate E_{th}^{env} from Eq. (1.1), we obtain:

$$\frac{d E_{grav}^{env}}{dt} = 2\Delta L . \quad (1.4)$$

Therefore, if $\Delta L > 0$ the envelope gains energy, its gravitational energy increases, that is, it becomes less gravitationally bound, and expands. Conversely, if $\Delta L < 0$ the envelope loses more energy from the surface than it receives from the bottom, its gravitational energy decreases, it becomes more bound, hence it contracts. This means that the star contracts in those evolutionary phases that lie in the upper half of Figure 1.2, and expands in those that lie in the lower half, and in both cases does so on a thermal timescale, if they are noticeably away from the $L_S = L_B$ TE line. All this is quite rigorous, but mere physical intuition should also suffice: if the envelope emits less energy than it receives from the bottom it inflates, if it emits more than it receives it deflates. Notice that only half of the energy needed to expand the envelope comes from the trapped energy flux, the other half comes from the thermal energy of the envelope itself.

This is indeed the case starting at point B: as the core is running out of fuel the envelope starts losing more energy than it receives, and contracts until at point C hydrogen is effectively exhausted over the central regions, and nuclear energy generation quickly shifts to a shell surrounding a hydrogen-exhausted helium core. The quick readjustment from core to shell burning leaves the star at point C′, somewhat out of TE, but then the structure tends to approach again TE, until at point D this tendency is reverted and a major excursion away from TE begins. In stars a little less massive than the one considered here, TE is actually well restored shortly after point C′, and yet – as here at point D – TE is broken and stars undergo an extensive loop in the $L_S - L_B$ diagram, as shown in Figure 1.2. The journey across the HRD from point D to point E takes place on a thermal timescale, and the star expands to red giant dimensions. Clearly, a *thermal instability* erupts at point D. This is indeed a quite severe thermal instability, suffice to say from Figure 1.2 that at the peak of the instability the core releases $\sim 7000\, L_\odot$ but the envelope radiates away only $\sim 4000\, L_\odot$, and $\sim 3000\, L_\odot$ are absorbed for its expansion.

The physical origin of this thermal instability is actually quite easy to understand. During phase C'D the luminosity provided by the hydrogen burning shell steadily increases as the shell moves out in the mass coordinate thanks to its own burning, and sits on a progressively more massive helium core. In response to the increasing luminosity impinging on its base (L_B) the envelope slowly expands. By expanding the envelope cools, and by cooling heavy metal ions begin to recombine. Besides being scattered by free electrons, photons now begin to be absorbed by such heavy ions via bound-bound and bound-free transitions: radiative opacity increases. This opacity increase is the key factor that determines the onset of the thermal instability. At any point within the star the luminosity transmitted outwards by the radiation field is:

$$L_r = 4\pi r^2 \frac{4acT^3}{3\kappa\rho}\frac{dT}{dr} = 4\pi r^2 F_r, \quad (1.5)$$

where r is the distance from the center, T the temperature, ρ the density, κ the opacity and F_r the radiative energy flux. During phase C'D the envelope slowly expands, hence r^2 increases while the flux F_r decreases, but their product still increases and the star is approaching TE. However, as this trend continues the increase in opacity accelerates and eventually the flux drops faster than r^2 increases, and their product L_r starts to decrease. The decrease happens first near the stellar surface, and then – very quickly – through the whole envelope. At this point the envelope is transferring outwards and radiating away less energy than it receives from the stellar core, that is, $\Delta L = L_B - L_S > 0$, and dE_{grav}^{env}/dt increases as demanded by Eq. (1.4). But as the envelope expands more ions recombine, opacity increases further, the flux drops even more, the thermal imbalance $L_B - L_S$ increases and expansion accelerates: the stellar envelope is in a thermal runaway, as it becomes more and more unable to radiate away the energy it receives from the stellar interior.

At point E the surface luminosity starts to rise again, $L_B - L_S$ begins to drop, and TE is rapidly restored. What relieves the instability and saves the star from literally falling apart is convection. As the expanding envelope cools, and opacity increases, eventually the radiative gradient ∇_{rad}[1] exceeds the adiabatic gradient ∇_{ad}, first near the photosphere, where hydrogen is only partly ionized, and then rapidly through the whole envelope. Thus, convection replaces radiative transfer in carrying out energy through the envelope, and the thermal instability is quenched since it was intimately related to the radiative mode of energy transfer. Most of the energy flux in the envelope being now carried by convective motions, the envelope ceases to absorb energy, and the surface luminosity L_S starts to increase again: the star now ascends the Hayashi line, until at point F the helium burning reactions ignite in the core.

1) Temperature gradients are defined as $\nabla(\rho, T, X_i) = (\partial \log T / \partial \log P)$.

Following helium ignition, the helium core initiates a very slow expansion, which will last through a major fraction of the helium burning phase. This is because the helium burning core works like a *breeder* reactor during this stage, that is, it produces more new fuel than it burns. As triple-α reactions produce fresh ^{12}C, the ^{12}C$(\alpha, \gamma)^{16}$O reactions release an increasing amount of energy, and the inner core is forced to expand slightly. Such a modest expansion of the core is sufficient to cause a decrease of temperature and density in the surrounding hydrogen burning shell and the stellar luminosity correspondingly starts to decrease. The star now evolves along the Hayashi line from F to G, burning helium in the core and hydrogen in the shell, until thermal stability is broken again at point G. As the envelope contracts it heats up gradually, and in particular near its base, heavy ions start losing electrons. Hence opacity decreases along with the radiative gradient. As ∇_{rad} drops below ∇_{ad} a radiative region appears at the base of the envelope and grows outwards in mass. As this growth progresses the envelope becomes more and more transparent to radiation, until it starts radiating away more energy than it receives from the core: $L_B - L_S$ turns increasingly negative, the envelope starts deflating, but the more it contracts the more "transparent" it becomes, and the more energy it loses into space. The thermal instability bringing the star in its envelope deflation from point G to point H is precisely the reverse analog of the thermal instability that causes the runaway expansion from point D to point E. As envelope inflation can be ascribed to a runway recombination of heavy elements in the envelope, envelope deflation is due to a runway ionization of such heavy elements. A comparison of the three figures (Figures 1.1–1.3) helps in visualizing the onset of the thermal instability and of its results.

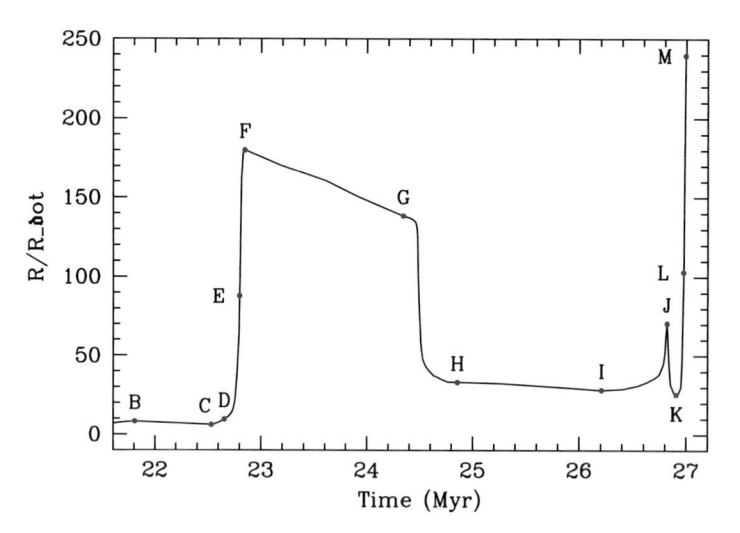

Figure 1.3 The stellar radius as a function of time for the same track shown in Figure 1.1. Labels of relevant points along the sequence are the same as in Figure 1.1 (source: Renzini *et al.* (1992, *Astrophys. J.*, 400, 280)) (Reproduced by permission of the AAS.).

Thus, in this star, the core helium burning phase in thermal equilibrium is spent in two distinct locations in the HRD: part on the Hayashi line as a red giant (F–G) and part as a blue giant (H–I) separated by a runaway contraction out of TE (G–H). This *first blue loop* (also called the Cepheid blue loop) continues after point I when TE is broken again. Here the envelope is slowly expanding in response to the slowly increasing luminosity from the core and shell burning, when the envelope turns unstable again due to the same physical process already described for the DE phase.

However, the journey towards the Hayashi line is suddenly interrupted and inverted at point J, which marks the start of the so-called *second blue loop*. What happens at point J and after is a rather complex series of interconnected events: helium is exhausted in the core, the core contracts and helium burning rapidly shifts from the helium-exhausted CO core to a shell surrounding it; helium shell ignition is quite violent and causes expansion of the helium buffer above this shell: when the expansion front breaks through the hydrogen burning shell temperature and density in the shell drop; burning in the hydrogen shell (which was still providing most of the stellar luminosity) is effectively shut off completely causing a sudden drop of the luminosity impinging on the base of the envelope: L_B drops below L_S and – following Eq. (1.4) – the envelope stops expanding and contracts. In the meantime the strength of the helium burning shell steadily increases until it leads L_B to exceed again L_S, contraction stops and the star resumes at point K its runaway expansion that was temporarily stopped at J. The rest, from K to L to M is quite similar to the DEF phase, with convection replacing radiative transfer in the envelope and TE being rapidly restored, shortly before point M. Figure 1.3 clearly illustrates the dramatic effects of the envelope thermal instabilities on the overall structure of the star. Those phases which are in TE are nearly flat in this plot, that is, the stellar radius changes quite slowly with time. With one exception, those phases which are out of TE are instead nearly vertical, that is, the radius varies very rapidly during such runaway inflations or deflations. The exception is phase BC, which is only modestly out of TE, and contraction is relatively slow.

What happens past point M is still an open issue. A 9 $M_\odot$ star lies between two domains: massive stars that eventually undergo core collapse and supernova explosion, and intermediate-mass stars, which shedding all their hydrogen-rich envelope die as a white dwarf. One believes that in a 9 $M_\odot$ star carbon is ignited in the central core under only mildly degenerate conditions, and the star keeps ascending along the Hayashi line as a *super asymptotic giant branch* star experiencing a few thermal pulses in its deeper regions, where hydrogen and helium are still burning in two separate shells. Thus, if the envelope is completely lost in a wind the star leaves an ONeMg white dwarf. If instead mass loss is less severe then the core keeps growing in mass thanks to the active burning shells until electron captures in the core trigger a core collapse and we have a supernova explosion. Clearly the fate critically depends on the strength of the mass loss process.

What does all this have to do with the global properties of stellar populations? An example is discussed next.

Since thermal instabilities in the envelope are driven by opacity variations, and opacity depends on the metal abundance Z, we should expect that stars with different metal content may or may not experience such instabilities. This is indeed the case for both the first inflation to red giant dimensions, and for the first blue loop.

For example, at low metallicity ($Z \sim Z_\odot/20 \simeq 0.001$) a 9 $M_\odot$ star ignites helium in the core before the envelope becomes thermally unstable, the star fails to become a red giant and the whole core helium burning phase is spent as a blue giant. This behavior extends to all metal-poor stars more massive than 9 $M_\odot$, and therefore a young (age $\lesssim$ 20 Myr), metal-poor stellar population fails to produce any sizable population of red giants and supergiants, and its integrated spectral energy distribution (SED) will correspondingly be appreciably stronger in the UV, and much fainter in the near-IR. Metallicity is also known to affect the extension of the first blue loop, to the extent that above solar metallicity the blue loops can be completely suppressed. This is because with the higher opacity provided by the metals, the envelope avoids the thermal instability that the solar abundance stars experience at point G, and after reaching a minimum luminosity, stars climb back along the Hayashi line without departing from it. Thus, in a metal-rich population the whole core helium burning phase is spent as a red giant (or supergiant) and its SED will correspondingly be weaker in the UV and stronger in the near-IR. In summary, low metallicity can suppress the thermal instability at point D and the ensuing runaway expansion, high metallicity can make this instability more severe with a deeper luminosity drop from D to E, but can also suppress the thermal instability at point G and the development of first blue loops.

1.1.2
The Evolution of Stars with Solar Composition

Figure 1.4 shows stellar evolutionary tracks of solar composition, covering a wide range of initial masses, from a 0.8 $M_\odot$ star, whose MS evolutionary lifetime is longer than one Hubble time, up to a 20 $M_\odot$ model. The evolutionary tracks of stars with mass greater than $\sim 20\ M_\odot$ are heavily affected by mass loss; those features which have an impact on the properties of stellar populations are discussed later.

Different gray shades in Figure 1.4 pertain to the different mass ranges as customarily distinguished in stellar evolution: low-mass stars, up to $\lesssim 2.2\ M_\odot$, intermediate-mass stars, up to $\lesssim 8\ M_\odot$, and high-mass stars. These ranges correspond to different physical behavior during the evolution of the stars, and the mass limits depend on chemical composition. Low-mass stars develop an electron degenerate helium core after their MS evolution; intermediate mass stars ignite helium under nondegenerate conditions, but develop a CO degenerate core after central helium burning; massive stars experience all successive nuclear burnings up to the production of an iron core. In the following we briefly point out those aspects of the stellar models which are more relevant in the context of the properties of stellar populations.

During their MS evolution, stars with mass below $\sim 1\ M_\odot$ burn hydrogen through the p-p chain, whose reaction rate is not extremely sensitive to tempera-

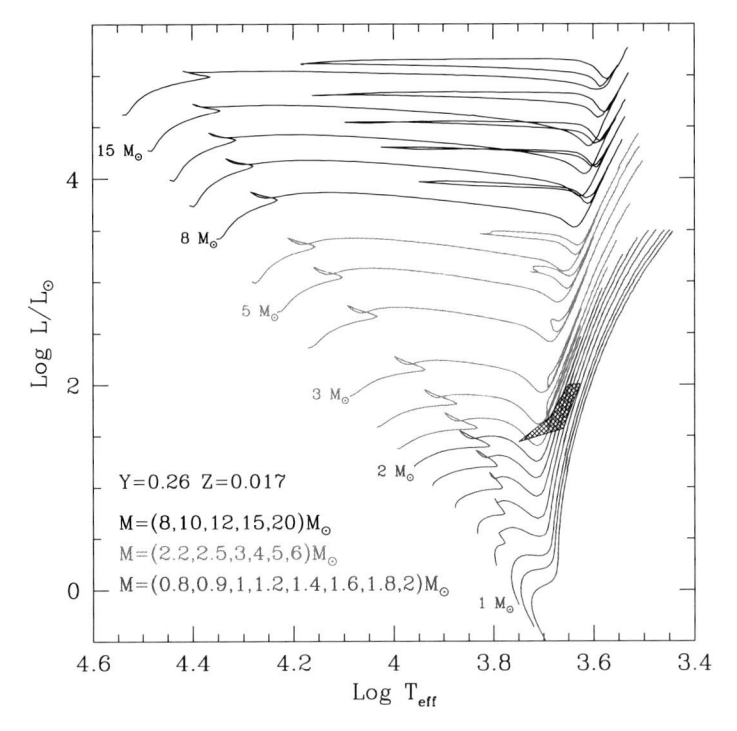

Figure 1.4 Evolutionary tracks of solar composition. The shaded area shows the location of low-mass ($0.55 \leq M/M_\odot \leq 2$) core helium burning models. Drawn using the YZVAR database (Bertelli, G. et al. 2008, Astron. Astrophys., 484, 815; 2009, Astron. Astrophys., 508, 355).

ture, so that these stars possess a radiative core. As evolution proceeds hydrogen is progressively depleted in the inner core, more in the center than in the periphery, leading to a smooth hydrogen profile from a central minimum up to the initial abundance in the outer layers. On the HRD, the stellar track climbs to higher luminosities and temperatures until the central hydrogen is severely depleted; at this point the effective temperature starts decreasing, producing the turnoff (TO), that is, the maximum temperature point on the MS, which is easily recognizable on the CMD of globular clusters. Shortly after the TO, hydrogen is completely exhausted in the center and the hydrogen shell burning phase starts as a natural progression from the previous core hydrogen burning phase. During shell hydrogen burning, the track forms the subgiant branch (SGB), evolving at (almost) constant luminosity and decreasing temperature until the external convection penetrates deeply inside, and the star becomes almost fully convective at the base of the red giant branch (RGB).

The MS stars (with $M \gtrsim 1\,M_\odot$) have a convective core, since at least part of the hydrogen burning occurs through the CNO cycle, whose energy generation rate is extremely sensitive to temperature. Because of central mixing, the hydrogen profile is characterized by an inner plateau; as evolution proceeds, the extension of the convective core progressively decreases, leaving behind a gradient of hydrogen abundance. In stars with a convective core, the fuel depletion affects a sizable central region, which starts rapid contraction when approaching fuel exhaustion. This happens at the local minimum effective temperature point during the MS evolution of stars with $M > 1\,M_\odot$ (Figure 1.4), which signals the beginning of the overall contraction phase. The evolutionary behavior across this phase and the following runaway expansion has been already described in detail in the previous section.

As already mentioned, the helium core of low-mass stars is (electron) degenerate: this implies that central helium ignition is delayed because core contraction does not lead to an efficient increase of the central temperature. The (almost) fully convective star climbs along the RGB, while the hydrogen burning shell progresses outward, thereby increasing the mass of the helium core. When the core reaches a critical limit, a helium flash occurs off-center, since neutrino losses induce a temperature inversion in the innermost layers. This event in not catastrophic for the star because local expansion removes the degeneracy; instead, a sequence of flashes occurring at progressively inner locations totally remove the core degeneracy. During this phase, which lasts ~ 1 Myr, the star moves downward along the Hayashi line to settle on the core helium burning locus, either the red clump, or the horizontal branch (HB). The maximum luminosity reached on the RGB (RGB Tip, or TRGB) is a very important feature in the color-magnitude diagram (CMD) of stellar populations because it is virtually independent of mass, and then of evolutionary lifetime. This is because on the one hand the critical mass for helium ignition under degenerate conditions is almost constant ($\sim 0.5\,M_\odot$), and on the other, along the RGB there exists a core mass-luminosity relation. The evolutionary lifetimes of low-mass stars range from ~ 1 Gyr up to the Hubble time, as seen in Figure 1.4, and all of them experience the helium flash at almost the same luminosity; therefore, the CMD of a stellar population with stars older than ~ 1 Gyr will show a prominent TRGB feature whose luminosity is known, thus allowing us to determine the distance of the stellar population. The TRGB luminosity depends slightly on metallicity, being higher for higher metal content. However, by a fortunate combination, the absolute magnitude in the I-band of the TRGB does not depend much on metallicity, for metal-poor populations. This is because the effective temperature at the TRGB also depends on metallicity: the higher Z, the cooler the TRGB stars, and the higher the bolometric correction (in absolute value). The trend with metallicity of the bolometric correction to the I-band largely compensates that of the tip luminosity, so that the I-band absolute magnitude of the TRGB ($M_{I,\mathrm{TRGB}}$) is almost independent of age and weakly dependent on metallicity of the parent stellar population. This makes $M_{I,\mathrm{TRGB}}$ a very effective distance indicator in galaxies. An empirical calibration yields (from Bellazzini M. et al. (2004, Astron. Astrophys. 424, 199)):

$$M_{I,\mathrm{TRGB}} = 0.258\,[\mathrm{M/H}]^2 + 0.676\,[\mathrm{M/H}] - 3.629 \qquad (1.6)$$

in the range $-2.2 \leq [\mathrm{Fe/H}] \leq -0.2$, with $M_{I,\mathrm{TRGB}} \simeq -4$. In this equation $[\mathrm{M/H}] = \log(n^M/n^H) - \log(n^M/n^H)_\odot$ and M indicates all metals. At higher metallicity $M_{I,\mathrm{TRGB}}$ starts to drop, due to the increasing strength of molecular bands, and can be replaced with the H-band magnitude ($M_{H,\mathrm{TRGB}}$) as a distance indicator.

The luminosity of core helium burning low-mass stars, being fixed by the mass of their hydrogen-exhausted core, is also largely independent of their total mass, thereby providing another distance indicator. However, evolution during the core helium burning phase spreads the red clump stars over ~ 0.6 magnitudes, as can be seen from the vertical size of the hatched region in Figure 1.4. In addition, the core mass at helium ignition is not a monotonic function of the total mass, and as the latter increases beyond the limit of the low-mass stars regime (M_{Hef}), it decreases, reaches a minimum of about 0.326 $M_\odot$, and then increases. As a result, stars with mass just above M_{Hef} start their core helium burning evolution at fainter luminosities, have the longest helium burning lifetimes, and cover a wider range of luminosities, compared to stars with $M < M_{\text{Hef}}$. Therefore, the core helium burners of a composite stellar population with a sizable component at ages just below 1 Gyr are spread over a wide magnitude range, which limits the use of this feature for distance determinations.

The effective temperature of low mass core helium burning stars depends on the mass of their envelope, a dependence which is very pronounced when the envelope is thinner than for example $\sim 0.2\ M_\odot$ for $Z \lesssim 0.1\ Z_\odot$. Below this threshold, the lower the envelope mass, the hotter the core helium burning stars. In a coeval and homogeneous stellar population the core helium burning stars all have virtually the same core mass and then luminosity; a spread of their envelope masses produces a feature on the HRD at about constant (bolometric) magnitude extending over a temperature range. The observational counterpart of this locus is the horizontal branch: a prominent feature in the HRD of globular clusters whose age is old enough to host core helium burning stars with low-mass envelopes. The existence of wide HBs in globular clusters was explained as due to a dispersion of mass lost during the RGB phase, but recently it became apparent that other effects are also at play (see Chapter 4). Evolutionary tracks during the core helium burning phase exhibit a wide variety of morphologies, depending on their mass, envelope mass, metallicity and helium abundance. At central helium exhaustion, a rapid core contraction leads to shell helium ignition; the model star expands and moves again towards the Hayashi line to start the asymptotic giant branch (AGB) evolutionary stage, which is discussed later. However, if the envelope mass is small enough, the shell helium burning phase is entirely spent at high effective temperatures, as an *AGB manqué star*. The hot HB stars and their AGB manqué progeny are likely responsible for the UV emission from elliptical galaxies which host old stellar populations.

The evolution of intermediate and high-mass stars during the hydrogen and helium burning stages is very similar to what has already been described for the 9 $M_\odot$ star. But, following helium exhaustion, intermediate-mass stars behave similarly to low-mass stars, as will be described in the next section. We only notice here that for stars with mass in the vicinity of M_{Hef} the core helium burning phase is spent in a red clump, which becomes a wider and wider blue loop as the model mass grows. Thus, the core helium burning phase of intermediate-mass stars is spent part in the blue and part in the red. The very occurrence of the loop, its extension and the fraction of lifetime spent on each side of the loop, are sensitive to a number of parameters describing the input physics of the models, like metallicity, opacity, convection and others. Therefore, the blue-to-red ratio in stellar populations with stars in this mass range is difficult to interpret. The luminosity of intermediate mass core helium burners is instead a more robust prediction of the models, and can be effectively used as an age indicator of stellar populations.

1.1.3
Dependence on Initial Chemical Composition

The evolutionary tracks of given mass are very sensitive to the (initial) helium content (Y) and metallicity (Z), which control the energy generation rates and the opacity. We briefly illustrate in this section some dependencies relevant to the interpretation of the HR diagram and of the spectral energy distribution of stellar populations.

At a given initial mass, evolutionary lifetimes become shorter as Y increases, because stars with higher molecular weight are more compact, hotter, brighter, hence faster in consuming the hydrogen fuel, of which less is available. Instead, lifetimes become longer as Z increases, because the hydrogen burning models get fainter due to the higher opacity, while the hydrogen fuel reservoir remains virtually unchanged. Table 1.2 lists the values of the coefficients of the following relation:

$$\log t = a \log^2 M_\odot + b \log M_\odot + c ,\qquad(1.7)$$

Table 1.2 Coefficients of Eq. (1.7) for various chemical compositions, resulting from the least square fit of the hydrogen burning and the total lifetimes as a function of $M_\odot$. The fit covers the range $0.6 \leq M_\odot/M_\odot \leq 20$; columns 5 and 9 report the average relative accuracy on the evolutionary lifetimes at the TO and at the first thermal pulse or central carbon ignition. We caution the reader that the accuracy of the fits at the low-mass end is substantially worse than what is reported in columns 5 and 9. The YZVAR database (Bertelli, G. et al. 2008, Astron. Astrophys., 484, 815; 2009, Astron. Astrophys., 508, 355). has been used to derive the coefficients.

Z, Y		H burning lifetime				Lifetime up to first pulse or C ignition			
	a	b	c	$\left(\frac{\Delta t_{TO}}{t_{TO}}\right)$		a	b	c	$\left(\frac{\Delta t_{tot}}{t_{tot}}\right)$
0.0001, 0.23	0.8751	−3.240	9.754	0.041		0.8130	−3.194	9.809	0.022
0.0001, 0.40	0.8248	−3.007	9.288	0.049		0.7784	−2.992	9.371	0.023
0.001, 0.26	0.8192	−3.168	9.705	0.040		0.7679	−3.147	9.776	0.031
0.001, 0.40	0.7590	−2.951	9.310	0.049		0.7363	−2.976	9.409	0.032
0.004, 0.26	0.8085	−3.230	9.796	0.040		0.7534	−3.211	9.875	0.042
0.017, 0.26	0.8071	−3.423	10.029	0.038		0.7260	−3.369	10.104	0.053
0.04, 0.26	0.7466	−3.516	10.213	0.049		0.6694	−3.457	10.281	0.063
0.04, 0.40	0.7649	−3.361	9.767	0.073		0.7324	−3.371	9.866	0.063
0.04, 0.46	0.7362	−3.243	9.554	0.082		0.7427	−3.303	9.669	0.068
0.07, 0.40	0.7322	−3.420	9.855	0.079		0.7065	−3.437	9.948	0.066

adopted to describe the evolutionary lifetimes (in years) as a function of initial mass (in $M_\odot$). The parabolic fit over the whole considered mass range is a rather drastic approximation; nevertheless these analytic relations can be useful to estimate evolutionary lifetimes and their dependence on composition.

As mentioned in the previous section, the values of mass defining the low, intermediate and high-mass range depend somewhat on chemical composition. For example M_{Hef} decreases with Y increasing and with Z decreasing. Therefore, an extended RGB on the HRD of a stellar population traces the presence of evolved stars with mass lower than $\sim 2.1\ M_\odot$, for solar composition, or with mass lower than $\sim 1.5\ M_\odot$ if $Z = 0.001$, $Y = 0.4$. However, evolutionary lifetimes at given mass also depend on composition, and, by and large, an extended and well populated RGB is developed in stellar populations older than ~ 1 Gyr almost irrespective of chemical composition.

At fixed initial mass, zero age main sequence models are hotter and brighter for a higher helium content, and/or a lower metallicity. Indeed, the locus of the corresponding stars on the HRD (ZAMS) is used to infer the composition of the target stellar population, and is virtually the only way to estimate the helium content of these stars, if Z is known from spectroscopy. The puzzling composition of multiple stellar populations in some galactic globular clusters has been derived from ZAMS fitting.

Figure 1.5 shows some evolutionary tracks of intermediate-mass stars for three different initial compositions. The described trend of the ZAMS with metal content is readily visible, together with some other properties already mentioned: at very low metallicity (Figure 1.5a) the entire core helium burning phase occurs in the blue part of the HRD, and the thermal runaway in the envelope which brings the model star to the Hayashi track is delayed to the very latest stages. As metallicity increases, the luminosity decrease associated with the thermal runaway gets more and more pronounced: this reflects the progressively higher opacity, and then radiative energy trapping in the envelope. At the same time, at higher metallicity it is more difficult to produce extended blue loops: the 3 and 4 $M_\odot$ tracks at $Z \sim 2 Z_\odot$ do not present a blue loop at all, and the loop of the 5 $M_\odot$ track is just alluded to. The models in Figure 1.5 are computed adopting classical recipes for the input physics, and even slight modifications of the assumptions lead to dramatic variations of the tracks shape. For example, intermediate-mass models with $Z = 0.04$, $Y = 0.40$ which adopt a modest overshooting from the convective core lack the loops completely, and the core helium burning phase is totally spent close to the Hayashi line.

Figure 1.6 illustrates the effect of chemical composition on core helium burners of low-mass. The gray tracks in Figure 1.6a show that at solar composition this evolutionary phase is completely spent in the red, even for masses as low as 0.55 $M_\odot$. Conversely, at low metallicity (black lines), low-mass core helium burners are blue, opening the possibility of producing extended HBs in old stellar populations. The tracks in Figure 1.6b show the effect of enhancing the helium abundance: at high metallicity the core helium burning phase is completely spent close to the Hayashi line for a solar helium abundance, but if the helium abundance is high, a blueward

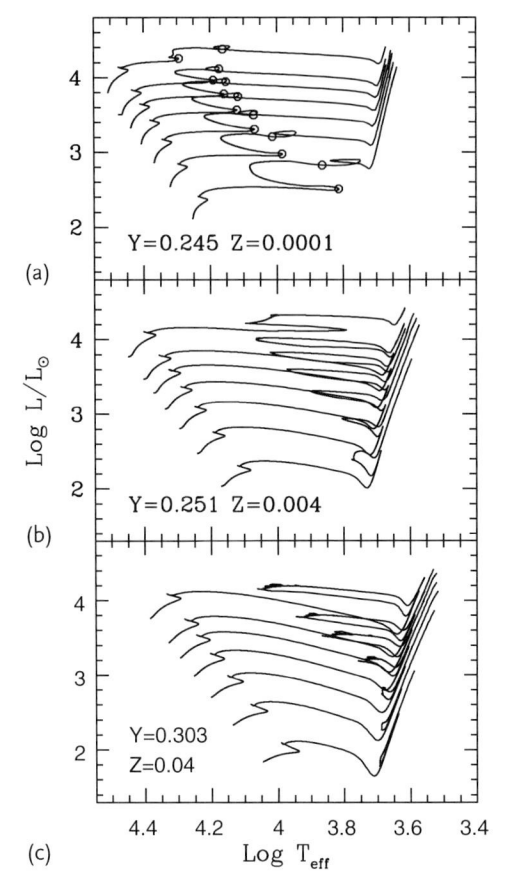

Figure 1.5 (a–c) Evolutionary tracks of intermediate-mass stars ($M/M_\odot$ = 3, 4, 5, 6, 7, 8, 10) for different compositions, as labeled. In (a), the open circles mark the start and the end of the core helium burning phase. Drawn using the BaSTI database (Pietrinferni, A. et al. 2004, *Astrophys. J.*, 612, 168).

excursion occurs during the core helium burning phase which is very wide for low-mass stars. Therefore the production of blue HB stars in old stellar populations can be achieved either assuming heavy mass loss on the RGB, or a high helium content, or both. Actually, the existence of multiple stellar populations with different helium content in NGC 2808 has been first suggested on the basis of its HB stars distribution, and confirmed later from the multiple MSs. Mass loss and helium abundance have indeed an important impact on the HRD of stellar populations, as well as on their spectral energy distribution, since stars in the core helium burning phase provide an important contribution to the total light.

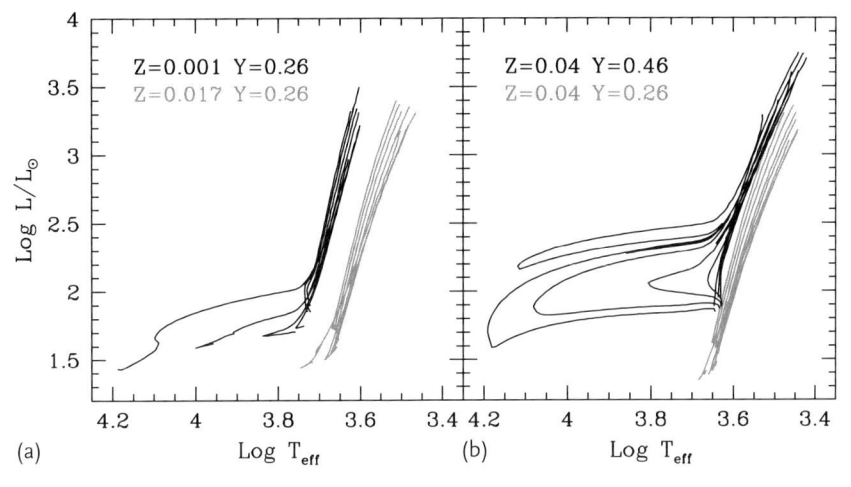

Figure 1.6 (a,b) Evolutionary tracks of low-mass stars during the core helium burning and early AGB phases. The chemical composition is labeled. Initial model masses are $M/M_\odot = 0.55, 0.6, 0.65, 0.7, 1, 1.2, 1.4, 1.6$ except for the ($Z = 0.04, Y = 0.46$) set, for which the 1.6 $M_\odot$ track ignites helium under nondegenerate conditions. Drawn using the YZVAR database (Bertelli, G. *et al.* 2008, *Astron. Astrophys.*, 484, 815; 2009, *Astron. Astrophys.*, 508, 355).

1.1.4
The Asymptotic Giant Branch Phase

Shortly after helium exhaustion in the core of stars less massive than $\sim 8\ M_\odot$, the carbon-oxygen (C-O) core has contracted to such high densities that electron degeneracy sets in. In the meantime, helium has ignited in a shell in a mildly violent fashion, causing expansion away from it in both directions, and hydrogen burning in its own shell is effectively switched off. The star now enters the asymptotic giant branch (AGB) phase, a name that comes from stars in this phase approaching asymptotically the RGB in the color-magnitude diagrams of globular clusters, as seen in Figure 1.7 for NGC 6752, with the main evolutionary stages being identified by their acronyms.

In the early part of this phase (E-AGB), the helium burning shell progressively eats through the helium buffer zone between the two shells, approaching in mass the inactive hydrogen shell. In low-mass stars this approach of the two shells takes place even if the hydrogen shell reignites.

This early fraction of the AGB phase comes to an end when the helium burning shell has consumed almost all the helium buffer zone, getting very close in mass to the hydrogen shell. Then helium burning in its shell almost dies out, and as a result matter above it starts to contract. As a consequence of this contraction, the hydrogen shell reignites, and starts moving outwards in mass: the mass of the helium buffer zone starts to increase as the hydrogen shell eats outward in mass, while the helium shell is inactive.

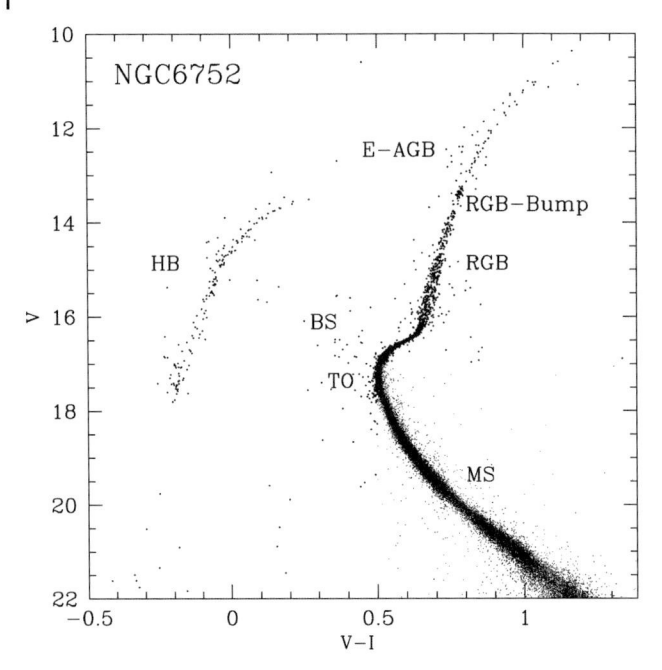

Figure 1.7 The $V-(V-I)$ CMD (HST magnitudes) of the globular cluster NGC 6752 with the main evolutionary phases labeled as follows: main sequence (MS), turnoff (TO), red giant branch (RGB), red giant branch bump (RGB-Bump), horizontal branch (HB), early asymptotic giant branch (E-AGB). A few blue stragglers (BS) are also indicated. Data to draw figure taken from Brown, T. et al. (2005, *Astron. J.*, 130, 1693).

This situation does not last for long. As the helium buffer zone grows, so do temperature and density in the helium shell, until this shell starts showing signs of reignition, first with some low-amplitude oscillations, until a first full scale helium shell flash develops. This is the beginning of the thermally pulsing phase of the AGB (TP-AGB), whose run for a 5 $M_\odot$ star is shown in Figures 1.8–1.9. The two shells cannot manage to burn together at the same time: when helium ignites, it does so so violently that the hydrogen shell is effectively switched off, and expansion of the helium buffer zone is so rapid that the helium burning itself is also quenched. During this sequence of events the star departs dramatically from thermal equilibrium until hydrogen can reignite, thermal equilibrium is restored, and another cycle begins. At the peak of the helium shell flash, the helium burning luminosity exceeds the surface luminosity by up to almost two orders of magnitude (cf. Figures 1.8–1.9).

In the specific TP-AGB sequence shown here, the temperature at the base of the convective envelope reaches very high values when hydrogen burning is active, to the extent that hydrogen burning proceeds vigorously near the inner boundary of the convective envelope. The situation, known as hot-bottom burning (HBB), can be described as the envelope convection starting within the outer part of the hydrogen burning shell. Meanwhile, mass is being lost from the surface, and the

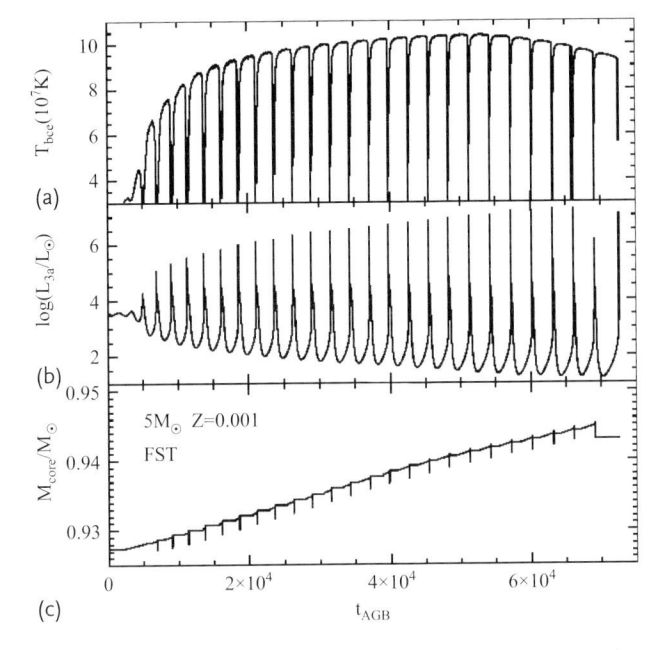

Figure 1.8 The time evolution of some relevant quantities during the thermally pulsing AGB phase of a 5 $M_\odot$ star: (a) the temperature at the base of the convective envelope, (b) the helium burning luminosity, and (c) the mass of the hydrogen-exhausted core. Notice the temporary decreases of this quantity, corresponding the third dredge-up episodes (source: Ventura, P. and D'Antona, F. (2005, *Astron. Astrophys.*, 431, 279)) (Reproduced with permission © ESO.).

stellar mass rapidly decreases. Whereas this thermal pulse behavior is generic to all TP-AGB stars, the onset and extent of the HBB in stellar models are extremely sensitive to how the mass loss process and the outer convection are described and parameterized. This specific aspect will be further discussed in the following.

Figure 1.8 also shows that shortly after each helium shell flash the mass of the hydrogen-exhausted core drops slightly. This is due to envelope convection being able to reach through the hydrogen-helium discontinuity, well into the buffer zone that had been just contaminated by the helium burning products during the previous helium shell flash. This event, which can repeat at each thermal pulse during the AGB phase, is called the third dredge-up, and brings to the surface fresh carbon and helium, as well as some neutron-capture elements. As carbon is dredged-up, its abundance can eventually exceed that of oxygen in the envelope and atmosphere of the AGB star, thus leading to the formation of a carbon star. Carbon (C-type) and oxygen-rich (M-type) stars have markedly different colors and spectra, and therefore their relative ratio can have important effects on the integrated spectrum and colors of stellar populations, given that stars in their TP-AGB phase can contribute a significantly large fraction of their total luminosity.

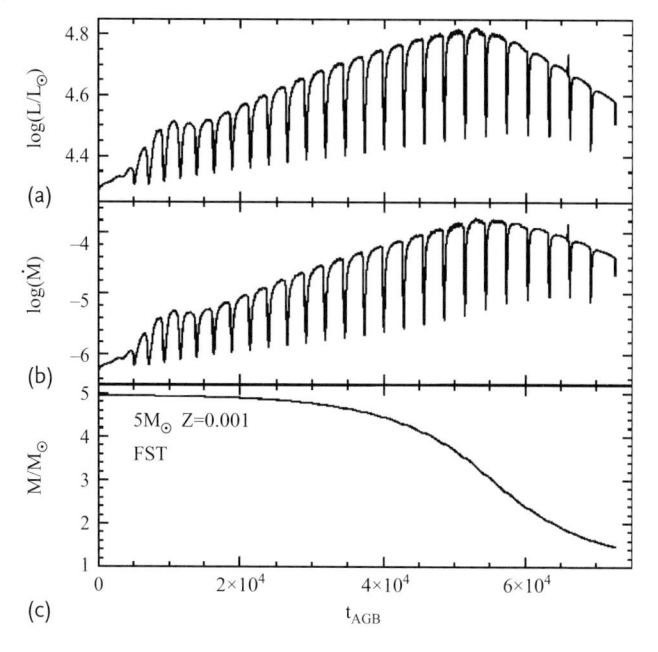

Figure 1.9 The same as Figure 1.8: (a) the surface (total) luminosity, (b) the mass loss rate (in $M_\odot$/yr), and (c) the steady decrease of the stellar mass as a result of mass loss by stellar wind. By comparing with Figure 1.8, notice that the surface luminosity increases by almost a factor of 3 while the mass of the core increases by only $\sim$ 0.01 $M_\odot$ (source: Ventura, P. and D'Antona, F. (2005, *Astron. Astrophys.*, 431, 279)) (Reproduced with permission © ESO.).

1.2
Strengths and Weaknesses of Stellar Evolutionary Models

The results of stellar evolutionary calculations depend on two main factors: (i) the sophistication of the stellar evolution code, and (ii) the input physics in model calculations. Virtually all active stellar model factories dispose of highly sophisticated codes, that are able to automatically adjust and optimize time steps and mass zoning ensuring that the outcome is independent of them. Thus, there is here no need to say more about the computational techniques of constructing stellar evolutionary sequences. More important is to examine the input physics, emphasizing those ingredients that are still subject to uncertainties, and highlight the possible consequences on the results of stellar population synthesis.

Stellar structure in quasistatic approximation is controlled by four differential equations, imposing hydrostatic equilibrium and mass continuity, and describing energy generation and energy transfer. Input physics appears in these equations as the (nuclear) energy generation coefficient $\epsilon(\rho, T, X_i)$, the radiative opacity $\kappa(\rho, T, X_i)$, the equation of state $P(\rho, T, X_i)$, and the adiabatic gradient $\nabla_{ad}(\rho, T, X_i)$, where P, ρ, and T are the pressure, density and temperature, respectively, and X_i

is the mass fraction of the *i* isotope, hence the X_i define the chemical composition. Collectively, these four functions represents the *microphysics* input, to which one can add the ionic diffusion coefficients when the diffusion equations for various ionic species are also included. Indeed, microphysics deals with elementary physical processes at the molecular, atomic, or nuclear levels, to be distinguished from a class of other physical processes that we call *macrophysics* as they all involve bulk motions of matter inside stars or out of them. Thus, macrophysics phenomena in stars include rotation, convection, mixing and mass loss. Mass transfer in binary systems can also be included in the macrophysics class of phenomena.

1.2.1
Microphysics

Nuclear Energy

The coefficient of nuclear energy generation results from the additive contribution of many kinds of nuclear reactions, involving up to several hundred different isotopes and thousands of nuclear reactions in the advanced evolutionary stages of massive stars. Energy sinks, via neutrino losses, are also part of this nuclear physics ingredient. From the point of view of stellar population synthesis, few major nuclear reactions have cross-sections with so high uncertainties that the emerging spectrum of a synthetic stellar population may be affected. Actually, only one reaction may be worth mentioning, that is, the $^{12}C(\alpha,\gamma)^{16}O$ reaction whose cross-section remains uncertain by a factor of ~ 2. The major consequence of this uncertainty concerns the nucleosynthesis yields of massive stars, which have no direct impact on synthetic stellar populations. On the other hand, by doubling the cross-section of this reaction the helium burning lifetime in the core is increased by $\sim 10\%$, and hence will have a very minor effect on the emergent flux of a synthetic stellar population.

Opacity

Opacity calculations for stellar mixtures have reached a high degree of sophistication and are generally considered extremely reliable. Publicly available codes allow opacities for *custom* mixtures to be calculated, including nonsolar proportions, in particular for α-elements with respect to iron-peak elements. Quite obviously, opacities tend to be more uncertain where it is more laborious to calculate them. This is the case in two different temperature-density regimes. Around 1 million degrees, most of the opacity comes from bound-bound and bound-free transitions of highly ionized metals, still with several bound electrons. Atomic physics approximations needed to deal with multielectron ions may affect the calculated opacities somewhat, but the net effect of these uncertainties on synthetic stellar populations should be modest. At much lower temperatures ($T \lesssim 3000$ K) molecules dominate the opacity, and its calculation becomes very complex and subject to several approximations. Molecular opacities then drive the stellar radius, hence determine the effective temperature, and can directly or indirectly influence the rate of mass loss during the red giant phases. Thus, in principle uncertainties in molecular opacities may have a strong effect on synthetic stellar populations. Yet, in practice uncertainties in molecular opacities are subsumed within the broader uncertainties affecting convection and mass loss, as cool stars with molecule-rich atmospheres have deep superadiabatic convection, and cool giants loose mass. Thus, uncertainties in this aspect of microphysics are overshadowed by uncertainties in macrophysics.

Equation of State and Thermodynamical Quantities
Through most of the stellar interior, and through most evolutionary stages of most stars, interactions between particles are weak enough for the perfect gas approximation to be perfectly adequate both in nondegenerate and degenerate conditions. This applies both to the equation of state $P(\rho, T, X_i)$, as well as to other thermodynamical quantities such as the thermal energy density and the adiabatic gradient. The exception is a domain in the temperature versus density diagram, at relatively low T and high ρ, where van der Waals forces are nonnegligible, electrons are partly degenerate, and yet ionization may still be incomplete. This regime requires approximations to be done to calculate the equation of state, $P(\rho, T, X_i)$, and the result is correspondingly somewhat uncertain. By good fortune (from the point of view of population synthesis), this $T - \rho$ regime is encountered only in low-mass stars ($M \lesssim 0.5\ M_\odot$), brown dwarfs and planets – all objects that contribute almost nothing to the energetics of stellar populations. Thus, within input physics ingredients, the equation of state does not introduce any important uncertainty in the energetic output of stellar populations.

In summary, uncertainties in the microphysics input in stellar model calculations are very minor, and do not appreciably affect the reliability of the models, and of the synthetic stellar populations built with them. As we shall see next, the story is different when it comes to macrophysics.

1.2.2
Macrophysics

Rotation
All stars rotate, yet all stellar models so far used in stellar population synthesis do not. Moreover, rotation is likely to generate magnetic fields, and also magnetic fields are ignored in the calculations of stellar evolutionary sequences. At first sight these limitations may appear to undermine the whole building of modern stellar evolution. However, Leo Mestel, one of the founders of stellar magnetism, used to say that magnetic fields make life simpler for stellar model makers. His provocative statement can actually be quite true.

One believes that stars are born with quite high angular momentum. Then, while settled on the main sequence stars also lose mass in a wind, even if at a relatively modest rate in most cases. Magnetic fields cause such winds to corotate up to a certain distance from the star, hence transfer much more angular momentum to the wind than in the case of no corotation. Moreover, inside stars magnetic fields dynamically connect layers in differential rotation, thus exchanging angular momentum among them. In essence, magnetic fields extract angular momentum from the stellar interior, transfer it to the wind, and the wind disperses it into space. The outcome is that stars rapidly lose most of their angular momentum, and the nonrotating, spherically symmetric approximation becomes quite viable. This should be particularly true for advanced stellar evolution stages, such as red giants, which shed a sizable fraction of their mass, and virtually all their residual angular momentum along with it.

This is not to say that rotation is never important in stellar evolution, but (perhaps with the exception of some fast rotating massive stars) it does not appear to play a leading role. For the time being, in dealing with stellar population synthesis we have no option other than using nonrotating stellar models. Still, we should be prepared to change this pragmatic attitude, if developments in theory and observations should dictate otherwise.

Convection

The Schwarzschild criterion is universally adopted to locate those regions where convection sets in, that is, all over $\nabla_{rad} > \nabla_{ad}$ convection develops. In the deep interior, convection is very efficient and to a very good approximation the actual temperature gradient that sets in in such regions (∇) is the adiabatic gradient, $\nabla = \nabla_{ad}$. Convective envelopes are present in stars with effective temperatures below ~ 9000 K, first very shallow and then deeper and deeper in stars close to the Hayashi line, for $T_e \lesssim 4000$ K. However, convection in the outer envelope is not so efficient, mostly because the thermal content of (cool) convective elements is lower than in the (hot) interior, and the actual gradient tends to be intermediate between the radiative and the adiabatic gradients ($\nabla_{ad} < \nabla < \nabla_{rad}$). In 1958 Erika Böhm-Vitense gave final form to the so-called *mixing length theory* (MLT) for superadiabatic convection, whereby the actual gradient ∇ can be calculated from local quantities and depending on an adjustable parameter: the mixing length.

The MLT of convection makes use of little more than dimensional arguments and definitely looks like a rather rough schematization of convective motions and the associated heat transfer. Shortly after the publication of the MLT, in 1963 Rudolf Kippenhahn in his lecture notes for a summer school on stellar evolution when coming to describe the MLT wrote: "…if one thinks about the state of the theory of convection then one should print parts of this chapter on a kind of paper which disintegrates already during the printing process." So strong was the perception of the inadequacy of the theory, and the expectation that it would soon be replaced by a more rigorous, if not exact theory. Ironically, half a century later everybody making stellar models is still using the infamous MLT, and the virtually full body of the stellar evolutionary sequences calculated in the intervening fifty years had the MLT as one of the key input physics ingredients. In the same span of time efforts where made at developing a more advanced theory, or modeling convection via numerical hydrodynamics, but these attempts did not pay dividends. The reasons for its success and resilience is that the MLT is simple to implement (just a few lines of code) and, after all, it works.

The adjustable parameter, the mixing length ℓ, is usually assumed to be proportional to the pressure scale height, that is, $\ell = \alpha_{ml} H_p$, with $H_p = -dr/d \ln P$, with α_{ml} expected to be on the order of unity. This parameter controls the efficiency of convection, which increases with it. Correspondingly, increasing α_{ml} the temperature gradient tends to be closer to the adiabatic gradient, with a strong effect only on the envelope itself. In particular, model stellar radii become smaller for higher values of α_{ml}. Conversely, the deep interior is almost completely unaffected by the value of α_{ml}, so the stellar luminosity and the time spent in various evolutionary phases are independent of the value of α_{ml}.

An empirical calibration is needed to fix the most appropriate value of this adjustable parameter. This is done by forcing models to reproduce the radius of stars whose radius is well known. Typically, one starts by reproducing the solar radius: a set of models for a 4.7 Gyr old $M_\odot$ star with $Z = Z_\odot$ are constructed for various values of the initial helium abundance Y and α_{ml}. Then helium is first forc-

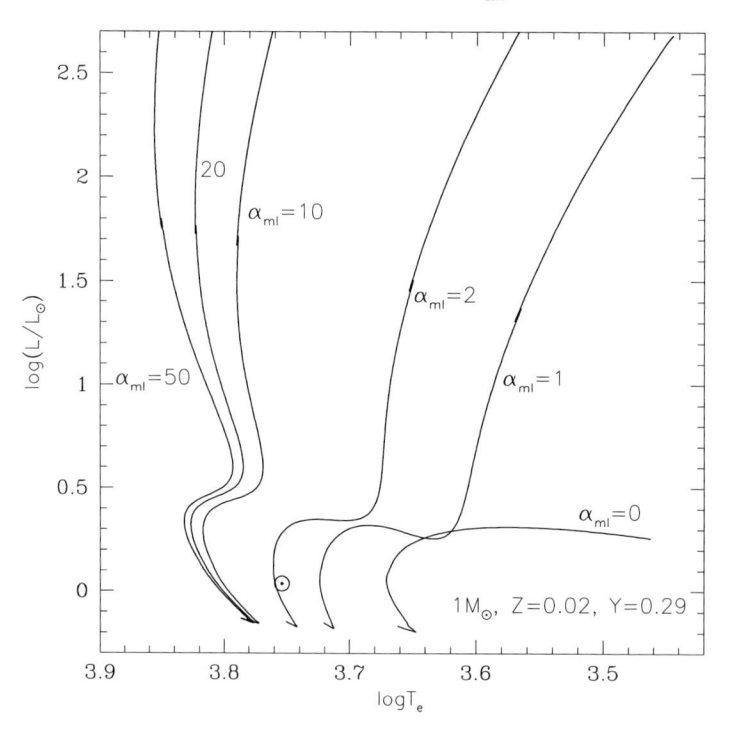

Figure 1.10 The evolution of a solar mass star from the main sequence to well into the red giant branch for several values of the mixing length parameter α_{ml}, and for solar composition. The position of the sun is also indicated, showing that a value of α_{ml} just a little smaller than 2 is required to reproduce the solar radius and effective temperature (source: Salaris, M. and Cassisi, S. (2005, Evolution of Stars and Stellar Populations, Wiley InterScience)).

ing the model to match the solar luminosity, that is, $L = L_\odot$, and having fixed Y the mixing length parameter is fixed by picking the value such that the model radius equals the solar radius. This simple calibrating procedure gives a solar helium $Y_\odot \simeq 0.28$, and $\alpha = 1.8$–2.0 (depending slightly on the adopted input physics). This is illustrated in Figure 1.10, showing how dramatic the changes are in the evolutionary path as the parameter α_{ml} is changed. Note that assuming $\alpha_{ml} = 0$ is equivalent to ignore convection altogether (i.e., $\nabla = \nabla_{rad}$) and $\alpha_{ml} \gg 1$ is equivalent to assume that convection is adiabatic (i.e., $\nabla = \nabla_{ad}$).

An immediate check that the calibration works as well for other stars can be done using local dwarfs and subdwarfs (low metallicity dwarfs) with precise trigonometric parallax, and the test works comfortably well. What is still surprising, is that having fixed α_{ml} by matching the solar radius, then evolutionary tracks from the main sequence all the way to the tip of the RGB give a virtually perfect match to the CMDs of globular clusters (see Figure 1.11). Now, red giants have much deeper convective envelopes compared to the sun, and radii several hundred times larger. It is its success over such a wide dynamical range that makes the MLT so dear to model makers, in spite of its simplicity contrasting so wildly with the complexity of the phenomena it is supposed to describe.

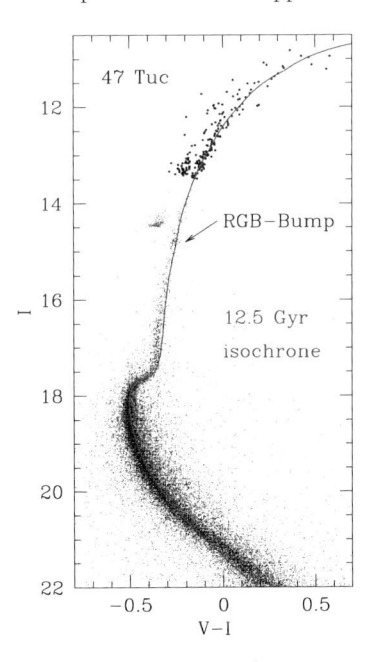

Figure 1.11 A 12.5 Gyr isochrone superimposed on the I–$(V-I)$ color-magnitude diagram of the globular cluster 47 Tuc. Magnitudes are in the HST system (F606W for V and F814W for I), and the isochrone refers to a metallicity $[Fe/H] = -0.70$ with an α-element enhancement $[\alpha/Fe] = +0.30$. The mixing length parameter has been calibrated to reproduce the solar radius, that is, $\alpha_{ml} = 1.9$. Data to draw figure taken from Brown, T. et al. (2005, *Astron. J.*, 130, 1693).

Mixing

The default assumption in making stellar models is that convective regions (with $\nabla_{rad} > \nabla_{ad}$) are fully mixed, hence chemically homogeneous, and radiative regions are unmixed. In some cases this simple prescription leads to embarrassments, such as a radiative region ($\nabla_{rad} < \nabla_{ad}$) may turn convective ($\nabla_{rad} > \nabla_{ad}$) if mixed, or vice versa. This situation is called *semiconvection* and there are physically sound algorithms to deal with it.

However, there is general agreement that the simple mixing criterion $\nabla_{rad} > \nabla_{ad}$ may not tell the whole story, even besides the cases when semiconvection develops. Indeed, it is the driving force that vanishes at the point where $\nabla_{rad} = \nabla_{ad}$, not the velocity of convective elements, which may *overshoot* this formal convective boundary. The mixing length theory is of no help to estimate the extent of such *overshooting* layers, hence of the chemically mixed region. This is because the MLT lacks the spatial resolution that would be required, and can actually produce physical nonsense if forced to make predictions anyway. The physics of the mixing process in the overshooting region is very complex, and the extent of this region cannot be predicted from first principles. This is likely to be an *erosion* process, in which much of the action is accomplished by extremely rare turbulent elements, in the extreme tail of the turbulence spectrum. Indeed, the situation bears some resemblance to the erosion of mountains on Earth, which is dominated by few, extremely rare but catastrophic storms, rather than by the day by day action of ordinary rains and winds. It may be worth quoting here the wisdom of Richard Feynman, when he said "a tremendous variety of behavior is hidden in the simple set of [hydrodynamics] equations ... the only difficulty is that we do not have the mathematical power today to analyze them except for very small Reynolds numbers." Unfortunately, stellar convection is in the high Reynolds number regime, hence fully turbulent, and we still fall short of mathematical power, in spite of the tremendous progress of our computers since Feynman gave his celebrated lectures.

Once more, lacking an adequate theory we are forced to parameterize our ignorance, and this is most often done by assuming that the extension of the overshoot region is proportional to the local pressure scale height, that is, $\ell_{ov} = \alpha_{ov} H_p$. Then the parameter α_{ov} can be fixed via an empirical calibration, but the effects are rather subtle, and no universally accepted value of α_{ov} currently exists. Estimated values range from vanishingly small $\alpha_{ov} \sim 0$ to ~ 0.2, thus favoring a relatively modest overshooting. Yet, with $\alpha_{ov} = 0.2$ the effects on stellar evolution are not negligible. By extending the size of the mixed convective cores, overshooting brings fresh fuel into the burning regions, and therefore prolongs the duration of the core hydrogen burning phase, compared to cases in which overshooting is neglected. For $\alpha_{ov} \simeq 0.2$ the duration of the hydrogen burning phase in the core is increased by $\sim 15\%$ compared to the no overshooting case, but the duration of the subsequent core helium burning phase is nearly halved, a result of the helium core being more massive, hence faster in consuming its fuel. Moreover, with bigger helium cores the luminosity during the power-down phase F-to-G (cf. Figure 1.1) does not drop enough to initiate the runaway deflation and first blue loops are suppressed. Thus, even a modest amount of overshooting may appreciably alter the mix of cool to hot stars, for those stars that develop convective cores during the main sequence phase ($M \gtrsim 1.2\ M_\odot$).

This is particularly important for massive stars, which are characterized by massive convective cores, hence for young stellar populations. Thus, with overshooting, models of young populations (say, younger than $\sim 10^8$ yr) have more hot main sequence stars per unit mass of the whole population, hence will produce more UV photons, compared to models that neglect overshooting. This will have an impact on the star formation rates (SFR) derived from either stellar counts or from fits to the spectral energy distribution, with overshooting models indicating lower SFRs compared to models without overshooting.

Estimates of the size of the overshooting region (usually quantified by the α_{ov} parameter), have been made by matching evolutionary tracks to the color-magnitude diagram (CMD) of star clusters, or to binary pairs. As such, the procedures find it difficult to disentangle overshooting effects from other effects that can alter the morphology of theoretical CMDs, for example color-temperature transformations and bolometric corrections. Thus, it would be of paramount importance to accurately measure by other means the size of mixed cores in as many bright main sequence stars as possible. Asteroseismology promises to do so, and this indeed should be its highest priority.

Overshooting can also happen at the lower boundary of the convective envelope, thus extending downwards the mixed region. This has less dramatic effects compared to overshooting from convective cores, but the surface composition could be seriously affected if mixing were to reach into the hydrogen burning shell, thus carrying up to the surface various products of advanced CNO processing and of proton-capture reactions. It is generally suspected that such an effect plays a role in developing some of the chemical anomalies observed among stars in globular clusters, but by and large this process is expected to have negligible effects on synthetic stellar populations.

Other forms of mixing may also take place in stars, being induced by stellar rotation. Meridional circulation is one possibility, another is mixing induced by turbulence originated by shear instability in differentially rotating stars. Another potential complication comes from rotation plus convection giving rise to magnetic fields. However, as already pointed out when examining the effect of rotation, stars are expected to lose a major fraction of their initial angular momentum, for the benefit of the spherically symmetric, nonrotating approximation.

A strong, empirical constraint on the extent of rotationally induced mixing comes from the so-called RGB bump, a sharp feature on the RGB luminosity function of globular clusters and other old stellar systems. This is produced by the temporary modest drop in luminosity that stars experience while ascending the RGB, when their hydrogen burning shell meets the chemical discontinuity left behind by the deepest penetration of envelope convection. This feature is clearly visible in Figure 1.10, as the small thickening of the tracks approximately midway in their RGB evolution, as well as in Figures 1.7 and 1.11. If rotational mixing were to smear out the mentioned discontinuity, or to locate it at different depths inside the star depending on star to star differences in rotation, then the RGB bump would disappear. Instead, the fact that it is observed to be so sharp demonstrates that no rotationally induced smearing takes place, and that all stars of given mass behave strictly in the same fashion. This suggests that rotational mixing does not play a major role in the evolution of near-solar mass stars.

Mass Loss

Mass loss was not generally recognized as a major factor in driving stellar evolution before the early 1970s, and the realization of its role in hot and cool stars had quite different histories. In hot stars its importance was recognized thanks to the early ultraviolet observations from space, that showed extremely strong, P-Cygni profiles for the resonant transitions of several abundant ions, with velocity widths of several thousand km s^{-1}. For red giants instead, optical spectra did not show much evidence for mass loss, apart from some quite weak absorption lines that were displaced some ~ 10 km s^{-1} with respect to photospheric lines. Thus, the fact that low-mass red giants had to lose a major fraction of their envelope was first inferred from the morphology of the CMDs of globular clusters.

Another major difference between mass loss in hot and cool stars is that the physical origin of the former is reasonably well understood and modeled, whereas that of the latter is still very poorly understood. For hot stars, it soon became clear that a great deal of momentum is transferred from the radiation field to the outflowing gas, thanks to the resonant absorptions. The theory of radiatively driven winds in hot stars was then developed, eventually incorporating huge numbers of transitions from several thousands energy levels. The resulting mass loss rate predictions are in fairly good agreement with the observations of massive hot stars and the nuclei of planetary nebulae. The mass loss rate from massive stars during their MS phase can be expressed as (from Vink, J.S. et al. (2001, Astron. Astrophys. 369, 574)):

$$\log \dot{M} = a + b \log \frac{L}{10^5} + c \log \frac{M}{30} + d \log T_{20} + d_1 \log^2 T_{20} + 0.85 \log \frac{Z}{Z_\odot}, \tag{1.8}$$

where $T_{20} = T_{\text{eff}}/20\,000$ K, $\dot{M}$ is in $M_\odot$/yr, and the star's luminosity and mass are in solar units. This relation describes the results of a grid of wind models computed for metallicities between 1/30 and 3 times the solar value and for effective temperatures between 12 500 and 50 000 K. The coefficients in Eq. (1.8) depend on the temperature regime: at high temperatures ($T_{\text{eff}} \gtrsim 25\,000$ K) $(a, b, c, d, d_1) = (-8.107, +2.194, -1.313, +7.507, -10.92)$, whereas at low temperatures $(a, b, c, d, d_1) = (-6.388, +2.210, -1.339, +1.07, 0.0)$. The mass loss rate increases with about the square power of the luminosity, and increases with metallicity, that is, the abundance of particles able to intercept such radiation. Mass loss in massive stars during their main sequence and blue supergiant phases has profound consequences on their evolution, to the extent that the whole hydrogen-rich envelope can be lost in the wind, with the formation of Wolf–Rayet (WR) stars.

WR stars expose their hydrogen-exhausted helium layers where almost all CNO elements have been converted to nitrogen (WN stars), or even layers once in the convective helium burning core, hence enriched in carbon and oxygen (WC stars). Even if radiatively driven winds may not completely peel stars of their hydrogen envelope, these stars have other chances to develop WR characteristics. When trying to cross the HRD on their way to the red supergiant region, the most massive stars enter the area occupied by the so-called luminous blue variables (LBV), which shed large amounts of mass through a series of ejection events whose prototype is offered by the famous η Carinae. One suspects that these mass ejections may signal that even convection is unable to carry out all the extremely high luminosities ($\gtrsim 10^5 L_\odot$) that are released by the burning core of these stars (suffice to say that large density inversions appear to be unavoidable in the outer layers of their shallow convective envelope). In this way, LBVs may well lose all their envelope and become WR stars finally exploding as supernovae of the Type Ib or Ic, or even producing γ-ray bursts (GRB) and/or hypernovae. For those that succeed to cross the HRD and become red supergiants, there is still another chance to become a WR star: losing part of their envelope in a red supergiant wind stars may evolve back to high temperatures, then losing the rest of it in a hot star wind. Finally, other massive stars can keep a hydrogen envelope till core collapse triggers a supernova explosion of Type II.[2]

Thus, several possible evolutionary paths can be followed by massive stars, such as:

- MS-BSG-LBV-WR-SNIb,c+GRB/Hypernova+GRB
- MS-BSG-WR–SNIb,c
- MS-BSG-RSG-WR–SNIb,c
- MS-BSG-RSG-BSG-SNII
- MS-BSG-RSG-SNII.

It is likely that these various options are followed in order of decreasing stellar mass, with for example, those producing GRBs being at the top end of the mass spectrum (say, above $\sim 60\ M_\odot$), and those ending as Type II supernovae being less massive than ~ 20–$30\ M_\odot$ (i.e., in the range 8–$10 \lesssim M \lesssim 20$–$30 M_\odot$). However, the precise mass ranges in which the various options are realized remain largely conjectural, together with their likely dependence on metallicity.

As mentioned above, the early recognition of mass loss in red giants as an important factor in stellar evolution was quite indirect, as it came from a comparison of globular cluster CMDs with stellar evolution models. Evolution at constant mass of stars in the appropriate mass range to fit the CMD of these clusters from the main sequence to the RGB tip ($0.8 \lesssim M \lesssim 1\ M_\odot$) would spend all of their core helium burning phase as *red clump* giants, failing to populate the full horizontal branch (HB) that in many clusters extends in the blue to temperatures

---

2) The distinction between supernovae of Type I (Ia, Ib and Ic) and Type II rests on the prominence of hydrogen (Balmer) lines in Type II SNe, and on their absence in Type I SNe, signaling that the progenitor still has an hydrogen envelope upon explosion, or lacks it completely.

well above 10 000 K. Only substantially less massive models could populate such extended HBs, which argued for some $\sim 0.2\,M_\odot$ being lost on the RGB before helium ignition in the core. Moreover, without additional mass loss during the subsequent AGB phase, $\sim 0.6\,M_\odot$ post-HB stars would reach over one magnitude brighter than the brightest stars known in galactic globular clusters. Such bright stars are simply not there. To prevent their formation, an additional $\sim 0.1\,M_\odot$ must be lost during the AGB phase, thus eventually leaving ~ 0.5–$0.55\,M_\odot$ white dwarf remnants.

On the observational side, in the early 1970s evidence accumulated for winds in red giants and supergiants, in the form of weak circumstellar absorption lines of neutral or singly ionized elements such as Ca, Na, Sr, K, Cr, Mn, and others. Some 40 years ago a collection of over 100 high-resolution red giant spectra from the Hale Observatories was analyzed, distilling from them a best fit formula relating the mass loss rate to the basic stellar parameters (from Reimers, D. (1975, Mem. Soc. R. Sci. Liege, 8, 369)):

$$-\dot{M} = 4 \cdot 10^{-13} \frac{L}{gR} \quad \left(\frac{M_\odot}{\text{yr}}\right), \qquad (1.9)$$

where luminosity, surface gravity and radius (L, g and R) are in solar units, and the numerical coefficient was said to be uncertain by a factor of ~ 3 either way. Taking this relation at face value, and integrating it along an RGB sequence it was soon realized that too much mass would be lost in the wind, globular cluster stars would lose the whole envelope before reaching the tip of the RGB, thus failing to ignite helium with the result that there would be no HB at all. Instead, placing an adjustable parameter (η) in front of this relation then model HBs resembling those of globular clusters could be realized by adopting $\eta \simeq 0.3$–0.4. As an additional bonus, the same value of η allows stars to lose another $\sim 0.1\,M_\odot$ during the AGB phase, then solving also the problem of the nonexisting very bright AGB stars in globular clusters. The final mass of the white dwarf remnants of $\sim 0.9\,M_\odot$ stars is then constrained to a very narrow range. On one side, the lowest possible mass of the WD remnant is set at ~ 0.51–$0.53\,M_\odot$ by the core mass at the helium flash, plus the advancement of the hydrogen shell during the subsequent HB and AGB phases. On the other side, an upper limit is set at ~ 0.52–$0.54\,M_\odot$ by the constraint that the AGB maximum luminosity in globular clusters does not exceed the RGB tip luminosity (except in metal-rich globulars). Thus, the mass of the white dwarf remnants of $\sim 0.85\,M_\odot$ stars must be $0.53 \pm 0.01\,M_\odot$, in agreement with the spectroscopic masses of white dwarfs in globular clusters.

The enduring success of this formula rests on its simplicity and on its double match to globular cluster HB and AGB extensions. But this is hardly the whole story. Indeed, serious difficulties arise if applying the formula to more massive AGB stars, all the way to $\sim 8\,M_\odot$, and using a core mass-luminosity relation that was believed to apply universally to all AGB stars (from Paczyński, B. (1971, Annu. Rev. Astron. Astrophys., 9, 183)):

$$L \simeq 59\,500\,M_c - 30\,900\,, \qquad (1.10)$$

with L and M_c in solar units, and M_c is the mass of the hydrogen-exhausted core. With $\eta \simeq 0.3$–0.4 stars less massive than $\sim 4\ M_\odot$ lose their whole envelope before the core reaches the Chandrasekhar limit of $\sim 1.4\ M_\odot$, and die as CO white dwarfs. Instead, this limit would be reached by $M \gtrsim 4\ M_\odot$ stars, carbon would be ignited in the core under extremely degenerate conditions, resulting in a carbon deflagration or detonation that would destroy the whole star in a supernova event. Moreover, with the core of so many stars growing up to $M_c = 1.4\ M_\odot$, Eq. (1.10) would imply the existence of very many bright AGB stars reaching up to $\sim 50\,000\ L_\odot$. Both these predictions are drastically contradicted by the observations.

First, in the now huge sample of supernova events there is no analog to a carbon deflagration/detonation event (the mechanism making Type Ia supernovae) taking place inside a still massive hydrogen-rich envelope, which then would be classified as a SN II event. If all stars with initial mass in the range $4 \lesssim M \lesssim 8\ M_\odot$ were to explode in this kind of SN I-1/2 (as it was dubbed), then they would instead dominate the SN statistics. Second, the first pioneering observations of the Magellanic Clouds in the near-IR showed that the brightest AGB stars are barely more luminous than $\sim 15\,000$–$20\,000\ L_\odot$, a factor ~ 3 fainter than predicted for AGB stars with a core approaching the Chandrasekhar limit. Clearly, intermediate-mass stars do not manage to grow a core as massive as $1.4\ M_\odot$, implying that either Eq. (1.9) does not apply to these stars, and/or other flaws affect the arguments leading to such grossly wrong prediction.

Before coming to clarify the origin of this discrepancy, it is worth noting that there are also other needs to entertain the notion of additional mass loss beyond the regime described by Eq. (1.9), hereafter referred to as the *normal wind*. Planetary nebulae are the immediate descendants of AGB stars, actually one of the *smoking guns* that such stars lose mass. The mass, radius and expansion velocity of planetary nebulae are typically $\sim 0.2\ M_\odot$, ~ 0.1 pc and ~ 20 km/s, respectively. Combining these values it is straightforward to infer that the mass loss rate during the event that produced the planetary nebula shell must have exceeded by several times $10^{-5}\ M_\odot/\text{yr}$, which is over a factor of ~ 10 higher than expected from Eq. (1.9). Thus, the AGB evolution must be terminated by a more dramatic mass loss episode, nicknamed *superwind*, to distinguish it from the normal wind.

Whereas a "superwind" may help to prevent AGB stars from reaching high luminosities, it does not appear to be the only actor at play. What happens during the AGB phase of stars initially more massive than $\sim 3\ M_\odot$ is indeed one of the most intricate situations encountered in stellar evolution. It was said about surface convection that the inner structure of stars, hence their luminosity, are virtually independent of the mixing length parameter α_{ml}. This is almost always true, with one notable exception: the TP-AGB phase of the more massive among the AGB stars. Here the temperature at the base of the convective envelope turns out to be almost a step function of α_{ml}. For low values of α_{ml} (say, around unity) such temperature is around a few 10^6 K, but above a threshold value this temperature increases steeply reaching over $\sim 10^8$ K for $\alpha_{ml} \sim 2$, in particular among the more massive AGB stars. Thus, the hot CNONe cycle efficiently operates at the base of the envelope, and the HBB process operates at its full regime as in the case shown in Figure 1.8.

Stars with active HBB (those more massive than $\sim 3\,M_\odot$ for $\alpha_{ml} \simeq 2$) do not follow the above core mass-luminosity relation anymore. Once HBB is ignited, the stellar luminosity increases steeply with time, the star expands and climbs quickly along the Hayashi line, and in a few 10^4 years runs into catastrophic mass loss leading to the ejection of the whole envelope.

Therefore, AGB stars with HBB is a case in which convection and mass loss together combine to determine the fate of the star. From the point of view of the model maker, there are at least three parameters to play with, α_{ml}, η, and some other describing the superwind. On top of this, lots of molecules form in the stellar atmosphere, and drastically affect the outer boundary conditions needed to start inward integrations. This is perhaps the most complex situation one encounters in making stellar models, and, on top of it, one has to say that we still do not know for sure even the basic physical processes driving the red giant wind and superwind.

Given the success of the theory of radiatively driven winds in hot stars, many believe that a similar process, radiation pressure on dust grains, drives the winds in cool giants. The circumstellar envelope and wind of many red giants abound indeed in dust grains, but it remains to be demonstrated whether grains drive winds or winds produce grains. If radiation pressure on dust grains were the sole mechanism driving winds, then one would expect a strong dependence of mass loss on metallicity, which does not appear to be the case. This is demonstrated by the trend with metallicity of the HB morphology of globular clusters. Very metal-poor globular clusters ([Fe/H] $\simeq -2.3$) have blue HBs, and to produce them $\sim 0.2\,M_\odot$ must be lost during the RGB phase. Instead, metal-rich globulars ([Fe/H] $\simeq 0$) have red HBs, hence their RGB stars cannot lose more than $\sim 0.3\,M_\odot$, otherwise their HB would be blue. Thus, mass loss rates during the RGB phase must scale with metallicity at a very, very small power.

This argument suggests that a physical mechanism other than radiation pressure on grains drives red giant winds, an obvious candidate being a *thermal wind*, similar to the solar wind. The extended convective envelope of red giants is indeed a powerful source of acoustic waves, and likely generates magnetic fields, hence all sorts of magnetohydrodynamical waves are likely to emerge from below the photosphere. All these waves must dissipate above the photosphere, thus heating a chromosphere or corona that eventually may evaporate in a wind. But obviously this kind of process is so complicated that it is currently impossible to predict the interesting quantity, that is, the mass loss rate, even in the case of our own sun, let alone for red giants.

In any event, AGB stars are likely to first support this kind of wind, that is, the same experienced by RGB stars, before running into a superwind regime. But then, what about the physical origin of the superwind? As AGB stars grow in luminosity and radius, their envelope becomes pulsationally unstable and stars appear as long period variables (LPV), or *Miras*, with periods of a few to several hundred days. Pulsation is likely to enhance mass loss, perhaps dramatically, perhaps favoring grain formation closer and closer to the photosphere. This may be the origin of the superwind, but we do not know for sure.

All these theoretical uncertainties lead to the conclusion that the whole AGB phase needs to be calibrated empirically. We know the physical processes that drive the evolution of AGB stars: besides thermal pulses and the dredge-up of carbon, AGB stars can experience HBB, envelope pulsation, winds and superwinds, but all these *macrophysics* ingredients cannot be usefully modeled from first principles. So, the duration of the AGB phase, its energetic output, and its contribution to the spectral energy distribution (SED) of stellar populations of different ages and metallicities all need to be calibrated empirically, using adequate observational datasets.

Binarism

A large fraction of stars, probably more than $\sim 50\%$, are born in binary or multiple systems, and a large fraction of them strongly interact with their companions in the course of their evolution. Mass transfer can take place, common-envelope stages in close binaries can result in envelope ejection, and the course of evolution is dramatically affected. This can prevent new fuels from being ignited, or make such fuels burn at radically different effective temperatures and luminosities compared to single stars of the same mass. This means that the binary fraction of a stellar population emits radiation with a different (perhaps radically different) SED compared to the single star component.

Mass transfer in binaries and common envelope phenomena are hydrodynamical processes which are difficult to model, and therefore a simple parameterization is commonly used to study the evolution of binary systems. An instructive example of how this can be done is discussed in detail in Chapter 7 on supernovae.

Existing libraries of synthetic stellar populations all ignore binaries. Including all sorts of binaries in a comprehensive and realistic fashion would be a monumental effort, and would bring in still more parameters. This is not likely to happen anytime soon, but binaries should not be completely forgotten when using synthetic stellar population tools. In particular, the SED of the binary component may extend in the UV and IR wings above that of the single star component, dominating the emission at those wavelengths. Some examples of how binaries can affect the SED of galaxies are discussed at the end of Chapter 5.

1.3
The Initial Mass-Final Mass Relation

Stellar populations form from the ISM and return mass to it via stellar winds and supernova explosions. In this metabolism, one important ingredient is the initial mass-final mass (IMFM) relation, $M_f = f(M_o)$, and the implied total mass loss $\Delta M = M_o - M_f$ suffered by stars in the course of their evolution. It has been mentioned above that a fairly precise value of the final mass can be set at $0.53 \pm 0.01\ M_\odot$ for $M_o = 0.85\ M_\odot$. The most massive single white dwarfs known, the likely remnants of $\sim 8\ M_\odot$ stars, have mass of $\sim 1.2\ M_\odot$. Joining linearly these two points, yields the IMFM relation:

$$M_f = 0.094\, M_o + 0.45, \tag{1.11}$$

which holds for the initial mass interval $0.85 \lesssim M_o \lesssim 8\, M_\odot$, and is in excellent agreement with the IMFM relation derived from the observational white dwarf mass-turnoff mass relation for open and globular clusters. As mentioned before, in a narrow mass range around $\sim 9 M_\odot$ (say, $8 \lesssim M_o \lesssim 11\, M_\odot$) stars may leave either an ONeMg white dwarf, or undergo core collapse depending on the strength of stellar winds. In both cases the mass of the remnant is ~ 1.3–$1.4\, M_\odot$. Above this range stars definitely undergo core collapse and leave either a neutron star, for $M_o < M_{BHR}$, or a black hole remnant for $M_o > M_{BHR}$. It is commonly assumed that the mass of the neutron star remnant is $\sim 1.4\, M_\odot$. The masses of the black hole remnants are poorly known, as in fact is M_{BHR} itself. Current reasonable choices are $M_{BHR} = 30$–$50\, M_\odot$, and $M_f = M_{BH} = (0.3$–$0.5)\, M_o$, but occasionally also lower values of M_{BH} are indicated. The truth is that this is a quantity quite poorly constrained by either theory or observations.

The theory of stellar structure and evolution is one of the most successful theories of complex astrophysical systems. It incorporates a major fraction of the accumulated knowledge of atomic and nuclear physics, as well as the behavior of matter in an extremely wide range of physical conditions, ranging from very diluted perfect gases to nuclear matter at extremely high densities. The outcome of evolutionary calculations consists of an impressive set of rich and detailed predictions, concerning the structure, evolution and outcome of stars in the whole range of masses and compositions that Nature has been able to produce. The resulting body of data represents the pedestal upon which rests the theory of stellar populations, its incarnation into synthetic spectral energy distributions, and ultimately their applications to galaxies all the way to the highest redshifts. Much of what we know today of such galaxies, their masses, star formation rates, ages and chemical enrichment, ultimately rests on the results of stellar evolution theory.

Ingredients in the evolutionary calculations that describe processes involving bulk motion of matter (such as convection, mixing, mass loss, and so on) contain most of the uncertainty affecting stellar evolutionary sequences, and need to be calibrated *vis à vis* relevant observations. These *macrophysics* processes are in most cases exceedingly complex to model, and progress over the last several decades has been extremely slow, if there has been progress at all (suffice to say the case of superadiabatic convection and the mixing length theory). Convection, mixing, and mass loss have a strong influence on the temperatures of stars in the course of their evolution, on stellar lifetimes in various evolutionary phases, and on the ultimate fate of stars. Thus, the algorithms and associated parameters used to describe these processes have profound effects on the SED of synthetic stellar populations.

When using on-line libraries of synthetic stellar populations it is important to check what assumptions were made in constructing the stellar evolutionary sequences and isochrones that have been used to calculate synthetic colors and spectra. In particular, the critical questions to ask are: how was the mixing length calibrated? Did the stellar models include overshooting, and if so how was it calibrated? What assumptions were made on mass loss, in particular for the red giant stages? How was the AGB contribution calibrated?

Further Reading

Books

Schwarzschild, M. (1958) *Structure and Evolution of the Stars*, Princeton University Press. The first two chapters of this classical book on structure and evolution of the stars still remain an excellent introduction to the fundamentals of stellar evolution theory.

More recent textbooks on the subject include:

de Boer, K.S. and Seggewiss, W. (2008) *Stars and Stellar Evolution*, EDP Sciences.

Iben Jr., I., Renzini, A., and Schramm, D. (1977) Saas-Fee Lectures on *Advanced Stages in Stellar Evolution*, Observatoire de Genéve.

Maeder, A. (2009) *Physics Formation and Evolution of Rotating Stars*, Springer.

Salaris, M. and Cassisi, S. (2005) *Evolution of Stars and Stellar Populations*, Wiley InterScience.

Classical Papers

Iben Jr., I. (1965) *Astrophys. J.*, **142**, 1447.

Iben Jr., I. (1966) *Astrophys. J.*, **143**, 505, 516 and 843).

Papers on More Specific Aspects of Evolution

AGB evolution:

Blöcker, T. (1998) *Astron. Astrophys.*, **297**, 727.
Iben Jr., I. and Renzini, A. (1983) *Annu. Rev. Astron. Astrophys.*, **21**, 271.
Marigo, P. and Girardi, L. (2007) *Astron. Astrophys.*, **469**, 239.
Renzini, A. and Voli, M. (1981) *Astron. Astrophys.*, **94**, 175.
Ventura, P. and D'Antona, F. (2005) *Astron. Astrophys.*, **439**, 1075.
Ventura, P. and D'Antona, F. (2009) *Astron. Astrophys.*, **499**, 835.

Thermal instabilities in stellar envelopes:

Renzini, A. *et al.* (1992) *Astrophys. J.*, **400**, 280.

Renzini, A. and Ritossa, C. (1994) *Astrophys. J.*, **433**, 293.

Low-mass stars:

Renzini, A., and Fusi-Pecci, F. (1998) *Annu. Rev. Astron. Astrophys.*, **26**, 199.
Rood, R.T. (1972) *Astrophys. J.*, **177**, 681.
Sweigart, A. and Gross, P. (1976) *Astrophys. J. Suppl.*, **32**, 367.
Sweigart, A. and Gross, P. (1978) *Astrophys. J. Suppl.*, **36**, 405.
Sweigart, A.V., *et al.* (1990) *Astrophys. J.*, **364**, 527.

Mass Loss

Castor, J.I. *et al.* (1975) *Astrophys. J.*, **195**, 157.
Chiosi, C. and Maeder, A. (1986) *Annu. Rev. Astron. Astrophys.*, **24**, 329.
Fusi Pecci, F. and Renzini, A. (1976) *Astron. Astrophys.*, **46**, 447.

Kudritzki, R. and Puls, J. (2000) *Annu. Rev. Astron. Astrophys.*, **38**, 613.
Reimers, D. (1975) *Mém. Soc. R. Sci. Liege*, **8**, 369.
van Loon, J.Th. *et al.* (2005) *Astron. Astrophys.*, **438**, 273.

Convective Overshooting and Its Influence on Stellar Evolution

Bertelli, G. *et al.* (1985) *Astron. Astrophys.*, **150**, 33.
Maeder (1976) *Astron. Astrophys.*, **47**, 389.

Testa, V. *et al.* (1999) *Astron. J.*, **118**, 2839.
Vandenberg, D.A. and Bergbusch, P.A. (2006) *Astrophys. J. Suppl.*, **162**, 375.

Advanced Stages in the Evolution of Massive Stars

Heger, A. *et al.* (2003) *Astrophys. J.*, **591**, 288.

Meynet, G. *et al.* (1994) *Astron. Astrophys. Suppl.*, **103**, 97.

Initial Mass-Final Mass Relation

Kalirai, J.S. *et al.* (2008) *Astrophys. J.*, **676**, 594.

Weidemann, V. (2000) *Astron. Astrophys.*, **363**, 647.

Publicly Available Sequences of Stellar Evolutionary Models and Isochrones

- Dartmouth: http://stellar.dartmouth.edu/~models/ (last accessed 2011-06-16)
- Geneva: http://obswww.unige.ch/Recherche/evol/Geneva-grids-of-stellar-evolution
- Padova: http://stev.oapd.inaf.it/ (last accessed 2011-06-16)
- Teramo: http://193.204.1.62/index.html (last accessed 2011-06-16)
- Victoria Regina: http://www3.cadc-ccda.hia-iha.nrc-cnrc.gc.ca/community/VictoriaReginaModels
- Yonsei-Yale: http://www.astro.yale.edu/demarque/yyiso.html (last accessed 2011-06-16)

2
The Fundamentals of Evolutionary Population Synthesis

In this chapter we present the fundamental properties of simple stellar populations (SSP), by which we mean an assembly of coeval single stars, all of the same chemical composition. Besides age and composition, only the initial mass function (IMF) needs to be specified in order to fully characterize the global properties of a SSP. Then, the results of the theory of stellar evolution – by describing the detailed evolutionary behavior of the constituent stars – can be used to predict several global properties of SSPs.

Galaxies are made of complex stellar populations, characterized by age and composition distributions, and can be regarded as collections of many SSPs. For this reason, it is worth familiarizing with simple populations before dealing with complex ones. Galaxies also contain binaries, and their possible contribution will be addressed separately.

2.1
The Stellar Evolution Clock

The lifetime of stars is a strong function of their initial mass $M_\circ$, and increases with decreasing $M_\circ$. For roughly 90% of their lifetime stars burn hydrogen at their center, and, on the HRD, they lie near the main sequence (MS). The MS turnoff (TO) is – at any age of the population – traditionally defined as the point of maximum effective temperature along the corresponding isochrone, which is the locus in the HRD occupied by stars of different mass but the same age. The TO is quite close to (but does not exactly coincide with) the point where stars exhaust hydrogen at their center, and beyond it they burn hydrogen in a shell.

Stellar evolution calculations provide the luminosity, effective temperature, and mass $M_{\rm TO}$ of stars at the TO, along with their colors and absolute magnitudes in whatever photometric system one is willing to adopt, all as a function of time (age) and composition. Some analytical approximations to tabular values of these quantities are particularly useful to grasp several basic properties of stellar populations. For example, the following relation provides a good fit to the mass of the star which

Stellar Populations, First Edition. Laura Greggio and Alivio Renzini.
© 2011 WILEY-VCH Verlag GmbH & Co. KGaA. Published 2011 by WILEY-VCH Verlag GmbH & Co. KGaA.

is exhausting hydrogen at its center and the age of the isochrone:

$$\log M_{TO}(t) = a \log^2 t + b \log t + c ,\qquad(2.1)$$

where M_{TO} is in $M_\odot$ units, t is in years, and a, b and c depend on the chemical composition. Table 2.1 reports the values of the coefficients which reproduce the characteristics of a specific database of tracks; for near-solar composition the relation is:

$$\log M_{TO}(t) = 0.0434 \log^2 t - 1.146 \log t + 7.119 .\qquad(2.2)$$

A similar relation can be fitted to the total evolutionary lifetime as a function of initial mass, which gives the mass of the dying star as a function of age, that is, of stars exploding as supernovae or about to become white dwarfs. Table 2.1 reports the coefficients also for this relation, which for near-solar composition becomes:

$$\log M_D(t) = 0.0379 \log^2 t - 1.048 \log t + 6.719 .\qquad(2.3)$$

Figure 2.1 shows M_D, M_{TO} and their time derivatives as functions of the age of the population. At any given time stars with $M_\circ < M_{TO}$ are still burning hydrogen at their center and lie close to the MS, stars with $M_\circ > M_D$ are already dead remnants, while stars in the narrow range $M_{TO} < M_\circ < M_D$ are venturing through their post-MS evolutionary stages. Hence, one does not commit a large error assuming that all post-MS stars have the same initial mass $M_\circ \simeq M_{TO} \simeq M_D$, and

Table 2.1 Coefficients of Eqs. (2.1) and (2.3) for various chemical compositions, resulting from the least square fit of the H burning and the total lifetime as a function of $M_\circ$. The fit covers the range $0.6 \leq M_\circ/M_\odot \leq 20$; columns 5 and 9 report the average relative accuracy on the evolutionary masses M_{TO} and M_D (the YZVAR (Bertelli, G. et al. 2008, Astron. Astrophys., 484, 815; 2009, Astron. Astrophys., 508, 355) database has been used to derive the coefficients listed).

Z, Y	Turnoff mass				Dying star mass			
	a	b	c	$\left(\frac{\Delta M_{TO}}{M_{TO}}\right)$	a	b	c	$\left(\frac{\Delta M_D}{M_D}\right)$
0.0001, 0.23	0.0637	−1.525	8.809	0.022	0.0582	−1.433	8.441	0.023
0.0001, 0.40	0.0758	−1.714	9.364	0.022	0.0691	−1.604	8.951	0.026
0.001, 0.26	0.0608	−1.472	8.553	0.021	0.0551	−1.374	8.159	0.023
0.001, 0.40	0.0692	−1.603	8.915	0.029	0.0631	−1.501	8.526	0.030
0.004, 0.26	0.0548	−1.362	8.077	0.023	0.0489	−1.260	7.660	0.026
0.017, 0.26	0.0434	−1.146	7.119	0.026	0.0379	−1.048	6.719	0.027
0.04, 0.26	0.0342	−0.969	6.325	0.030	0.0300	−0.895	6.026	0.029
0.04, 0.40	0.0419	−1.100	6.739	0.040	0.0380	−1.033	6.488	0.037
0.04, 0.46	0.0445	−1.143	6.850	0.046	0.0420	−1.099	6.699	0.042
0.07, 0.40	0.0361	−0.990	6.247	0.043	0.0330	−0.936	6.042	0.036

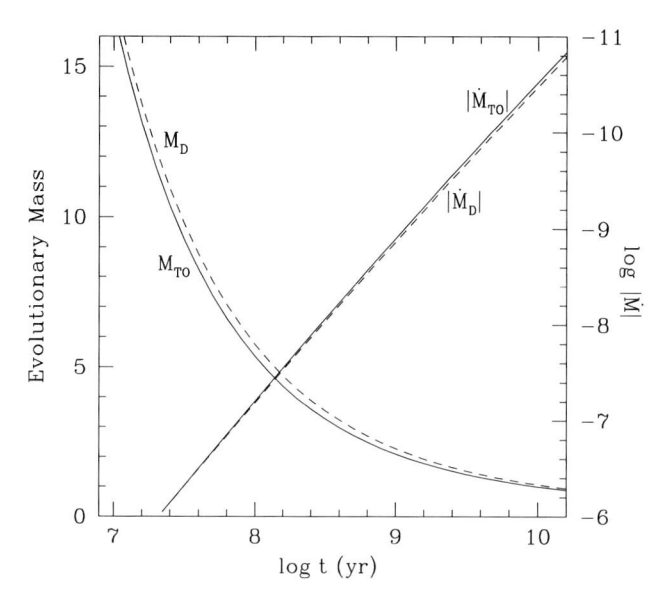

Figure 2.1 The time variation of the turnoff mass M_{TO} and its derivative (solid lines), and of the mass of dying stars M_D and its derivative (dashed lines) for solar composition stellar models, as derived from Eqs. (2.1) and (2.3). The small difference between M_D and M_{TO} at any age reflects the short duration of post-MS evolutionary phases relative to the MS lifetimes.

that the corresponding evolutionary line on the HRD for $M_\circ = M_{TO}$ nearly coincides with the isochrone. For the following sections it is assumed that $M_\circ \simeq M_{TO}$ for all stars in their active post-MS phases.

Analytical approximations can also be derived for the TO luminosity and temperature, or for their more directly observable quantities: the TO magnitude and color. For example, for an old population ($t \gtrsim 6$ Gyr) one can use (derived using the models of Dotter, A. et al. (2007, Astron. J., 134, 376)):

$$\log t \simeq (0.45 + 0.05\,[\text{Fe/H}])\, M_V^{TO} - 0.322\,[\text{Fe/H}] - 0.11\,[\alpha/\text{Fe}] + 8.155\,, \quad (2.4)$$

where t is the age in years, M_V^{TO} is the absolute visual magnitude of the MS turnoff, and [Fe/H] and [α/Fe] are respectively the iron abundance and the α-elements to Fe ratio in standard notations.[1] In the same age range the $(V - I)$ color at the turnoff can be approximated by the relation (derived using the models of Dotter, A. et al. (2007, Astron. J., 134, 376)):

$$(V - I)_{TO} = A \log t + B - 2.253\,, \quad (2.5)$$

where $A = 0.0473\,[\text{Fe/H}]^2 + 0.0023\,[\text{Fe/H}] + 0.2814$ and $B = -0.453\,[\text{Fe/H}]^2 + 0.1231\,[\text{Fe/H}] + 0.075\,[\alpha/\text{Fe}]$. These approximate relations assume $Y = 0.245 +$

---

1) $[\text{Fe/H}] = \log(n^{Fe}/n^H) - \log(n^{Fe}/n^H)_\odot$ and similarly for [α/Fe]. Note that the total metal abundance [M/H] is equal to [Fe/H] if elemental proportions are solar.

1.54 $Z_\odot$, are optimized for old populations and used here merely for illustrative purposes. Similar relations can be constructed for any desired age range, depending on the specific needs.

2.2
The Evolutionary Flux

The aging process of a SSP can be visualized as a flow of stars leaving the main sequence at progressively lower TO luminosities as time goes by, and moving through the various post-MS evolutionary stages, before turning into dead remnants. The rate of such flow is the *evolutionary flux* of the population:

$$b(t) = \phi(M_{TO})|\dot{M}_{TO}| = A M_{TO}^{-(1+x)}|\dot{M}_{TO}|, \quad \text{yr}^{-1} \qquad (2.6)$$

which is measured in stars per year, and ϕ is the IMF $= A M_\circ^{-(1+x)}$. With $x = 1.35$ over the entire mass range one has the classical Salpeter's IMF, whereas a *bottom-light* IMF with $x = 0.3$ for $M < 0.5\,M_\odot$ and $x = 1.35$ above is now adopted quite extensively, and this and following chapters refer to it as the *Salpeter-diet* IMF. The coefficient A is the scale factor fixing the size of the population; the other two factors have a different rôle in controlling the evolutionary flux. The IMF represents the reservoir of stars that is progressively depleted at a rate that is established by the stellar evolution *clock* of the population, that is, $|\dot{M}_{TO}|$. At young ages, that is, for large M_{TO} values, the clock is very fast, but its rate drops dramatically as time goes by and M_{TO} decreases. Indeed, $|\dot{M}_{TO}|$ drops by about five orders of magnitudes as M_{TO} decreases from $\sim 15\,M_\odot$ ($t \simeq 10^7$ yr) to $\sim 1\,M_\odot$ ($t \simeq 10^{10}$ yr), as shown in Figure 2.1. This trend is partly compensated by the modulating effect of the IMF, which increases for decreasing mass. The result of this interplay is illustrated in Figure 2.2, together with the death rate of the population, the latter being equal to the evolutionary flux computed for $M_D(t)$. Figure 2.2 shows that there is little difference between the rate at which stars leave the main sequence, and the rate at which they die: one can state that at any given age the evolutionary flux of an SSP is nearly constant along the post-MS evolutionary path. This is to say that to a fairly good approximation the flux $b(t)$ gives the rate at which stars enter or leave any particular post-MS evolutionary stage.

It immediately follows that the number N_j of stars in a generic jth post-MS stage of duration t_j is given by the number of stars entering (or leaving) that stage during a time t_j, hence:

$$N_j = b(t) t_j, \qquad (2.7)$$

a relation that states that N_j is proportional to the duration t_j and the coefficient of proportionality is just the evolutionary flux $b(t)$.

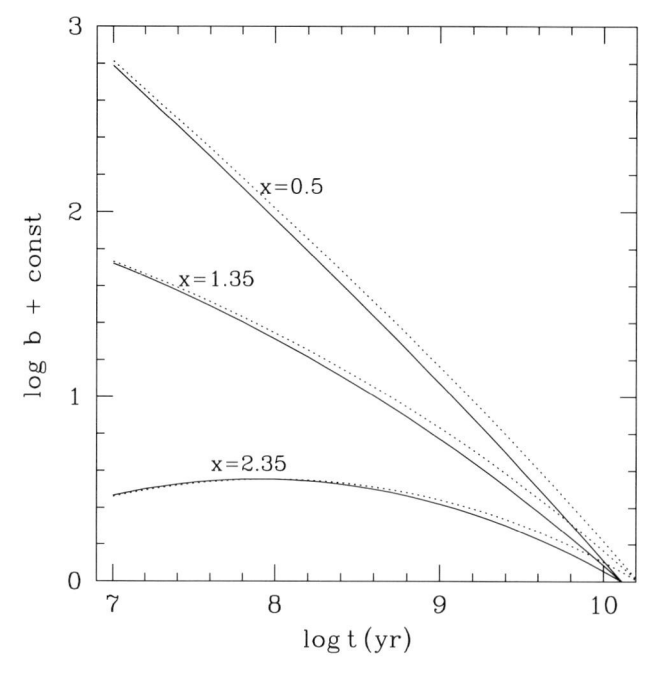

Figure 2.2 Time variation of the evolutionary flux $b(t)$ for different choices of the IMF slope. The evolutionary flux has been normalized to its value at 12 Gyr. The flatter the IMF, the smaller its effect in contrasting the drop of $\dot{M}_{\rm TO}$, which results in a relatively steep trend of the evolutionary flux. For each value of the slope x the solid lines show $b(t)$ computed at the MS TO, while the dotted lines show the death rates.

2.3
The Fuel Consumption Theorem

One of the most conspicuous property of a SSP is certainly its total bolometric luminosity $L_{\rm T}$. One can split $L_{\rm T}$ into two terms, one for the contribution of stars still burning hydrogen in their core (those still in their MS phase) and one for stars in their post-MS phases:

$$L_{\rm T} = L_{\rm MS} + L_{\rm PMS} = \sum_i^{\rm MS} L_i + \sum_j N_j L_j , \qquad (2.8)$$

where the first sum extends to all MS stars, and the second to all post-MS evolutionary stages (labeled by j) with L_j being the average luminosity of stars on their jth stage. Given the adopted approximation ($M_\circ = M_{\rm TO}$ for all post-MS stages) all L_j are relative to stars of mass $\sim M_{\rm TO}$. The various post-MS stages can be subdivided as it may turn out most convenient for specific applications.

2 The Fundamentals of Evolutionary Population Synthesis

The contribution of MS stars can be obtained from the integral

$$\sum_i^{MS} L_i = \int_{M_{inf}}^{M_{TO}} L(M_\circ)\phi(M_\circ)\,dM_\circ , \qquad (2.9)$$

where M_{inf} is the lower mass cutoff of the IMF (typically 0.1 $M_\odot$) and $L(M_\circ)$ is the luminosity-mass relation for MS stars. To a first approximation one can neglect that stars just below M_{TO} have somewhat increased their luminosity compared to their initial luminosity on the zero age main sequence, and use just a power-law fit to the ZAMS luminosity-mass relation: $L_{ZAMS} \simeq kM_\circ^a$, where k and a depend on composition and on the mass range. Thus, integrating Eq. (2.9) one gets:

$$L_{MS}(t) \simeq \frac{k\phi(M_{TO})M_{TO}^{a+1}}{a - x} , \qquad (2.10)$$

where the time dependence comes from $M_{TO}(t)$ as given by Eq. (2.1).

For the sake of a more accurate approximation one can take into account that stars have evolved somewhat in luminosity compared to the ZAMS, and use an evolutionary correction factor in front of L_{ZAMS}:

$$L(M_\circ, t) = \left\{1 + c\left[\frac{M_\circ}{M_{TO}(t)}\right]^\gamma\right\} L_{ZAMS} , \qquad (2.11)$$

where $c = L_{TO}(M_{TO})/L_{ZAMS}(M_{TO}) - 1$ and γ varies between $\simeq 3$ and $\simeq 5$ as the population ages from 10 Myr to 10 Gyr. At a given age, the correction factor is maximum for $M_\circ = M_{TO}$ and rapidly vanishes towards lower masses. Including these evolutionary corrections Eq. (2.10) becomes:

$$L_{MS}(t) \simeq k\phi(M_{TO})M_{TO}^{a+1}\left(\frac{1}{a-x} + \frac{c}{a-x+\gamma}\right) = KAM_{TO}^{a-x} , \qquad (2.12)$$

where K lumps together all the constants (k, a, c, x and γ).

Turning now to the contribution of post-MS stars, from Eqs. (2.7) and (2.8) one obtains:

$$L_{PMS} = \sum_j N_j L_j = b(t)\sum_j t_j L_j . \qquad (2.13)$$

The product $t_j L_j$, that is, the average luminosity during phase j times the duration of this phase, is clearly the total radiative energy emitted by the star during phase j. For those stages during which stars are in thermal equilibrium the virtual totality of this energy comes from nuclear burning, while gravitational energy release is a tiny fraction of it. Moreover, phases out of thermal equilibrium are so short that they contribute a negligible amount to the energetics of stellar populations. Thus, the product $t_j L_j$ is proportional to the amount of nuclear fuel burned during phase j.

The conversion of 1 g of hydrogen into helium via the CNO cycle (which dominates during all post-MS phases) releases 6.04×10^{18} erg, not including neutrino losses. Similarly, the conversion of 1 g of helium into carbon releases just $\sim 10\%$ of this energy, which increases to $\sim 12\%$ when including the conversion of $\sim 80\%$ of carbon into oxygen via the $^{12}C(\alpha,\gamma)^{16}O$ reaction (of course, the precise fraction of this conversion depends on the cross-section of the reaction). Therefore, when measuring the evolutionary flux in stars per year, and luminosities and masses in solar units, Eq. (2.13) becomes:

$$L_{PMS} = 9.8 \times 10^{10}\, b(t) \sum_j F_j(M_{TO}), \qquad (2.14)$$

where F_j is the amount (mass) of fuel burned during phase j. In turn

$$F_j = M_j^H + 0.12\, M_j^{He}, \qquad (2.15)$$

where M_j^H and M_j^{He} are the mass of hydrogen and of helium burned during phase j, respectively. This neglects carbon burning and more advanced evolutionary stages, but on the one hand the vast majority of stars do not ignite carbon at all (for $M_\circ \lesssim 8\, M_\odot$), and on the other hand, for massive stars, the contribution of such stages to the total amount of emitted radiation is negligible.

The construction of synthetic spectra of stellar populations is a rather complex endeavor, and before embarking in such a demanding task it is worth having an idea of what is important and should be taken into account, and what is less important, and can be neglected to make the effort easier. The integrated light of a population results from the overlapping contributions of stars in the various evolutionary stages: giants and dwarfs, horizontal branch and asymptotic giant branch stars, all have their role to play. Thus, it is useful to know, before hand, such relative contributions, hence the relative importance of the various evolutionary stages. This is established by Eq. (2.14), that gives formal shape to the *fuel consumption theorem* (in essence, not much more than the principle of energy conservation): *the contribution to the bolometric light of a SSP by stars in any post-MS stage is proportional to the amount of fuel burned during that stage*.

After these developments, the total luminosity of a SSP can be written as:

$$L_T(t) = KAM_{TO}^{a-x} + 9.8 \times 10^{10}\, b(t) \sum_j F_j(M_{TO}), \qquad (2.16)$$

and be further approximated by

$$L_T(t) \simeq 10^{11}\, k'\, b(t)\, F_T(M_{TO}), \qquad (2.17)$$

where

$$F_T(M_{TO}) = \sum_j F_j(M_{TO}) \qquad (2.18)$$

is the total fuel consumption of stars with $M_\circ \simeq M_{TO}$ in their post-MS phases, and $k' = 1 + L_{MS}/L_{PMS}$ is a measure of the *dwarf-to-giant* light ratio. This is quite a slow function of age, as we shall see later, decreasing from ~ 7.7 for a 10^7 yr old SSP, down to ~ 1.3 for a 10 Gyr old SSP.

2.4
Fuel Consumptions

The F_j appearing in Eq. (2.14) are among the most natural products of stellar evolutionary calculations, although most often are left unrecorded and are not included in stellar evolution databases. This unfortunate circumstance makes it quite laborious to reconstruct how much fuel is actually burned in the various post-MS stages as a function of stellar mass and composition.

Figure 2.3 shows two results of such attempts, using different stellar evolution databases, having divided the post-MS evolution into main phases according to the following convention: subgiant branch (SGB), from central hydrogen exhaustion to the arrival on the Hayashi line; red giant branch (RGB), from such arrival to helium ignition in the core; horizontal branch (HB), from helium ignition to central helium exhaustion; early-AGB (E-AGB), from central helium exhaustion to the first thermal pulse (helium shell flash) or up to central carbon ignition for high-mass stars; thermally pulsing AGB, TP-AGB, from the first thermal pulse to the end of the AGB superwind. This nomenclature is most appropriate for low- and intermediate-mass stars (i. e., $M_\circ \lesssim 8\text{–}10\,M_\odot$). More massive stars ($M_\circ > 10\,M_\odot$, not shown in the figure) burn an increasing amount of fuel during their core helium burning phase, up to many solar masses. The additional contribution of more advanced stages (from carbon burning to core collapse) is negligibly small in comparison.

Solid and dashed lines refer to evolutionary sequences computed assuming a modest amount of overshooting and no overshooting at all, respectively, and both are for solar composition. Overall, the two datasets appear to yield fairly similar fuel consumptions. Some main features of Figure 2.3a–f are worth noting:

- The smaller SGB fuel for models with overshooting, compared to models without overshooting, is due to the former ones being brighter at central hydrogen exhaustion, and therefore the stellar envelope is already close to the thermal runaway which brings it to red giant dimension. Correspondingly, phase C-D (cf. Chapter 1) is shorter in overshooting models, and less fuel is burned. On the other hand, this phase contributes a relatively minor fraction of the total post-MS fuel consumption.
- The fuel burned during the RGB phase dominates the total post-MS fuel for low mass stars, that is, for those which develop an electron degenerate helium core and experience the core helium flash ($M < M_{\text{HeF}} \simeq 2.1\,M_\odot$). As the mass approaches M_{HeF} the RGB shortens dramatically, and the fuel burned during this stage drops to become negligibly small. At the same time the fuel burned in the HB phase rises to a (local) shallow maximum, which is due to the small mass of the helium core of stars just above M_{HeF} resulting in a longer HB lifetime, with more hydrogen being processed by the hydrogen burning shell. This evolutionary property has been described in Chapter 1. Note that M_{HeF} is slightly lower in the case of overshooting models.

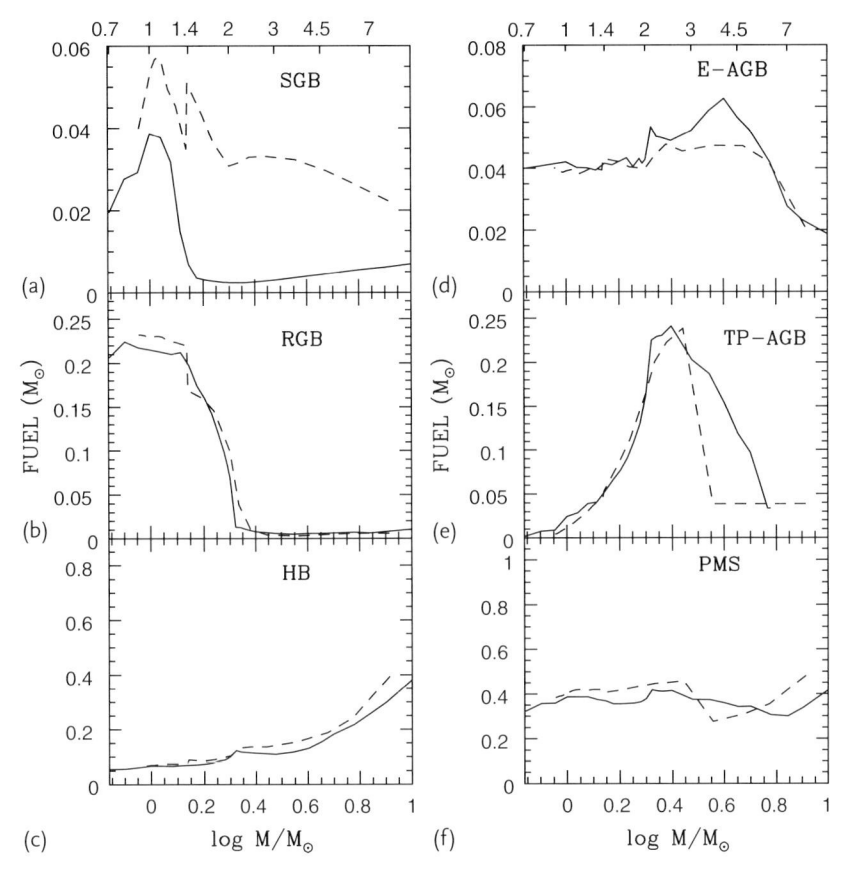

Figure 2.3 The amount of nuclear fuel burned in the various evolutionary stages (in $M_\odot$ units) as a function of stellar mass (labels shown on top x-axis), for the evolutionary stages indicated in each panel (a–f). The total post-MS fuel consumption (i.e., the sum over all post-MS phases) is shown in (f). Two different stellar evolution datasets have been used, based on tracks with (solid lines) and without convective overshooting (dashed lines) (Solid lines are derived using the YZ-VAR (Bertelli, G. et al. 2008, Astron. Astrophys., 484, 815; 2009, Astron. Astrophys., 508, 355) database of stellar tracks; fuel consumptions for the no-overshooting case are from Maraston (1998, Mon. Not. R. Astron. Soc., 300, 872).

- The fuel burned during the TP-AGB phase dominates the total post-MS fuel for masses between ~ 1.5 and $\sim 3.5\ M_\odot$. As discussed in Chapter 1, the behavior of stellar models in this phase is extremely sensitive to assumptions concerning mass loss, convection and mixing, and therefore an empirical calibration is indispensable. No-overshooting models in Figure 2.3 have used Large Magellanic Cloud (LMC) globular clustersfor such calibration, measuring the relative contribution of TP-AGB stars to the total bolometric light of clusters in several age bins lumped together, so to ensure statistical completeness. This is illustrated in Figure 2.4, showing the relative contributions of TP-AGB stars to the total

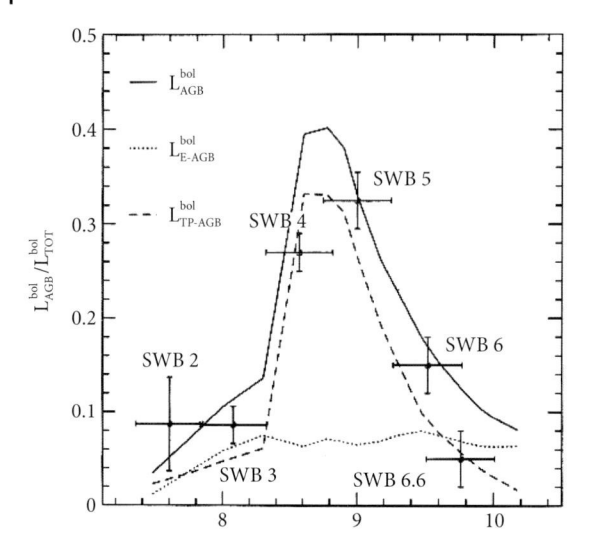

Figure 2.4 The relative contributions of AGB stars to the total bolometric luminosity of Large Magellanic Cloud clusters in six age bins (data points). The adopted calibration is shown separately for TP-AGB (dashed line) and E-AGB (dotted line) stars, as well as for their cumulative contribution (solid line) (source: Maraston (1998, *Mon. Not. R. Astron. Soc.*, 300, 872)) (Reproduced by permission of the Royal Astronomical Society.).

bolometric light of LMC clusters in several age bins. In the case of overshooting models, the calibration was achieved by reproducing the luminosity function of TP-AGB stars in the LMC field, having assumed a star formation history. The resulting TP-AGB fuel consumptions agree fairly well up to $\sim 2.7\ M_\odot$, when the fuel drops more rapidly for the no-overshooting models compared to the overshooting ones, simply because different calibrators have been used. These calibrations refer to the LMC metallicity ($\sim Z_\odot/2$), and it is assumed to hold also at other metallicities as Nature does not provide, within our reach, other suitable calibrators. In any event, the luminosity functions of bright AGB stars in globular clusters and fields of the LMC support the notion that the AGB evolution of stars more massive than ~ 3–$3.5\ M_\odot$ is severely affected by mass loss as hot bottom burning quickly drives them to high luminosities (see Chapter 1). The residual TP-AGB fuel consumption for $M > 3.5$ reported in Figure 2.3 is therefore quite tentative, as is its sudden drop at $M_{TO} = 7\ M_\odot$ that was assumed to correspond to the transition between intermediate and high-mass stars.

- An additional complication comes from the fact that TP-AGB stars come in two flavors: carbon-rich (C-type) stars and oxygen-rich (M-type) stars, with distinctly different spectra. Their proportion at other than LMC cluster metallicities is fixed using TP-AGB stellar models, whose predictions in this specific case are reasonably robust: as metallicity increases, more carbon needs to be dredged-up from the interior before carbon exceeds oxygen in the atmosphere then turning stars from M- to C-type. Thus, at intermediate ages the TP-AGB phase is dominated by carbon stars at low metallicity, and by M-type stars at high metallicity.

Given the overall similarity of the fuel consumptions derived from the two datasets and shown in Figure 2.3, in the following we adopt the fuels now shown together in Figure 2.5 to illustrate the luminosity evolution of an SSP. The interplay of the fuels burned in the various post-MS phases determines the upper curve in Figure 2.5 which shows that the total post-MS fuel consumption is a rather weak function of stellar mass in the range $1 \lesssim M_\circ \lesssim 10\ M_\odot$, being confined between ~ 0.3 and $\sim 0.4\ M_\odot$. Instead, above $\sim 10\ M_\odot$ the fuel consumption increases steeply with mass reaching many solar masses towards the top end of the mass range at $M_\circ \simeq 100\ M_\odot$.

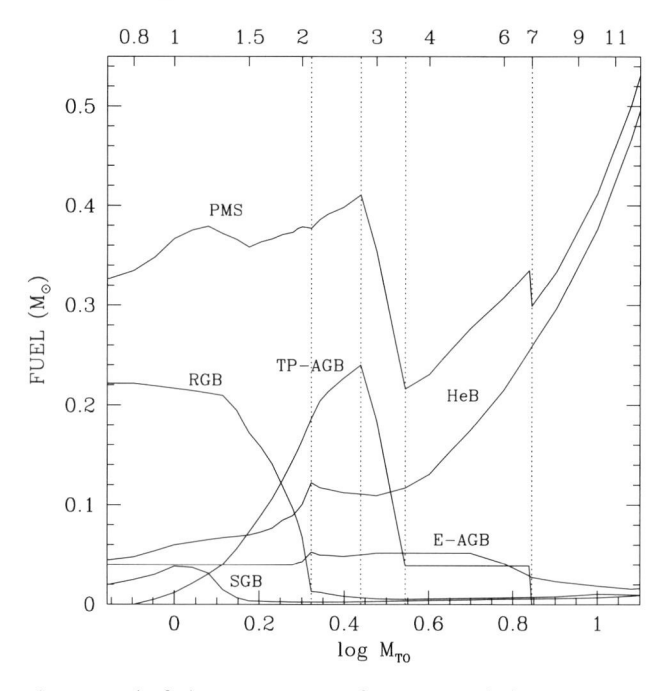

Figure 2.5 The fuel consumption as a function of the turnoff mass in $M_\odot$ units (labeled on the top axis) adopted here to derive the luminosity evolution of a SSP. The disappearance of the extended RGB (at $M \sim 2.1\ M_\odot$) does not have an appreciable impact on the total post-MS fuel consumption because it is balanced by the growth of the TP-AGB and HB fuel consumptions. Notice the peak in TP-AGB fuel at $M_{TO} \sim 2.75\ M_\odot$, and the two discontinuities at $M \sim 3.5$ and $7\ M_\odot$, a consequence of the TP-AGB phase becoming very short and eventually disappearing. These critical masses are highlighted by the vertical lines.

2.5
Population Synthesis Using Isochrones

For given age t, theoretical isochrones give the luminosity and temperature of stars that populate the isochrone as a function of their initial mass, that is, $L(t, M_\circ)$ and $T_e(t, M_\circ)$. The total, bolometric, luminosity of a SSP can then be obtained as:

$$L_T(t) = \int_{M_{\rm inf}}^{M_D} L(t, M_\circ)\phi(M_\circ)\,dM_\circ , \qquad (2.19)$$

where the integration extends from the minimum mass up to $M_D(t)$, the mass of dying stars at time t. Similar integrals can be written for the luminosity in any particular band, then deriving synthetic colors, or for a set of wavelengths, hence deriving synthetic spectra.

At first sight this approach looks more straightforward than that using fuel consumptions. In principle, the two approaches should be equivalent, but in practice results may differ because of flaws in their concrete implementation. This is because a nontrivial fraction of L_T may come from stars in very short evolutionary phases, hence in an extremely narrow range of masses. This would dictate using extremely small $dM_\circ$ steps in the integral along the isochrone, but in a different interval of masses at the various ages. For example, for an age $t = 1$ Gyr, from Eq. (2.2) one has $dM_{\rm TO}/dt \simeq 10^{-9}\,M_\odot\,{\rm yr}^{-1}$, hence a phase with a duration of a few million years (such as for example the TP-AGB phase) is populated by stars in a mass range of just a few $10^{-3}\,M_\odot$. Failures at properly handling this problem may result in spurious large amplitude oscillations in the resulting synthetic spectral energy distribution (SED), and examples of this effect can be found in the literature.

The fuel consumption approach is designed also to bypass this difficulty, as the summation is made over the fuels themselves. On the other hand, it requires an approximation to be made in treating the stars still on the MS band, where instead integrating along the isochrone is no problem. Thus, a quite natural compromise is to integrate along the isochrone up to the MS turnoff (or thereabout), and switch to the fuel consumption approach afterwards. Then the two *pieces* (the MS and the post-MS contributions) need to be properly matched. This is ensured by the scale factor A that appears in the IMF. Therefore, the total luminosity can be evaluated as:

$$L_T(t) = A \int_{M_{\rm inf}}^{M_{\rm TO}} L(t, M_\circ) M_\circ^{-(1+x)}\,dM_\circ$$
$$+ 9.8 \times 10^{10} A M_\circ^{-(1+x)} |\dot{M}_{\rm TO}| F_T(M_{\rm TO}) . \qquad (2.20)$$

This is probably the most robust and flexible way of calculating the evolution of the luminosity of a SSP. It exploits the advantage of fuel being a better variable in post-MS phases, which allows us to include uncertain evolutionary stages in a mod-

2.6
The Luminosity Evolution of Stellar Populations

Figure 2.6 shows the evolution of the bolometric luminosity of a SSP as derived from Eq. (2.20), for three different slopes of the IMF, as adopted for Figure 2.2. As a SSP grows old, it becomes fainter and fainter, and the drop is much faster the flatter the IMF. Indeed, both the MS and the PMS contributions decrease faster for a flatter IMF: from Eq. (2.12), the former scales as $\propto M_{TO}^{a-x}$, and smaller values of x imply a steeper exponent (note that always $a > x$, except for an unrealistically steep IMF). At the same time, most of the time variation of the PMS luminosity is due to the drop of the evolutionary flux, which is faster for a flatter IMF (see Figure 2.2).

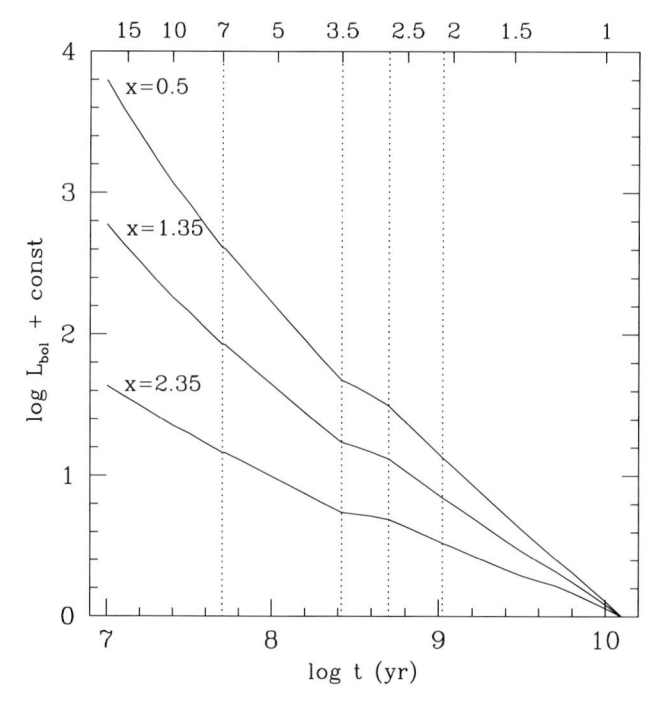

Figure 2.6 The time variation of the bolometric luminosity of a SSP for three different choices of the IMF slope. Each curve is normalized to its value at 12 Gyr. The upper horizontal axis is labeled with the values of the turnoff mass corresponding to the age on the bottom axis. Vertical dotted lines mark the same critical epochs (initial masses) as in Figure 2.5 (the YZVAR (Bertelli, G. et al. 2008, Astron. Astrophys., 484, 815; 2009, Astron. Astrophys., 508, 355) database of stellar tracks has been used to compute the MS contribution to the luminosity).

The upper horizontal axis in Figure 2.6 is labeled with the values of the TO mass at the corresponding ages on the lower *x*-axis. The vertical dotted lines mark the discontinuities in the fuel consumption due to the appearance of the TP-AGB phase. It can be noticed that the rapid development of the TP-AGB contribution has the effect of temporarily slowing down the drop of the bolometric luminosity at intermediate ages. Of course, the exact ages at which this contribution is present, and its size, are still subject to some uncertainty.

Finally, Figure 2.7 shows the contributions of the different phases to the total bolometric light of SSPs of different ages, that is, it illustrates the result of the *fuel consumption theorem*. The MS contribution dominates in young populations, and then steadily declines with time thus dropping from over ~ 80 to $\sim 25\%$. The contribution of core helium burning stars (labeled HB) is instead roughly constant at the $\sim 15\%$ level, with a local maximum at $M = M_{\rm HeF} \simeq 2.1 M_\odot$, reflecting the local maximum in the HB fuel consumption. As the TP-AGB develops, it reaches a maximum at the $\sim 40\%$ level for $t \simeq 1$ Gyr, rivaling the MS contribution. At nearly the same time the RGB begins its fast development, and beyond ~ 3 Gyr it surpasses the MS contribution until it levels off at a $\sim 45\%$ contribution. Last to become significant is the contribution of SGB stars, that is, those in between the

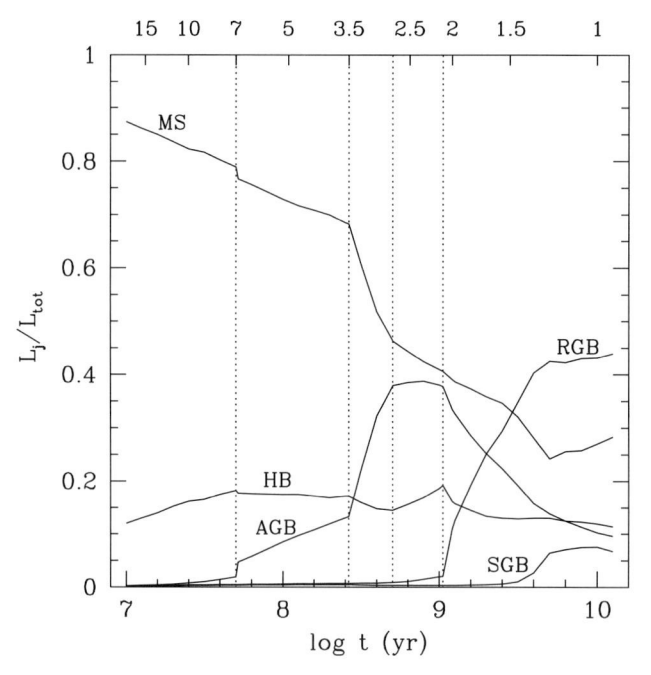

Figure 2.7 The time variation of the contributions of the various evolutionary phases to the total bolometric light of a SSP with a Salpeter-diet IMF ($s = 1 + x = 2.35$). The upper axis is labeled as in Figure 2.6. Vertical dotted lines mark the same critical epochs (initial masses) as in Figures 2.5 and 2.6 (the YZVAR (Bertelli, G. et al. 2008, Astron. Astrophys., 484, 815; 2009, Astron. Astrophys., 508, 355) database of stellar tracks has been used to compute the MS contribution in this figure).

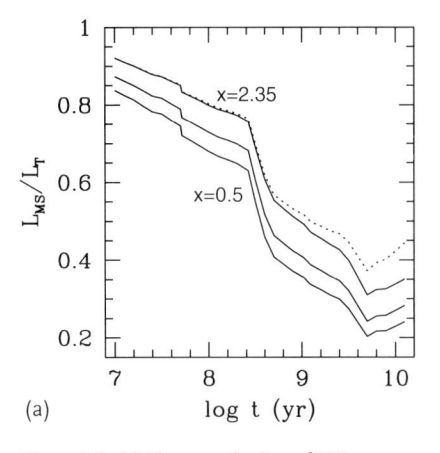

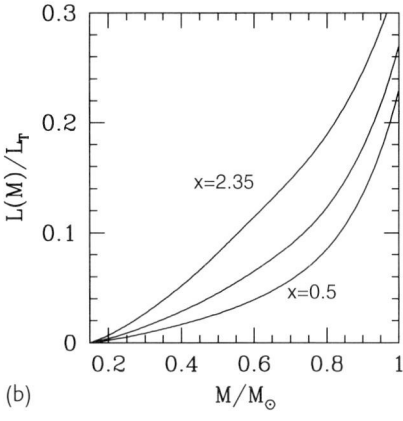

Figure 2.8 (a) The contribution of MS stars to the total bolometric light of a SSP as a function of age for three values of the IMF slope above 0.5 $M_\odot$: $x = 0.5, 1.35, 2.35$ (the three solid curves, respectively), and $x = 0.3$ in all three cases below 0.5 $M_\odot$. The dotted line shows the ratio for a single slope IMFs down to 0.1 $M_\odot$ with $x = 2.35$. (b) The build up of bolometric light along the MS of a 10 Gyr old SSP, for the same IMFs as in (a) (the YZ-VAR (Bertelli, G. et al. 2008, Astron. Astrophys., 484, 815; 2009, Astron. Astrophys., 508, 355) database of stellar tracks has been used to compute the ingredients to draw this figure).

MS turnoff and the base of the RGB. Only stars less massive than $\sim 1.5\,M_\odot$ burn an appreciable amount of fuel during this stage.

Changing the IMF slope does not affect the relative contributions of the post-MS phases (as such phases span a narrow mass interval), but affects slightly the relative weight of the MS and post-MS stages. This is illustrated in Figure 2.8a, showing that the MS contribution increases with increasing IMF slope, as more low-mass stars are added. This (relatively minor) sensitivity follows from the increased contribution of low-mass stars for steeper IMFs, as illustrated in Figure 2.8b: the steeper the IMF, the broader the mass range of MS stars that contribute to the bolometric light, and the higher the overall contribution of stars still on the MS.

2.7
The Specific Evolutionary Flux

A quantity with manifold useful applications is the *specific evolutionary flux*, that is, the evolutionary flux per unit light of the stellar population:

$$B(t) = \frac{b(t)}{L_T(t)}, \qquad (2.21)$$

where $b(t)$ is given by Eq. (2.6) and $L_T(t)$ by either Eqs. (2.16) or (2.20). One can easily realize that numerator and denominator have several quantities in common, especially if one uses the approximation given by Eq. (2.17). Hence, one expects $B(t)$ to be a very weak function of both time and IMF slope. This is indeed illus-

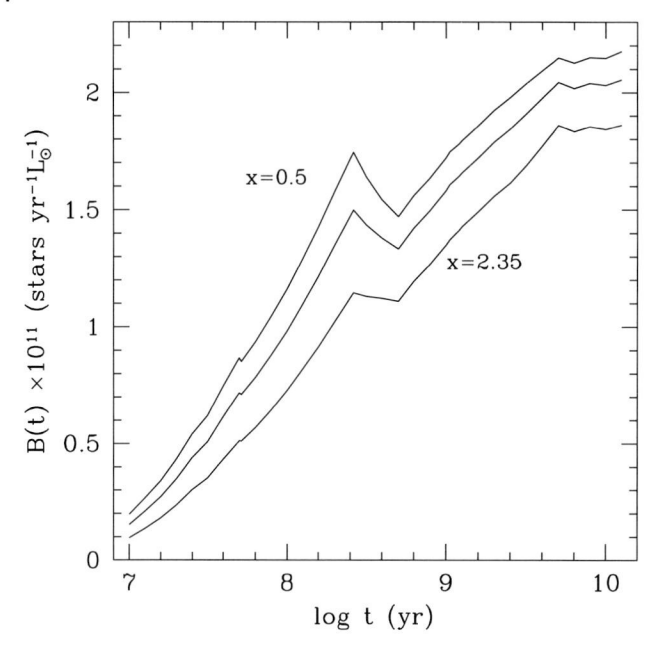

Figure 2.9 The specific evolutionary flux $B(t)$ as a function of age for three values of the IMF slope $x = 0.5$, 1.35, and 2.35 for the three curves. The sharp discontinuity at $t \sim 2.5 \times 10^8$ yr is an artifact of the appearance of an extended TP-AGB phase.

trated by Figure 2.9, showing that $B(t)$ is virtually independent of the IMF, and depends very mildly on age, varying by just a factor ~ 2 between an age of 10^8 yr and 10^{10} yr.

By using the specific evolutionary flux, Eq. (2.7) becomes:

$$N_j = B(t) L_T t_j , \qquad (2.22)$$

a notable relation which is extremely useful in a number of situations. Having even a rough idea of the age of a population, this relation allows us to estimate the number of stars in any given evolutionary phase of fairly known duration, once the total luminosity is measured for the sampled population, for example, a stellar cluster, or an elliptical galaxy, or just a portion of them. For example, a globular cluster with $L_T = 10^5 \, L_\odot$ contains $\sim 2 \times 10^{-11} \times 10^5 \times 10^8 = 200$ HB stars, being $\sim 10^8$ yr the duration of the HB phase. This relation can also be used to infer the duration of an evolutionary stage, once stars in that stage are counted and the sampled luminosity is measured. For example, the old globular cluster 47 Tuc contains 4 long period variables (LPV), likely in their AGB stage. With a total cluster luminosity of $\sim 10^6 \, L_\odot$, Eq. (2.22) then indicates $\sim 2 \times 10^5$ yr for the duration of the LPV phase in 47 Tuc.

Note that the death rate of a SSP is given by $b(t) = B(t) L_T$: even without making any guess about the age of the population (or the stellar age distribution within a galaxy), by simply adopting $b \sim 10^{-11} \, L_T$ the death rate of the population is obtained within a factor of a few once its total luminosity is known. Similarly, if one knows the age within a factor ~ 2, then one can pick a more precise value of $B(t)$ and the resulting death rate can be obtained with an accuracy of ~ 10–20%. The specific evolutionary flux is one of the most robust tools provided by stellar evolution theory for the study of stellar populations. Some applications of this relation are presented in Chapter 3.

As the luminosity L^j provided by all the stars in any given post-MS stage is given by $L^j = 9.8 \times 10^{10} b(t) F_j(M_{TO})$, by dividing both sides of this relation by L_T one gets the fractional contribution of each post-MS stage to the total luminosity of the SSP, that is:

$$\frac{L^j}{L_T} = 9.8 \times 10^{10} B(t) F_j(M_{TO}) . \tag{2.23}$$

For an old SSP, say ~ 10 Gyr or older, $B(t) \simeq 2 \times 10^{-11}$ (stars per year per solar luminosity), and therefore the fractional contributions are very simply given by:

$$\frac{L^j}{L_T} \simeq 2 \times F_j(M_{TO}) . \tag{2.24}$$

For example, reading the fuel consumption from Figure 2.5 one gets that the amount of fuel burned by $\sim 1 M_\odot$ stars during their HB phase is $F_{HB} \sim 0.06\, M_\odot$, hence HB stars contribute $\sim 12\%$ of the total bolometric luminosity of such a SSP. Similarly, $F_{RGB} \sim 0.20$–$0.25\, M_\odot$, and therefore RGB stars contribute ~ 40–50% of the total luminosity.

2.8
The IMF Scale Factor

The scale factor A appearing in the IMF is clearly a measure of the size of a SSP, as the number of stars in the population is proportional to it. It is convenient to express this scale factor in terms of some global property of the population which is directly observable, such as its total luminosity. Thus, one can solve for A the relation $b(t) = A\phi(M_{TO})|\dot{M}_{TO}| = B(t) L_T$ and get:

$$A = B(t) L_T(t) M_{TO}^{1+x} |\dot{M}_{TO}|^{-1} . \tag{2.25}$$

With A being a constant, the r.h.s. of this equation can be evaluated for any arbitrary age. In particular, for an age $t \simeq 12$ Gyr one derives $A \simeq 0.9\, L_T(12\,\text{Gyr})$, as can be appreciated from Figure 2.10, which shows the A/L_T ratio as a function of age. Note that for old SSPs this ratio is almost independent of the adopted IMF, and therefore their IMF can be written as $\phi(M) \simeq 0.9\, L_T M^{-(1+x)}$, which allows us to estimate the number of stars formed in the various mass intervals, once the present luminosity of an old population is known, and adopting a suitable IMF slope.

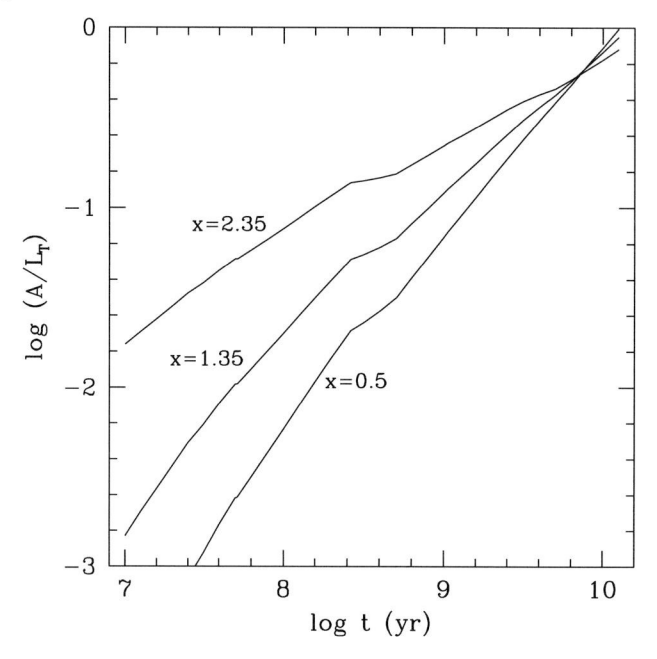

Figure 2.10 The ratio of the IMF scale factor A to the total bolometric luminosity of a SSP as a function of age, and for three IMFs as in Figure 2.6.

2.9
Total and Specific Rates of Mass Return

The mass return to the interstellar medium (ISM) is one of the major metabolic functions of stellar populations. Each star sheds a large fraction of its mass before concluding its evolution as a dead remnant. In a SSP of mass $\mathcal{M}$ the rate of this process is given by the product of the death rate (i. e., the evolutionary flux) times the total mass loss suffered by one evolving star:

$$|\dot{\mathcal{M}}| = b(t)\Delta M = \phi(M_{\rm TO})|\dot{M}_{\rm TO}|\Delta M , \qquad (2.26)$$

where $\Delta M = M_{\rm TO} - M_{\rm f}(M_{\rm TO})$, and $M_{\rm f}$ is the final mass, that is, the mass of the remnant, and is a direct implication of the initial mass–final mass relation discussed in Chapter 1. Figure 2.11a shows the temporal behavior of the rate of mass return for SSPs with different IMF slopes. Following the evolutionary flux, the rate of mass return is a rapidly decreasing function of time, and is steeper for relatively flat IMFs.

It is worth noting the remarkable similarity of this equation for the rate of mass return with Eq. (2.17) for the evolution of the total luminosity. In both equations the evolutionary flux $b(t)$ provides most of the time dependence, and multiplies either the mass of burned fuel, or the mass lost by one evolving star. Indeed, the initial mass $M_\circ = M_{\rm TO}$ of an evolving star is partly burned (the remnant is composed

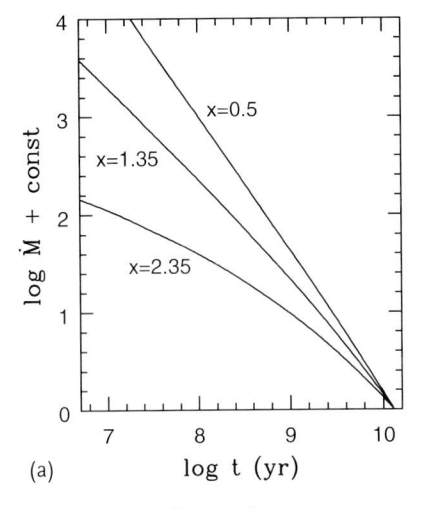

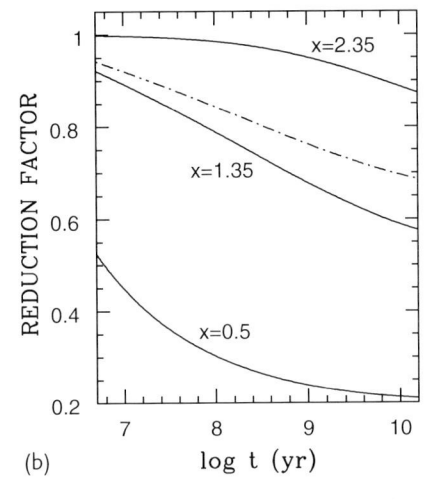

Figure 2.11 (a) The rate of mass return from a SSP for various slopes of the IMF, as labeled. The three cases have been normalized to provide the same value at an age of ∼ 12 Gyr. The adopted initial-final mass relation includes Eq. (1.11) up to 8 $M_\odot$, extended with a piecewise linear relation through the points $M_f/M_\odot$ = 1.4, 1.4, 20 for $M_\circ/M_\odot$ = 10, 25, 100. (b) The ratio between the current mass in stars and remnants and the initial mass of the SSP. The three considered IMFs are flattened below 0.5 $M_\odot$, where they share a common slope of $(1 + x) = 1.3$. The dot-dashed line shows the case of a single slope, Salpeter IMF, which, compared to the corresponding Salpeter diet, has a larger fraction of low-mass stars.

of thermonuclear ashes) and partly lost to the ISM. The former activity inevitably produces light (L_T), the latter produces exhaust ($\dot{\mathcal{M}}$). By varying mass loss prescriptions one cannot reduce the exhaust without increasing the ashes, and therefore without getting a brighter population (i.e., what is not lost is burned). In practice, this connection allows one to put firmer constraints on the stellar mass loss than direct observational estimates of stellar mass loss rates.

Note that some ashes may be engulfed in the exhaust, as a result of either dredge-up processes, hot bottom burning, or explosions, the rest remaining locked into the dead remnants, that is, white dwarfs, neutron stars or black holes. Thus, Eq. (2.26) can be generalized to give the rate of mass return for each individual element, isotope, or type of dust grains. For this purpose it is sufficient to replace ΔM with the mass of the particular species that is lost in the course of its evolution by one star of mass $M_\circ = M_{\rm TO}$. For example, in the case of dust grains, a SSP initially returns only oxygen-rich species such as silicate grains, until $M_{\rm TO}$ drops below ∼ 3$M_\odot$ (at $t \simeq 3 \times 10^8$ yr), when carbon stars first appear on the AGB, and the particulate exhaust suddenly switches to graphite, soot, or other sorts of carbon-rich grains.

As a consequence of stellar mass loss, the mass of a SSP decreases with time at the rate $-\dot{\mathcal{M}}$, and therefore its current mass at time t after formation is given by:

$$\mathcal{M}(t) = \mathcal{M}_\circ - \int_0^t b(t) \Delta M \, dt , \quad (2.27)$$

where $\mathcal{M}_\circ$ is the initial mass of the SSP. Figure 2.11b shows the mass reduction factor $\mathcal{M}(t)/\mathcal{M}_\circ$ as a function of time, for various choices of the IMF slope. Note that, unless the IMF is very steep, much of the reduction takes place during the first Gyr, and then the rate of decrease slows down considerably, approaching $\sim 60\%$ at $t = 12\,\mathrm{Gyr}$ for a Salpeter-diet IMF.

For a system with a continuous star formation rate (SFR) $\psi(t)$ the rate of mass return is given by:

$$|\dot{\mathcal{M}}(t)| = \int_{M_{\mathrm{TO}}}^{M_{\mathrm{up}}} \phi(M_\circ)(M_\circ - M_{\mathrm{f}}) \psi(t - \tau_M) \, dM_\circ , \quad (2.28)$$

where M_{up} is the upper limit of the stellar masses, and τ_M is the evolutionary lifetime of a star with mass $M_\circ$, as given for example, by Eq. (1.7). For a constant SFR, the rate of mass return secularly increases with time as more and more stellar generations add their contribution to the total mass loss, but again most of the variation occurs within the first Gyr as shown in Figure 2.12a. Correspondingly, the reduction factor levels off quite soon (see Figure 2.12b): for a Salpeter IMF the reduction factor ranges between 0.7 and 0.6 as the stellar population evolves

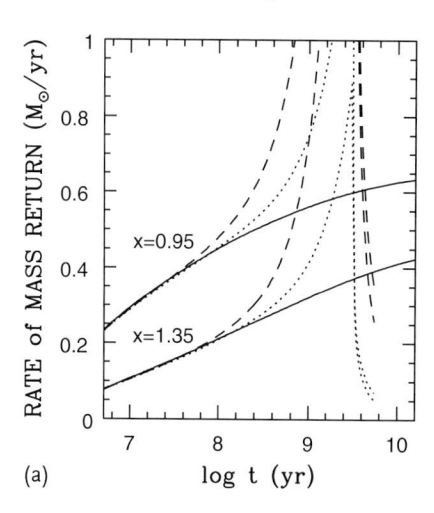

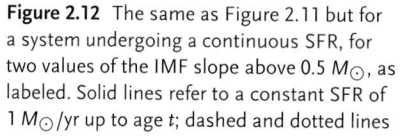

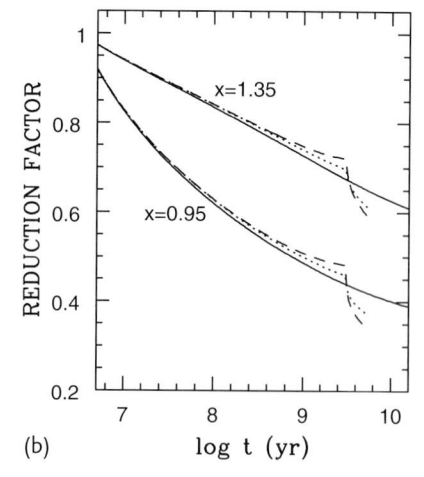

Figure 2.12 The same as Figure 2.11 but for a system undergoing a continuous SFR, for two values of the IMF slope above 0.5 $M_\odot$, as labeled. Solid lines refer to a constant SFR of 1 $M_\odot$/yr up to age t; dashed and dotted lines refer to a SFR which, starting from a value of 1 $M_\odot$/yr, increases exponentially for the first 3 Gyr, and then drops to zero. The e-folding time for the increasing SFR is of 1 and 3 Gyr for the dashed and dotted lines, respectively.

from 1.5 to 12 Gyr. In the case of star-forming galaxies part of the exhaust can be lost to the intergalactic medium, part can be incorporated into successive stellar generations, thus driving their chemical evolution.

By dividing the rate of mass return by the total luminosity of the parent population one gets the specific rate of mass return:

$$\dot{m} = \frac{|\dot{\mathcal{M}}|}{L_T} = B(t)\Delta M \propto \sim \frac{\Delta M}{\sum_j F_j}, \qquad (2.29)$$

which is measured in $M_\odot\,\mathrm{yr}^{-1}\,L_\odot^{-1}$. This relation says that the specific rate of mass return is roughly proportional to the *exhaust-to-ash* ratio for $M_\circ \simeq M_{\mathrm{TO}}$, and therefore it must be almost independent of the IMF and only weakly evolving with time, as indeed shown in Figure 2.13. This follows from the very basic fact that both most of the light and most of the mass shed by an evolving SSP comes from evolved stars, which at any time span a fairly narrow range of initial masses.

A straightforward way of estimating the rate of mass return follows from the above equation:

$$|\dot{\mathcal{M}}| = B(t)\Delta M L_T, \qquad (2.30)$$

which provides a direct relation between the rate of mass return and the luminosity of the SSP. For example, for a 12 Gyr old, solar metallicity population one has

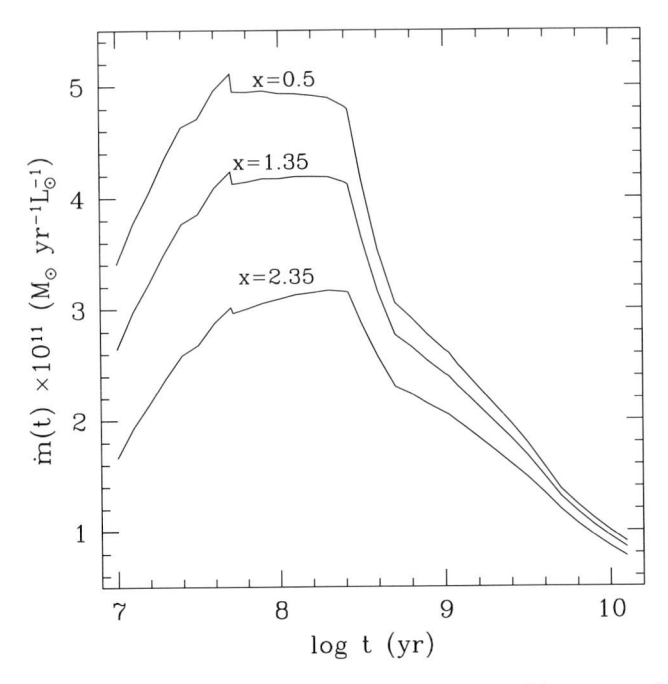

Figure 2.13 The rate of mass return per unit luminosity of the parent stellar population for a SSP as a function of age, also called specific rate of mass return.

$B(t) \simeq 2 \times 10^{-11}$, $M_{TO} \simeq 0.95\ M_\odot$, $M_f \simeq 0.55\ M_\odot$ ($\Delta M \simeq 0.4\ M_\odot$), and therefore $\dot{M} \simeq 8 \times 10^{-12}\ L_T$. Thus, the current metabolism of an old and passive elliptical galaxy with $L_T = 2 \times 10^{11}\ L_\odot$ generates an exhaust of 1.6 $M_\odot$ yr^{-1}, an estimate whose accuracy is better than $\sim 10\%$. Figure 2.11 shows the rate of mass return that such a galaxy was supporting at earlier times, scaling up from this value. Thus, at $t = 10^7$ yr, in the infancy of this SSP, the rate of mass return was ~ 3000 times higher than at $t = 12$ Gyr.

2.10
Mass and Mass-to-Light Ratio

The mass-to-light ratio is a fundamental property of stellar populations, which allows us to estimate the mass of stellar systems from their light, and it is widely used to derive the mass of star clusters and galaxies. Since galaxy photometry yields magnitudes in specific wavelength ranges, the mass-to-bolometric light ratio is not of direct interest; nevertheless it is instructive to examine its secular evolution and its dependence on the star formation history and on the IMF. This is shown in Figure 2.14a,b.

Since most of the light of SSPs is provided by stars with mass $\simeq M_{TO}$, the mass-to-light ratio is proportional to the inverse of the number of stars at the TO. For the same total mass of formed stars ($\mathcal{M}_\circ$), SSPs with steep IMFs are most populated in the low-mass range, so that at young ages they are fainter than SSPs with a flatter IMF. Conversely, at old ages, when $M_{TO} \simeq 1 M_\odot$, SSPs with steep IMFs become brighter than SSPs with flat IMFs, as shown in Figure 2.14a. The effect of

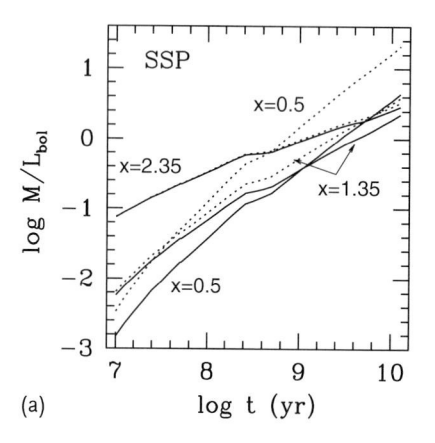

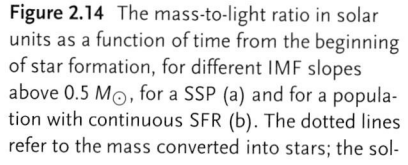

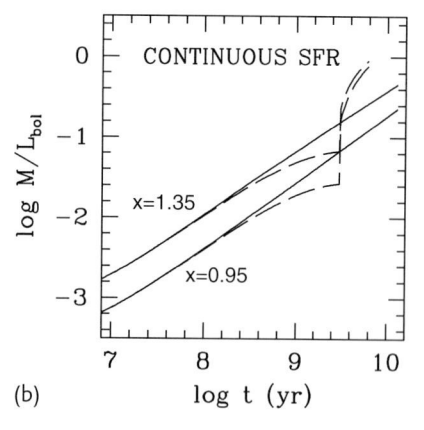

Figure 2.14 The mass-to-light ratio in solar units as a function of time from the beginning of star formation, for different IMF slopes above 0.5 $M_\odot$, for a SSP (a) and for a population with continuous SFR (b). The dotted lines refer to the mass converted into stars; the solid and dashed lines refer to the current mass of the stellar population, having taken the mass return into account. In panel b the different line types refer to different assumptions of the SFR law with the same line encoding as in Figure 2.12b.

the mass return (see the dotted lines) is to decrease the mass-to-light ratio below $\mathcal{M}_\odot/L(t)$, and do so progressively more as the SSP ages. At old ages, for a Salpeter-diet slope, the stellar mass-to-light ratio is $\sim 0.6\,\mathcal{M}_\odot/L$, while for $x = 2.35$ the effect of the mass return is virtually negligible. At the other extreme, for a very flat IMF ($x = 0.5$), a 10 Gyr old SSP has a mass-to-bolometric light ratio of only ~ 3.5, since the mass return is very large. A straight Salpeter IMF is often used in the literature: with respect to the Salpeter-diet case its mass-to-light ratio is enhanced by 0.13 dex almost independent of the age of the SSP, since the additional low-mass stars provide only mass without affecting the light.

The temporal variation of the mass-to-light ratio for a constant and exponentially increasing SFR are shown in Figure 2.14b. Notice that in the latter case, the exponential increase only lasts for the first 3 Gyr, before the SFR drops to zero. For a continuous SFR the mass-to-light ratio at an epoch t is obtained as:

$$\frac{\mathcal{M}_\odot}{L} = \frac{\int_0^t \psi(\tau)d\tau}{\int_0^t L^{SSP}(\tau)\psi(t-\tau)d\tau}, \qquad (2.31)$$

where L^{SSP} is the luminosity of a unit mass SSP, and τ is its age. The light-to-mass ratio in Eq. (2.31) is the average of the light-to-mass ratios of the SSPs born up to epoch t, weighted by the star formation history. It follows that the mass-to-light ratio for continuous star formation histories at any epoch t will assume values between the minimum and maximum of the SSP mass-to-light ratio in the range of ages between 0 and t. Therefore, its age dependence is milder than in the SSP case. For the exponentially increasing SFR, the mass-to-light ratio at 3 Gyr is ~ 0.37 dex below the constant SFR case and ~ 1.1 dex below that of an SSP with the same age, equivalent to that of an SSP of only 100 Myr old.

2.11
IMF-Dependent and IMF-Independent Quantities

Besides age and composition, the IMF is the third fundamental property of a SSP. It is therefore important to have clear which observable properties of stellar populations strongly depend on the IMF and which do not, if at all.

The evolutionary flux $b(t)$ is a strong function of the IMF slope, as illustrated in Figure 2.2, and therefore so are all those quantities that are (roughly) proportional to it. This is indeed the case for the luminosity evolution of a SSP (Figure 2.6) and for the evolution of the rate of mass return (Figure 2.11).

Specific quantities, that is, those per unit luminosity, are almost independent of the IMF slope, because having the luminosity at the denominator they contain the evolutionary flux $b(t)$, and therefore the IMF, both at the numerator and at the denominator. Thus, the specific evolutionary flux and the specific rate of mass return are both almost independent (or weakly dependent) on the IMF, see Figures 2.9 and 2.13.

Perhaps the most important observable which is almost independent of the IMF is the spectrum of a SSP. This depends on the fact that at any given time the bulk of the light of a SSP comes from stars in a fairly narrow range of masses. As far as the integrated spectrum of the SSP is concerned, what is the slope of the IMF in that particular range does not quite matter. For the contribution of post-MS stars this is documented by Figure 2.1, whereas for the contribution of MS stars, that is, those with $M < M_{TO}$ this is illustrated by Figure 2.8.

For the same reason, the luminosity of the SSP in any particular band and at any particular time is fairly insensitive to the IMF slope. What matters is the absolute value of the IMF at the main sequence turnoff, that is, $\phi(M_{TO})$, see Eq. (2.16).

Still, being *almost* independent of the IMF it means that a weak dependence exists, after all. But use such weak dependence to infer the IMF itself would be an extremely dangerous exercise. Indeed, colors and spectra are far more sensitive to many other parameters than they are to the IMF, and trying to infer from them something about the IMF can produce nonsense.

Moreover, the IMF slope in different intervals of the stellar mass has various specific effects on the integral properties of stellar populations and their evolution. This will be quite extensively explored in Chapter 8.

2.12
The Age-Metallicity Degeneracy

The distribution of stars across the HRD is a function of age and metallicity, and therefore the SED of a stellar population is sensitive to all details of their distribution. Still, there are parts of the HRD which have a stronger effect on the derived ages than others, and for this reason deserve special attention. It is the main sequence turnoff temperature (or color) that has the strongest effect on the part of the SED of stellar populations that is most sensitive to age, hence on ages derived from fitting synthetic spectra to the observed SED of galaxies.

However, besides from age, the turnoff temperatures and colors depend also on metallicity, and disentangling one from the other is one of the main troubles of the stellar population synthesis approach to galaxy diagnostics. This is illustrated by Eq. (2.5); a fit to the same models used to construct this equation yields:

$$\left(\frac{\partial \log t}{\partial [Fe/H]}\right)_{(V-I)^{TO}} \simeq -0.52 - 0.19\,[Fe/H] \ . \tag{2.32}$$

For example, for near-solar metallicity $\delta \log t \sim -0.5\delta[Fe/H]$, and a 0.3 dex error in the estimated metallicity translates into an age error of $\sim 40\%$. Moreover, the RGB color is far more sensitive to metallicity than to age, and a 0.3 dex uncertainty in metallicity results into a factor of ~ 2 total uncertainty in age. For young populations an error of this size may not be a major concern, but when it comes to very old stellar populations, a factor of two error in age implies that a galaxy may be found as old as the Universe, or just half of that, thus leading to dramatically different scenarios of galaxy formation.

This *age-metallicity degeneracy* has plagued the population synthesis approach to galaxy evolution from the very beginning, and needs to be either solved or circumvented in order to achieve robust conclusions. In the following chapters this problem will be encountered over and over again. In particular, error circulation will obviously manifest itself as spurious age-metallicity anticorrelations, as an overestimate of metallicity immediately implies an underestimate of age, and vice versa.

Evolving stars are the site of many physical processes and phenomena, and some stages are dominated by especially complex (hydrodynamical) processes that make their modeling particularly uncertain. Yet, when one considers whole populations of stars, one finds that their evolution is subject to quite simple laws. Such laws control the overall energy production of a population, its evolution as a function of time, the contribution of the various evolutionary stages to the integrated light, as well as the rate at which part of the population's mass is shed back to the ISM. These laws then provide the backbone for the calculation of synthetic spectra of stellar populations, their evolution with time, and their dependence on chemical composition. *Evolutionary population synthesis* is indeed the tool commonly used to measure the stellar masses, ages, and star formation rates of galaxies all the way to the distant Universe, at extremely high redshifts. It is then good to know that its foundations are fairly solid and well understood.

Further Reading

Classical papers on Synthetic Stellar Populations

Tinsley, B.M. (1975) *Mem. Soc. Astron. Ital.* **46**, 3.

Tinsley, B.M. (1980) *Fund. Cosm. Phys.*, **5**, 287.

Concepts in this Chapter

have been introduced and developped in:

Renzini, A. (1981) *Ann. Phys. (France)*, **6**, 87.
Renzini, A. and Buzzoni, A. (1986) *Spectral Evolution of Galaxies* (eds C. Chiosi and A. Renzini), Dordrecht, Reidel, p. 195.

Concrete applications can be found in:

Buzzoni, A. (1989) *Astrophys. J. Suppl.*, **71**, 8170.
Greggio, L. and Renzini, A. (1990), *Astrophys. J.*, **364**, 35.
Maraston, C. (1998) *Mon. Not. R. Astron. Soc.*, **300**, 872.
Maraston, C. (2005) *Mon. Not. R. Astron. Soc.*, **362**, 799.

3
Resolving Stellar Populations

3.1
The Stellar Populations of Pixels and Frames

The use of sophisticated software packages for the individual star photometry of stellar systems is a current, routine activity. CCD/IR-array frames are fed into a package, which automatically flat-fields them, subtracts dark, bias and sky, removes cosmic ray hits, restores bad pixels, extracts a point spread function (PSF), and uses it to match all count peaks found in the frame. The instrumental magnitudes, that is, the integration of the counts under individual PSFs, are then converted into true magnitudes via calibration with standard stars. Finally, a catalog of star magnitudes and positions is delivered. All the astronomer has to do is to plot color-magnitude diagrams (CMD), sort out an astrophysical interpretation of them, and write the paper.

The use of these packages has resulted in tremendous progress in the study of resolved stellar populations, including for example, those in galactic and extragalactic globular clusters, the Galactic bulge, the Magellanic Clouds, the bulge, disk and halo of M31, irregular galaxies within and beyond the Local Group, and stellar fields out to the Virgo cluster. However, relying entirely on automated procedures can also produce nonsense. Here a few simple tools are offered that can be useful to most rapidly and efficiently plan for specific observational projects, given the properties of the target field and the available telescope facilities. These tools can also help to gather a deeper understanding of the scientific content of the data, once they have been obtained, and have two straightforward applications, namely: (i) to estimate to which extent a given frame contains a fairly complete sample of the stellar population of the targeted stellar system, and (ii) to estimate down to which flux limit stellar photometry can be undertaken without being significantly degraded by crowding.

3.1.1
The Stellar Population of a Frame

Given a target stellar system and the field of view F_{oV} (in arcsec^2) of the imaging camera with which it is observed, the bolometric luminosity of the population sampled by the detector is given by:

$$L_T = BC_\lambda F_{oV} 10^{-0.4(\mu_\lambda - A_\lambda - \mathrm{mod} - M_{\lambda,\odot})} ,\qquad(3.1)$$

where μ_λ, A_λ and mod are, respectively, the surface brightness of the target in magnitudes arcsec^{-2}, its extinction and distance modulus, $M_{\lambda,\odot}$ is the absolute magnitude of the sun in the λ-band, and BC_λ is the bolometric correction (BC) factor (i. e., L_T/L_λ), given by:

$$BC_\lambda = 10^{-0.4(M_{\mathrm{bol}} - M_\lambda - M_{\mathrm{bol},\odot} + M_{\lambda,\odot})} ,\qquad(3.2)$$

where M_{bol} and M_λ are the bolometric and λ-band magnitudes of the target system, and $M_{\mathrm{bol},\odot}$ and $M_{\lambda,\odot}$ are the corresponding absolute magnitudes of the sun.

The BC factor can be evaluated from synthetic stellar populations, and Figure 3.1 shows the $BC_\lambda(t, Z)$ for a SSP of solar metallicity. One can notice that for ages in excess of ~ 0.3 Gyr the BC factors for the V and redder bands are almost constant. Therefore, typical values readable from Figure 3.1 can be used for all stellar systems if one has reasons to believe that the light of the target in those bands is dominated by stars older than 0.3 Gyr, irrespective of the actual age distribution, a welcome simplification.

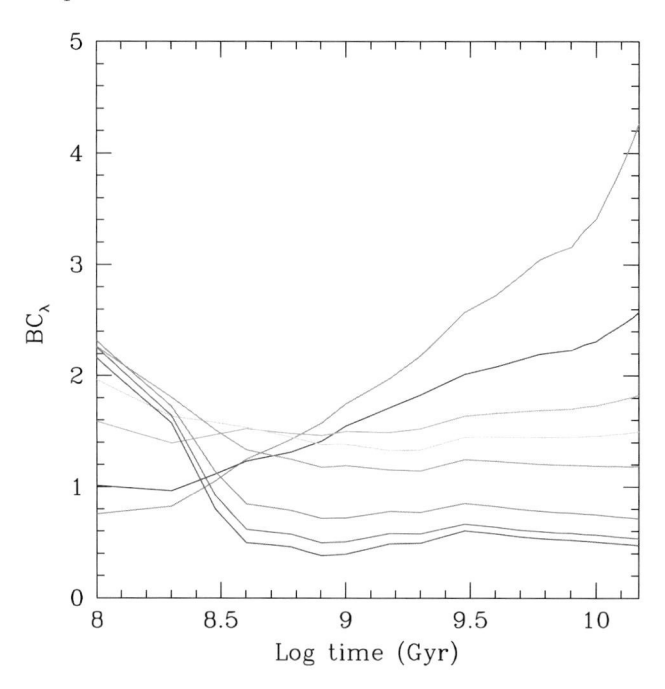

Figure 3.1 The bolometric correction for a SSP of solar metallicity as a function of age, for various broad bands in the Johnson–Cousin photometric system. On the intercept with the right, vertical axis one can identify the bolometric correction for the bands U, B, V, R, I, J, H, and K, from top to bottom, respectively (figure is based on the synthetic populations by Maraston, C. (2005, *Mon. Not. R. Astron. Soc.*, 362, 799)). For a color version of this figure please see in the front matter.

For the sake of concreteness, let us suppose that the camera samples a luminosity $L_K = 2 \times 10^5 \, L_\odot$, and that the stellar population is very old, say ~ 12 Gyr or so. Then, $BC_K \simeq 0.5$, and $L_T = 10^5 \, L_\odot$, corresponding to the luminosity of a typical globular cluster. Then from the relation

$$N_j = B(t) L_T t_j \tag{3.3}$$

one can estimate the number of stars in any of the post-MS evolutionary stages, once their duration is known from the stellar evolutionary models. To estimate the number of main sequence stars one has instead to integrate the IMF from its lower mass limit to the turnoff mass, which for $t = 12$ Gyr is $\sim 0.9 \, M_\odot$. Adopting a straight Salpeter IMF:

$$N_{MS} = A \int_{0.1}^{0.9} M^{-2.35} dM \simeq 14 \, L_T, \tag{3.4}$$

having inserted $A = 0.9 \, L_T$ following Eq. (2.25).

The resulting number of stars in the various evolutionary phases is reported in Table 3.1, which illustrates that for the given size of the population ($L_T = 10^5 \, L_\odot$) several evolutionary phases are well represented, that is, include a statistically significant number of stars, while for others (of very short duration) there is a clear dearth of objects. When such a number is smaller than unity, it is an excellent proxy for the probability to find one such star in the sampled population. For example, with such a small sampled luminosity the probability to find an AGB star in the superwind stage is only $\sim 0.2\%$, thus a luminosity of order of $\sim 5 \times 10^7$ should be explored in order to a have a reasonable chance to pick just one such star.

Table 3.1, for a very old population, or similar tables that can be constructed for other ages using Eqs. (3.1), (3.3) and (3.4), can be used to properly tailor specific observations if the goal is to collect the appropriate number of stars in a particular evolutionary phase, depending on the science goals of a specific project. Indeed, one can plan the observations in such a way that the telescope points to regions of the target where the surface brightness is appropriate to ensure an adequate sampling, or plan for multiple pointings in order to achieve the desired result. For example, one way of calibrating the TP-AGB contribution to the integrated light of stellar population for ages around 1 Gyr is to resort to Magellanic Cloud globular clusters. From evolutionary models one can expect a duration of the TP-AGB phase of the order of ~ 1 Myr, but with substantial uncertainty. This is indeed why one needs an empirical calibration, hence we need to assemble clusters in age bins around 1 Gyr such that a sufficient total luminosity is sampled in each bin. From Eq. (3.3), with $B(1 \, \text{Gyr}) \simeq 1.5 \times 10^{-11}$ one can see that one would need to sample a population of $\sim 10^6 \, L_\odot$ to have a chance to include ~ 15 TP-AGB stars, which would give roughly a $\sim 25\%$ uncertainty in the calibration. The Clouds have not been so prolific of bright globulars in this age range, which ensures an enduring controversy on the effective contribution of the TP-AGB phase of stellar evolution.

Equation (3.3) can also be used in the reverse direction, that is, to infer the duration of a specific evolutionary phase by counting the number of stars in that phase,

Table 3.1 Time spent in the various evolutionary phases and the corresponding number of stars for a ~ 12 Gyr old, solar metallicity population with $L_T = 10^5 \, L_\odot$. All star numbers for post-MS phases are derived from Eq. (3.3) (this table is an update of a similar table in Renzini, A. and Fusi Pecci, F. (1988, *Annu. Rev. Astron. Astrophys.*, 26, 199)).

Evolutionary phase (j)	t_j (yr)	N_j	Evolutionary phase (j)	t_j (yr)	N_j
MS	$> 10^{10}$	1.4×10^6	TP-AGB	$< 10^6$	< 2
SGB	3×10^9	6000	LPV	2.5×10^5	0.5
RGB	6×10^8	1200	SW-AGB	10^3	0.002
RGB-bump	4×10^7	80	P-AGB	3×10^5	0.6
TRGB	5×10^6	10	PN	2.5×10^3	0.005
HB	10^8	200	WD	10^9	2000
E-AGB	1.5×10^7	30	BS TP-AGB	3×10^6	0.3

MS: main sequence, N_j from Eq. (3.4); SGB: subgiant branch, from the MS turnoff to the base of the RGB; RGB: the red giant branch, ending with helium ignition at the RGB tip; RGB-Bump: the RGB feature (cf. Chapter 1) and the time spent within such a feature; TRGB: the upper portion of the RGB, within one bolometric magnitude from the RGB tip; HB: horizontal branch; E-AGB: early asymptotic giant branch; TP-AGB: thermally pulsing AGB; LPV: part of the AGB phase during which stars are long period variables; SW-AGB: final AGB phase during which the superwind operates and the star is enshrouded in an optically thick circumstellar envelope; P-AGB: post-AGB, from the AGB tip down to a luminosity 10 times lower, joining the white dwarf cooling sequence; PN: portion of the P-AGB phase during which the star is surrounded by an observable planetary nebula; WD: first part of white dwarf cooling sequence, down to a luminosity $10^{-3} \, L_\odot$; BS TP-AGB: TP-AGB progeny of blue stragglers, that is, merged binaries, assuming their production rate is ~ 5% of the evolutionary flux $b(t)$ (as the average frequency in galactic globular clusters).

and dividing it by $B(t) L_T$. This is actually the correct procedure to estimate the duration of phases such as the TP-AGB, the Post-AGB, and so on, that are controlled by mass loss processes, hence hardly predictable from first principles. For example, the duration of the LPV phase reported in Table 3.1 was inferred in this way, from the 4 LPVs present in the globular cluster 47 Tuc.

For a giant elliptical galaxy, with $L_T = 10^{11} \, L_\odot$, all numbers in Table 3.1 must be scaled up by a factor 10^6, hence all mentioned evolutionary phases are very well represented in such a galaxy, with exquisite statistical completeness. Yet, another problem is encountered: there are too many stars for the limited capabilities of our detectors, as they may crowd inextricably together. How to deal with such cases is discussed next.

3.1.2
The Stellar Population of a Pixel

To sample an adequate number of stars, one may be tempted to observe a field close to the center of the stellar system under study, where the surface brightness, hence the sampled luminosity, are higher. However, by moving to regions of higher surface brightness crowding will inevitably degrade the photometric accuracy, to the extent that meaningful stellar photometry may even become impossible. Clearly, an optimization is needed, finding a compromise between the conflicting needs of collecting an adequate sample of the stellar population, and that of being able to keep the photometric accuracy within an acceptable limit.

Having from Eq. (3.1) the luminosity L_T sampled by the whole CCD/IR-Array frame, the average luminosity sampled by each pixel of the detector is:

$$L_1^{pix} = \frac{L_T}{N^{pix}}, \qquad (3.5)$$

where N^{pix} is the number of pixels in the detector. In general, the actual resolution element exceeds the size of one pixel, and in ground-based observations this depends on seeing. The average luminosity sampled by one resolution element is then the product of L_T^{pix} times the number of pixels in one resolution element. In the following, *pixel* is used here to denote the actual *resolution element*, which is generally several detector's pixels. Equivalently, L_T^{pix} can be evaluated from Eq. (3.1) where for F_{oV} one takes the effective area of the PSF (in arcsec^2).

Crowding can affect the results of automatic reductions of detector frames in two main ways, respectively, in a moderate crowding and an extreme crowding regime. One effect is the reduced photometric accuracy, as the PSF fit to individual stars is degraded by the presence of other stars nearby, and the photometric error exceeds that expected from the mere Poisson photon noise. The less obvious effect arises when two or more stars fall on the same pixel, the photometric package is not able to de-blend them and mistakes them for a single star. The former effect broadens the stellar loci in the CMD and smears out features in the luminosity function, but the scientific content of the data, although somewhat reduced, remains useful. The latter effect is instead much more insidious, as it generates false stars that have no counterpart in the real stellar population, but mimic objects with attractive characteristics that may appear as plausible members of the stellar system being investigated. Figure 4.9 in Chapter 4 nicely illustrates this issue, showing how individual *stellar* sources on the ground-based image break up into small *star clusters* on the NICMOS2 image, thanks to the superior resolution of the latter.

Let

$$N_{j,pix} = B(t) L_T^{pix} t_j \qquad (3.6)$$

be the number per pixel of stars in the post-MS phase j; if $N_{j,pix} < 1$, this is also close to the probability that a pixel contains one star in phase j. Hence, the probability for a pixel to contain two such stars is $\sim N_{j,pix}^2$, and the number N_{2j} of such events over the whole frame is

$$N_{2j} = N_{j,pix}^2 N^{pix} = \left[B(t) L_T^{pix} t_j \right]^2 N^{pix} = \frac{[B(t) L_T t_j]^2}{N^{pix}}. \qquad (3.7)$$

Since L_T is proportional to the surface brightness μ of the target, the result is that the number of 2-j-star blends is proportional to the square of the surface brightness expressed in $L_\odot$ arcsec^{-2} units. This *square law* can be generalized to any kind of star pair (e.g., the blend of an HB and a RGB star), with the number N_{jk} of jk blends in the frame being

$$N_{jk} = N_{j,\text{pix}} N_{k,\text{pix}} N^{\text{pix}} = \frac{[B(t) L_T]^2 \, t_j t_k}{N^{\text{pix}}}. \tag{3.8}$$

In summary, the number of any kind of 2-star blends in the frame scales as the square of the surface brightness, that of triplets as the cube, that of quartets as the fourth power, and so on.

Blends which are not recognized as such by the photometric package are counted as single stars, and are assigned the cumulative luminosity of all the blended stars. This is clearly a way of producing bright stars that do not exist, doing so preferentially in the high surface brightness regions of a stellar system. Indeed, the crowding square law has another insidious effect: the surface brightness of a stellar system is usually highest at the center, and therefore the number of pairs sharing the same pixel will increase towards the center as the square of the surface brightness. As a consequence, the fake bright stars will appear much more centrally concentrated than the underlying light distribution, mimicking for example, the residue of a central burst of star formation that did not actually occur.

A safe criterion for the accurate photometry of stellar fields states that *meaningful magnitudes can be obtained only for stars brighter than* L_T^{pix}, that is, for $L > L_T^{\text{pix}}$. Astronomers are used to measure stars much fainter than the *sky*, that is, stars for which $L \ll L_{\text{sky}}^{\text{pix}}$, where $L_{\text{sky}}^{\text{pix}}$ is the luminosity of the sky sampled by the PSF: it is sufficient to integrate long enough. Albeit the above criterion seems to contradict this common practice, in reality it does not. The point is that the sky background is smooth and a stellar field is not smooth at all. The sky fluctuations are Poisson photon-noise dominated, while the fluctuations of a stellar field come in *quanta* made by whole stars(!). This makes it impossible to extract meaningful photometry for stars fainter than the above limit, i.e., for $L < L_T^{\text{pix}}$: that is, below such limits the frame is *burnt out*, almost like an old overexposed photographic plate.

Of course, the photometric accuracy degrades in a continuous fashion as the luminosity of stars to be measured approaches the L_T^{pix} limit. There will be stars measured with fairly good accuracy next to others whose luminosity is overestimated by a factor of 2 or so, and there is no way of distinguishing which one is right and which one is wrong. In such mined terrain, Eqs. (3.7) and (3.8) can be used to estimate the number of contaminants in a given magnitude bin, in such a way to beware of the result when the number of contaminants becomes a sizable fraction of the total star counts in that bin.

In any stellar population (and in most if not all photometric bands) the luminosity function steeply increases with decreasing luminosity, apart from a few bumps here and there. For example, this is illustrated by Figure 3.2, which shows the *I*-band luminosity function of the globular cluster 47 Tuc. The brightest stars, those at the RGB tip, are always accompanied by more stars in the immediately adjacent bin at fainter luminosity, and the same applies to stars all over the luminosity function apart from the two peaks due to the RGB bump and the HB. Ideally, it would be nice to deal with stars which are much brighter than any other in the field, such

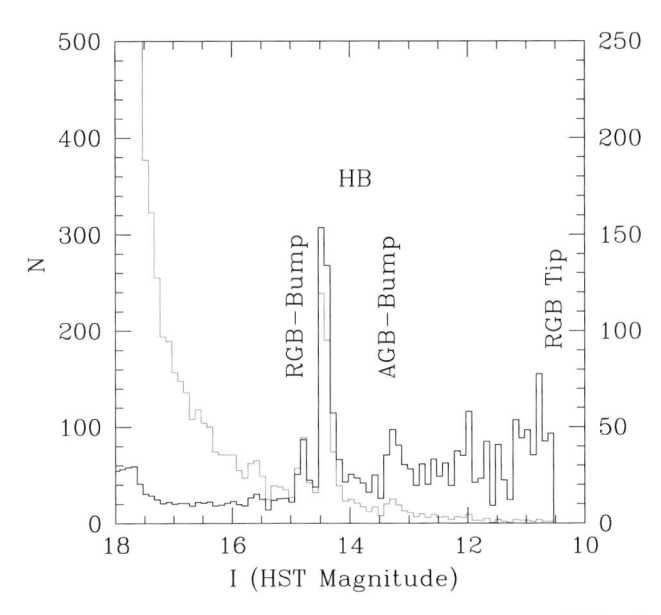

Figure 3.2 The *I*-band luminosity function of the globular cluster 47 Tuc (gray line) with HB, RGB-bump, AGB-bump and RGB tip labeling the corresponding features. The black line shows the contribution of each magnitude bin to the total *I*-band luminosity of the cluster ($\propto N \times 10^{-0.4I + \text{const.}}$, where N is the number of stars in the bin) (data from Brown, T. et al. (2005, Astron. J., 130, 1693)). For a color version of this figure please see in the front matter.

that the fainter stars would make a continuum background. Unfortunately this is almost never the case: most of the crowding problems in one magnitude bin arise from the fainter bins immediately adjacent to it. Figure 3.2 offers the opportunity to show the contribution to the total cluster light of stars in the various *I*-band bins. Although stars become very rare near the RGB tip, still their contribution to the total light is very large, larger indeed than that of the much more numerous stars at fainter magnitudes.

A concrete example is useful in order to see how these tools work in practice. Suppose one is willing to do stellar photometry of stars in elliptical galaxies in the Virgo cluster and use a 40 m extremely large telescope (ELT) assisted with high-Strehl adaptive optics, with a resolution of 0.01 arcsec. Thus, one has mod = 31.3 and a pixel area of $\sim 10^{-4}$ arcsec^2. The typical central surface brightness of a Virgo elliptical is $\mu_I \simeq 16$ mag/arcsec^2, and therefore Eq. (3.1) indicates that each pixel will collect $\sim 5.7 \times 10^3 \, L_\odot$ in the *I*-band (using $M_{\odot,I} = 4.1$), or $\sim 7 \times 10^3 \, L_\odot$ in bolometric (BC$_I \simeq 1.2$, cf. Figure 3.1).

The light of ellipticals is dominated by very old stars, so $B(t) \simeq 2 \times 10^{-11}$, and therefore from Eq. (3.6), with $L_T^{\text{pix}} = 7000 \, L_\odot$ and $t_{\text{TRGB}} = 5 \times 10^6$ yr from Table 3.1, one gets $N_{\text{TRGB,pix}} = 0.7$, or that on average each pixel has a $\sim 70\%$ probability to include one RGB star within one magnitude from the RGB tip (i.e., one TRGB star). By the same token, each pixel has a $\sim 50\%$ probability to contain 2 TRGB stars (i.e., ~ half the RGB tip stars will be measured ~ 0.75 mag brighter than they are), a ~ 35% probability to contain 3 TRGB stars, and so on. Clearly the crowding is way too severe for meaningful individual star photometry to be attempted. Either one improves the resolution by a factor of several, increasing the size of the telescope, or one investigates regions of lower surface brightness.

3.2
Simulated Observations and Their Reduction

The criterion for safe photometry formulated in the previous section is still qualitative. To make it more quantitative, we rely on simulations with the aim of assessing how the photometric accuracy degrades and how many fake stars appear as one approaches the L_T^{pix} limit.

A PSF photometry package runs through the image to identify peaks over a local background, fits a PSF to them, measures the local background, and assigns an instrumental magnitude, as a function of the excess counts associated to the stellar source. Moreover, if deviations from a single PSF are perceived (e.g., some stellar images appear elongated) the corresponding count peaks are modeled with multiple, partially overlapping PSFs in the attempt of de-blending two or even more stars crowded together. Thus, at least in principle, the PSF photometry can recognize sources fainter than the luminosity of the resolution element, but in practice the measurement is quite risky. This is so because the stellar background light, rather than smoothly distributed, is the result of the fluctuating overlap of many stars of luminosity just slightly fainter than the object one would like to measure.

Several sources of error affect the PSF magnitude of a given star, including the Poisson photon noise associated with the source, with the sky, and with the stellar background. One must also consider the noise due to the wings of the PSFs of the nearby stars and their overlapping portions, and, last but not least, the blend of sources within the same pixel mentioned earlier. The accuracy of the PSF photometry will further depend on how well the PSF is described, on its spatial sampling by the detector (i.e., one needs at least three physical pixels to sample the PSF adequately), and on the performance of the fitting algorithm. On top of this, the procedure is not completely automatic, requiring stacking multiple images and tuning the package parameters, especially to force the photometry at the faint magnitudes. There is then some dependence of the results on how much effort the astronomer dedicates to the measurement. In the following, the behavior of the photometric error is described for an idealized situation by analyzing the results of simulated observations, so to have full control of the experiment. Not all sources of error will be included, and in real frames the photometric accuracy may be worse. Still, these numerical experiments can illustrate the effect of crowding and provide useful guidance.

The simulations described below are meant to reproduce observations conducted with a 42 m telescope equipped with an adaptive optics module, which enables imaging very close to the diffraction limit. The adopted PSF includes a diffraction limited central core and an extended halo due to a natural seeing of $0''.6$, with a Strehl ratio 0.6 in the K band and 0.22 in the J band. In the K band the core of the PSF has a FWHM of 13 mas, and includes 40% of the total energy, whereas in the J band the core has a FWHM of 7 mas and includes 12% of the total energy. A detector pixel scale of 3 mas is assumed, along with an overall throughput (telescope + detector) of 30%.

Figure 3.3 shows (in gray) the theoretical K-band luminosity function for an old stellar population located at a distance of 18 Mpc, with a total $L_K = 3.84 \times 10^7 L_{K,\odot}$ (and a total mass of formed stars of $7.53 \times 10^7 M_\odot$). This luminosity function has been used as the input stellar list to generate a synthetic $3'' \times 3''$ frame, with a random spatial distribution of the stellar sources. The resulting surface brightness of the simulated target is therefore $\mu_K = 18\,\text{mag}/\square''$. The synthetic frames are produced for an integration time of 5 h, including the sky and thermal background, as well as the light produced from all stars, down to the minimum stellar mass of $0.1 M_\odot$.

The area of one resolution element is set equal to that of a circular aperture whose diameter is the FWHM of the PSF and therefore it is $\sigma_{RE} \simeq 1.3 \times 10^{-4} \square''$

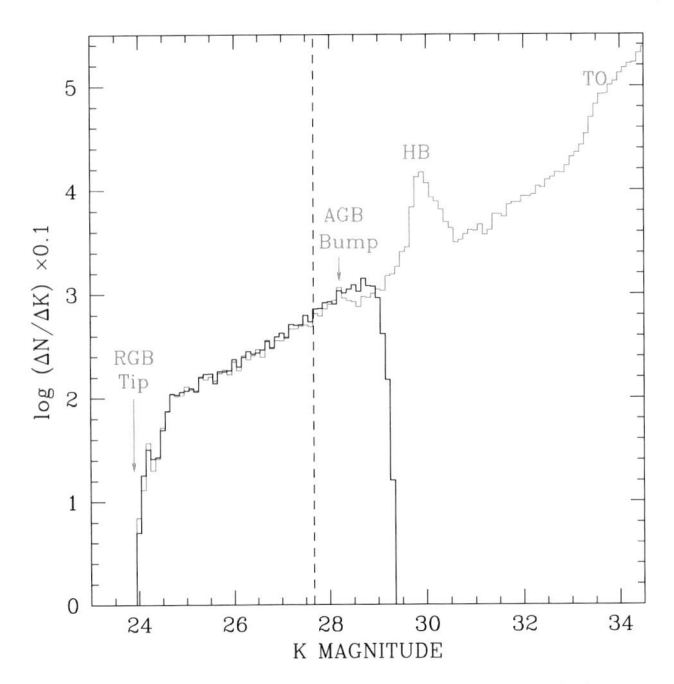

Figure 3.3 Input and recovered luminosity function. The gray line shows the K-band luminosity function of a simulated stellar population with $m - M = 31.3$, a flat age distribution between 10 and 12 Gyr and a wide metallicity distribution (from 0.02 to 1.5 times solar). The location of the typical features on the CMD are labeled. The black line shows the luminosity function measured with DAOPHOT (Stetson, P. (1987, *PASP*, 99, 191)) on a synthetic frame generated with the theoretical luminosity function as input, for an observation of 5 h exposure with a 42 m telescope. See text for other parameters. The dashed line marks the magnitude sampled by the resolution element.

in the K band. Given the surface brightness of the target, each resolution element is then sampling on average a luminosity of $\sim 560 L_{K,\odot}$, corresponding to an apparent magnitude of $m_{RE} \simeq 27.7$. This level is shown as a vertical dashed line in Figure 3.3, as well as in Figures 3.4 and 3.5.

The simulated image has been measured with a suitable package for PSF photometry to yield the *observed* luminosity function shown in Figure 3.3 as a black line. The luminosity function is well recovered up to $K \simeq 29$, where *incompleteness* sets in and gets rapidly worse with increasing magnitude. This abrupt drop depends on the detection threshold imposed when measuring the K image, which corresponds to $K \sim 29$. Besides this macroscopic effect, a more subtle discrepancy between the input and output luminosity function is visible in Figure 3.3, namely the excess of recovered stars, present virtually at all magnitudes, and particularly strong just above m_{RE}. This is due to blending, which causes the migration of stars to brighter magnitudes.

A more detailed view of the photometric performance and its dependence on magnitude is shown in Figure 3.4, where the input magnitudes m_{in} of individual stars are compared to their measured magnitudes m_{out}. Matching the detected sources to their input counterparts is not a trivial matter. Different criteria can be outlined, that tend to give somewhat different results, and all inevitably fail when reaching too faint magnitudes. In this experiment, a search radius of 1 pixel around each output star has been adopted to identify candidate mates in the input list, and

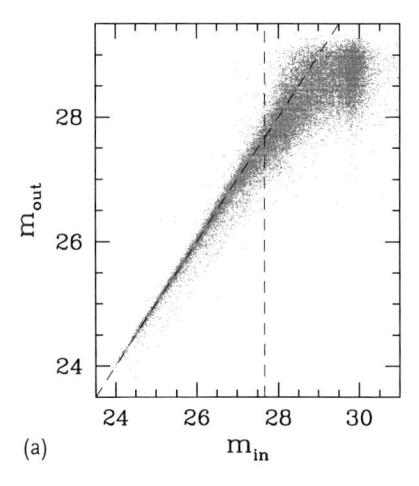

Figure 3.4 Comparison between the output and the input magnitudes as a function of the latter for the simulated stellar population shown in Figure 3.3. The dashed lines show the m_{RE} level (vertical) and the $m_{in} = m_{out}$ locus. In panel (b) the dotted lines show the theoretical error associated with photon noise from the stellar source plus the sky and instrument background; the solid lines show the 1σ levels of the Δmag distribution, computed separately for the positive (upper) and negative (lower) Δmag. These loci are drawn up to the value of m_{in} for which the full Δmag distribution is sampled: fainter than $m_{in} \simeq 28$ the detection threshold prevents the measurement of large and positive errors; for the same reason a fraction of the negative Δmag is missed fainter than $m_{in} \simeq 29$.

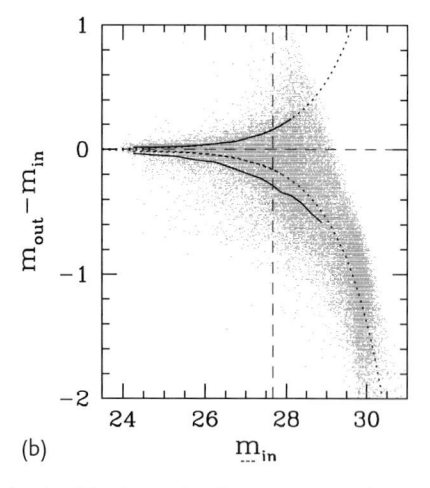

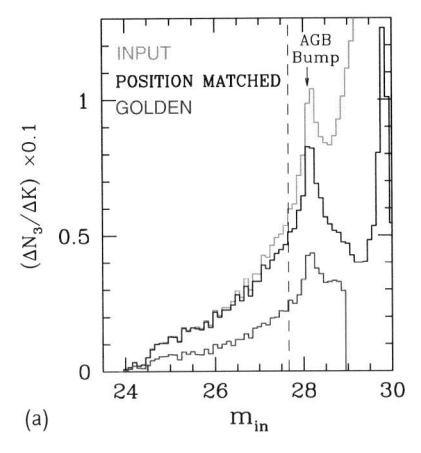

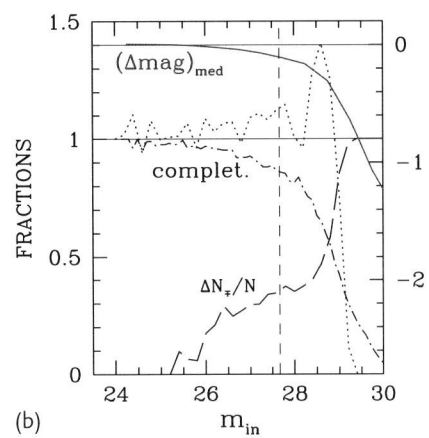

Figure 3.5 Further illustration of the same K-band simulation shown in Figure 3.3. (a) Input luminosity function (gray); distribution of those input stars which have been position-matched in the data reduction (black); and distribution of stars which have been position-matched and are measured with $|\Delta\mathrm{mag}| \leq \sigma_+$, where σ_+ is given by the upper line on Figure 3.4b. This criterion assumes that all sources of error but blending are symmetric, so that the distribution of the positive $\Delta\mathrm{mag}$ can be used to estimate empirically the photometric uncertainty associated with the photon noise and the stellar background. (b) Photometric quality as a function of input magnitude as quantified by the median $\Delta\mathrm{mag}$ (solid, to be read on the right axis); the ratio between the output and input luminosity functions (dotted); the ratio between the position-matched and the input luminosity functions (dot-dashed); and difference between the fraction of objects recovered brighter and those recovered fainter than the input magnitude (dashed). The vertical dashed lines show the location of m_{RE}, the magnitude sampled, on average, by one resolution element, while the two horizontal lines in (b) show the $\Delta\mathrm{mag} = 0$ (upper) and 100% completeness (lower) loci.

the brightest of them has been chosen as the counterpart. This criterion works well at bright magnitudes, but mate matching becomes progressively more ambiguous towards faint magnitudes. At these faint magnitudes, the surface density of candidate mates increases, photometric errors increase (at least by mere Poisson photon noise) and several input sources of similar magnitudes are present within the search area. When this happens it means that the *confusion limit* is being reached and the above mating criterion can produce spurious associations, like any other reasonable criterion one may consider.

The brightest stars are clearly recovered with a small error, but as the input magnitude increases, the distribution of the discrepancies $\Delta\mathrm{mag} = m_{\mathrm{out}} - m_{\mathrm{in}}$ widens as an increasing number of stars is measured either fainter or brighter than its true magnitude. The upper envelope of the $\Delta\mathrm{mag}$ distribution in Figure 3.4b increases up to $m_{\mathrm{in}} \simeq 28$; brighter than this limit, sources are measured also when the photon noise is unfavorable or the local background is high, albeit with output magnitudes fainter than the true ones (and vice versa when the local background is low). Fainter than $m_{\mathrm{in}} \simeq 28$ sources are recovered and measured *only* if at least one of two conditions apply: (i) the local background has fluctuated to a minimum and/or

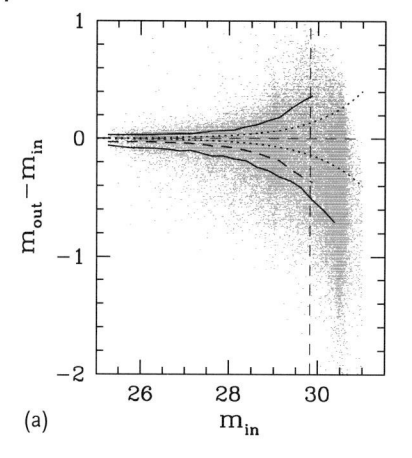

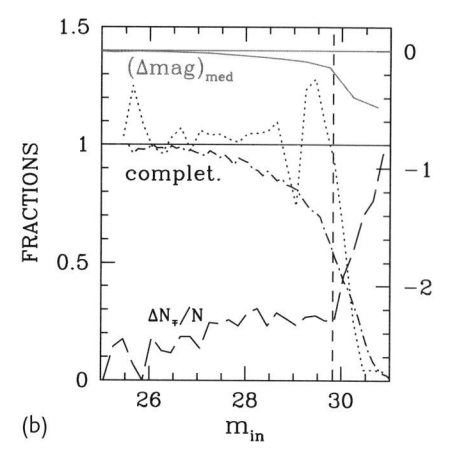

Figure 3.6 (a) Same as in Figure 3.4b, (b) same as in Figure 3.5b, but for the J-band image produced for the same synthetic stellar population. In the F_{oV} of $9\square''$ the total sampled luminosity is of $L_J = 2.6 \times 10^7 \, L_{J,\odot}$, corresponding to a surface brightness $\mu_J = 18.8$ at the adopted distance. The resolution element has an area of $\sigma_{RE} = 3.85 \times 10^{-5}\square''$, so that each resolution element samples $\sim 100 \, L_{J,\odot}$, on average. The corresponding magnitude is shown as a dashed vertical line in both panels.

the photon noise to a maximum, or (ii) sources are blended with other sources of comparable luminosity. These indeed are the conditions needed for the software to recognize that there is a peak above the estimated noise in the photon counts, and to start a PSF fitting. Thus, the resulting values of Δmag become increasingly negative with decreasing luminosity as sources are identified only if their magnitude is boosted by blending. The detection threshold used in this experiment corresponds to $m_{out} \simeq 29$ and generates the slanted cut-off in Figure 3.4b. Clearly, below $m_{in} \sim 29$, sources are found only when their magnitude is estimated incorrectly, as visible in Figure 3.4a. For example, simulated stars with $m_{in} = 29.5$ are recovered, but only as objects at least half a magnitude brighter. From this experiment it appears that incompleteness (i. e., some stars are not recovered at all) is only one aspect of faint photometry in crowded fields. What happens is much more insidious than mere drowning of sources below an ocean of background light. Actually, the more disturbing effect has the opposite result: stars from below a reasonable detection limit are artificially propelled to a higher luminosity by being lumped together with others and mistaken for a single star.

The upper 1σ level of the Δmag distribution in Figure 3.4b matches very well to the error expected for isolated stars, while the distribution of the negative Δmag is wider, showing an excess of sources measured brighter than their true magnitude. This *asymmetry* is due to blending, and is detected over the whole range of input magnitudes. The maximum Δmag that can be obtained from summing the light of two stars is -0.75, and corresponds to the case in which the blended objects have the same luminosity. Such cases are present in the simulation virtually over the whole magnitude range, even near the tip of the luminosity function at $K \sim 24$, and

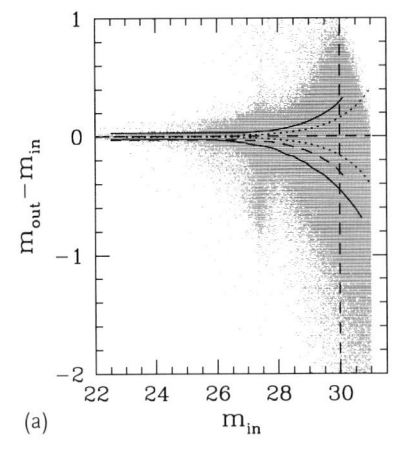

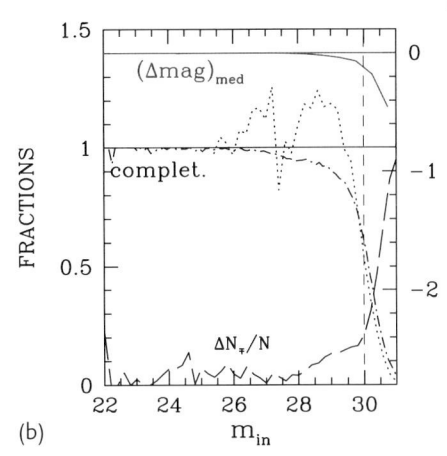

Figure 3.7 Same as Figure 3.6 but for a stellar population akin to late-type galaxies, generated assuming a constant SFR over 12 Gyr, and an age-metallicity relation similar to what characterizes the disk of the Milky Way. A distance modulus of $m - M = 28.3$, and a surface brightness of $\mu = 18.95$ in the J band have been adopted. The instrumental set up is the same as for the simulation in Figure 3.6 except for the F_{oV}, which in this case is of $12'' \times 12''$. With these parameters, the total luminosity sampled is $L_J = 2.27 \times 10^7\ L_{J,\odot}$ and the resolution element samples just $6\ L_{J,\odot}$. The corresponding magnitude is shown as a dashed line in both panels. Compared to the case in Figure 3.6, this field is 3 magnitudes closer, so that the red clump is well detected at $J \simeq 27.5$. Notice that the RGB tip is located at $J \simeq 21$, brighter than the plotted magnitude range.

have no counterpart in the positive Δmag region of the diagram. In spite of being rare at bright magnitudes, these fake stars may lead to the wrong interpretations, as discussed earlier, and their number can be evaluated for specific evolutionary phases using Eq. (3.7). At faint magnitudes however this blending effect dominates, and the majority of stars are measured brighter than they are.

From what was discussed thus far it follows that a more complete way of describing an artificial star experiment such as the present simulation would be via a matrix n_{ij} giving the number of stars with input magnitude in the *i*th bin, which are recovered in the *j*th bin. In the absence of crowding effects the matrix is symmetrical, and the population of non-diagonal elements increases with magnitude, dominated by Poisson photon noise. When crowding is important, blending has the effect of boosting the non-diagonal side of the matrix where stars are recovered in brighter bins.

Figure 3.5 further illustrates the photometric outcome of this experiment. The *position-matched* luminosity function plotted as a black line in Figure 3.5a shows the portion of the input luminosity function which is recovered taking into account only the positional coincidence of the input and output sources, irrespective of their Δmag. Different from the output luminosity function shown in Figure 3.3, this input matched function has no excess in any bin, as only a fraction of the stars are associated with an output mate according to this matching criterion. The *golden* luminosity function includes only those stars which are positionally matched as above, and also measured with a photometric error corresponding to a $\sim \pm 1\sigma$ photon noise. Comparing the three luminosity functions it appears that the true photometric quality is somewhat worse than grasped from just Figure 3.3, where the star counts brighter than the AGB bump appear to be well recovered. Actually, to a good extent unrecovered stars are replaced by other intrinsically fainter stars that are boosted by blending.

Figure 3.5b shows several quantities meant to further illustrate the completeness and photometric errors in this experiment. The dotted and dot-dashed lines show the completeness computed in two distinct ways: the former is the ratio between the output and input luminosity functions as seen in Figure 3.3; the latter is the ratio between the input matched and the input luminosity functions. Counterintuitively, completeness defined in the former way is higher than 1 as blending results in an excess of measured stars at almost all magnitudes. Following instead the other definition, completeness drops monotonically with increasing magnitude, and this is the meaning of completeness taken in most photometric studies. Clearly, the effect of blending is forgotten. The dashed line shows the *asymmetry* of the photometric errors, quantified as the excess of stars measured with negative Δmag with respect to the number of stars measured with positive Δmag, $(N^- - N^+)/N$. Finally, the solid line shows how the median Δmag (to be read on the right axis) becomes increasingly negative as the input magnitude gets fainter. The vertical dashed line shows the magnitude m_{RE} sampled on average by one resolution elements: at this level the recovered luminosity function exceeds the input function by $\sim 15\%$, and this excess explodes below m_{RE}; whereas a significative asymmetry is present up to ~ 2 magnitudes brighter than m_{RE}.

Figure 3.6 refers to the same stellar populations and observing conditions illustrated in the Figures 3.3, 3.4, and 3.5, but now observed in the J band. The cutoff at faint magnitudes is less abrupt compared to what happens in the K band, partly because of the smaller resolution element and lower background favoring the detection of faint stars. The 1σ width of the positive Δmag distribution now exceeds the expected theoretical uncertainty for isolated sources (dotted lines), showing the effect of the background stellar light which increases Δmag at given m_{in}. This contribution is not appreciated in Figure 3.4 because of the large contribution of the photon noise associated with the sky and thermal background in the K band, which dominates the error in this band. The dashed line in Figure 3.6a mirrors the upper solid line, representing the expected negative Δmag due to all sources of photon noise. The measured 1σ width of the negative Δmag distribution is clearly wider than this, again signalling blending at work. As for the K band, occasionally at all magnitudes stars are measured much brighter than they are and, in spite of the smaller resolution element, the asymmetry is noteworthy up to the top of the luminosity function (see Figure 3.6b). Fainter than m_{RE} the asymmetry boosts, and the median Δmag rapidly becomes worse. In this case photometry below the magnitude of the resolution element is meaningless, but also above m_{RE} there is a persistent, albeit sparse, population of objects measured to be substantially brighter.

Figure 3.7 shows the same plots as Figure 3.6 but for a different stellar population. In this case a constant star formation rate over 12 Gyr has been adopted, so that the color magnitude diagram also contains massive young stars. Other adopted parameters for this simulation include a distance of 4.6 Mpc, a surface brightness of $\mu_J = 18.9$, and a F_{oV} of $12'' \times 12''\square''$. The detected stars span a range of more than 10 magnitudes, including the well populated red clump at $m_{in} \sim 27.5$ (J band). Figure 3.7a shows that some of the clump stars are measured brighter than normal up to a $\Delta mag \simeq -0.75$, which corresponds to the complete blending of two such clump stars. Overall, the photometric quality of this example is better than that shown in Figure 3.6: at the same input magnitude the 1σ width of the Δmag values is smaller, completeness is better and the asymmetry remains smaller than 10% down to $J \simeq 28.5$. This reflects the lower crowding conditions of this case with respect to the previous one, with a 16 times larger FoV sampling about the same luminosity. Still, the effect of blending is clearly visible in Figure 3.7a as a wider distribution of the negative Δmag with respect to the positive Δmag (i.e., the difference between the solid and the dashed lines), as an excess of objects in the output luminosity function for magnitudes approaching m_{RE} (dotted line in Figure 3.7b), and as a rapid increase of the asymmetry setting in right at $m_{in} \simeq m_{RE}$.

The average magnitude sampled by the resolution element is shown as a vertical dashed line in Figures 3.3–3.7: it can be noticed that, remarkably, the photometric accuracy gets abruptly worse in the vicinity of this level as do both asymmetry and incompleteness. On the basis of the theoretical considerations developed in the previous section, and of the experiments just illustrated, one can then re-iterate the following simple criterion for safe photometry in crowded fields: *meaningful photometry over stellar fields can be attempted only for stars brighter than the confusion limit represented by the average luminosity sampled by one resolution element.* As one approaches this limit, stars are measured sensibly brighter than they truly are, for two reasons: (i) it becomes increasingly difficult to detect these stars against the background, so that the peaks get identified preferentially in regions with negative fluctuations of the background, and, especially, (ii) multiple sources within the same resolution element become very frequent.

The simulations presented above do not refer to situations of extreme crowding, that is, a substantial portion of the luminosity function is much brighter than the magnitude sampled by the resolution element. For this reason it has been possible to discuss several sources of error. Blended stars are present basically at all magnitudes, but their number grows tremendously only when approaching the magnitude of the resolution element. Therefore there is always a certain number of stars that are measured brighter than they are, and tools to estimate this number are provided in Section 3.1.2. Yet, one can easily imagine what would happen to the resulting photometry if the same experiments were conducted on a target with a much higher surface brightness, or placed at a larger distance, such that, for example, the average luminosity sampled by one resolution element is comparable, or even brighter, than the brightest stars in the field.

Based on stellar evolution concepts and tools, it is relatively easy to figure out what to expect from the reduction and photometry of CCD frames of stellar fields, that is, what one can expect to gain from the process and what should not be attempted. This results in a very simple, easy to implement *rule of thumb* for safe photometry in crowded fields that can have a variety of useful applications, for example helping an insightful interpretation of observed color-magnitude diagrams, or providing tools to properly design observational projects. Numerous examples of macroscopic violations of this simple rule can be found in the literature, along with their extravagant results.

Acknowledgments

The simulations presented in this chapter were tailored as part of the Science Cases for the Phase-A study of the MICADO camera for the European Extremely Large Telescope (E-ELT). Synthetic color-magnitude diagrams have been calculated by G.P. Bertelli, synthetic frames were produced by R. Falomo and M. Uslenghi, and reductions were done using the DAOPHOT photometric package (Stetson, P. (1987) *Publ. Astron. Soc. Pac.*, **99**, 191) by S. Zaggia. We are indebted to these collaborators for their contribution and scientific discussions.

Further Reading

Renzini, A. (1998) *Astron. J.*, **115**, 2459.
  Presents concepts and ideas over which this chapter expands.

Drukier, G.A., *et al.* (1988) *Astron. J.*, 95, 1415
  Presents a matrix approach to photometry in crowded fields.

4
Age Dating Resolved Stellar Populations

Measuring the age of stellar aggregates, from local star clusters all the way to galaxies at the highest redshifts, is one of the most prominent application of stellar evolution theory and of its incarnation in stellar population synthesis. Age dating stellar populations allows us to establish a cosmic chronology of the formation and evolution of galaxies and their constituents that is independent of the cosmological model. Yet, it can be inscribed in the cosmological context so to achieve a comprehensive view of the evolution of the Universe as a whole, as well as of galaxies and cosmic structures.

Similar to the indicators of the familiar *distance ladder,* also in age-dating methods we may speak of primary, secondary and tertiary age indicators, with the aim of tracing how errors can propagate from one indicator to the next, in an *age ladder*. In the case of distances, the starting point are the trigonometric parallaxes, then used to calibrate the main sequence of stellar clusters, and from them the RGB tip and Cepheids, and finally supernovae, so to bypass local peculiar motions and enter the undisturbed Hubble flow. How errors propagate from primary, to secondary, to tertiary distance indicators is rigorously traced, using a strictly empirical approach, and avoid relying on models that may bring systematic errors that are difficult to quantify.

In the case of the age ladder, a similar approach is in order. Although it is impossible to avoid using the stellar evolution clock as given by theory, the use of models needs to be restricted to the indispensable minimum, and for the rest proceed empirically keeping rigorous track of the errors and how they propagate.

This chapter deals with current dating methods of resolved stellar populations, starting with globular clusters, and using their ages to calibrate other age indicators. This also offers the opportunity to show some of the most representative results of recent photometric studies of resolved populations.

4.1
Globular Cluster Ages

Historically, measuring globular cluster (GC) ages has been a major driver to refine stellar evolution models and bring them to the best possible state of the art.

Stellar Populations, First Edition. Laura Greggio and Alivio Renzini.
© 2011 WILEY-VCH Verlag GmbH & Co. KGaA. Published 2011 by WILEY-VCH Verlag GmbH & Co. KGaA.

The same applies to GC observations, both photometrically (especially with HST) and spectroscopically, thus submitting stellar models to more and more stringent observational checks. Indeed, with GC ages a strict lower limit on the age of the Universe was at stake, and for quite some time in the 1980s they have been an embarrassment for the young, matter dominated Universe many were advocating at the time. Indeed, GC ages contributed to keeping open the door to a cosmological constant (or some equivalent to it) prior to the unambiguous discovery of the universal acceleration. Now, at the time of the *precision/concordance cosmology*, the age of the Universe issue appears to be solved, and as a *side benefit* we can be quite confident on the reliability of stellar evolution models, that never indicated GC ages significantly below ~ 12 Gyr.

4.1.1
Absolute and Relative Globular Cluster Ages

The mainstream method to measure absolute globular cluster ages relies of the relation between the luminosity of the main sequence turnoff of theoretical isochrones, the age, and chemical composition, such as that given in Eq. (2.4). In turn, the absolute visual magnitude of the turnoff is given by $M_V^{TO} = V_o^{TO} - \text{mod}$, where V_o^{TO} (the turnoff apparent magnitude corrected for extinction) is the directly observable quantity, and mod is the distance modulus of the cluster. This equation allows one to estimate the relative importance of the uncertainty in each of the input quantities (IQ, namely V_o^{TO}, mod, Y, [Fe/H], and [α/Fe]) in determining the total uncertainty in the age determination.

This is reported in Table 4.1, which illustrates how errors in the IQs – σ(IQ) – propagate into age errors. On purpose, the σ(IQ) given in this table refer to the situation in the pre-*HST* and pre-*Hipparcos* era; this allows one to appreciate the progress that has been achieved thanks to these two space observatories. Clearly, the error budget used to be dominated by distance errors, which with a typical one quarter of magnitude uncertainty in distance modulus, $\sigma(\text{mod}) \simeq 0^m.25$ mag, implied a relative error in age of $\sim 22\%$, or some ± 3 Gyr. All other input quantities convey substantially smaller errors, and therefore it was on determining more accurate distances that the attention has focused ever since.

Table 4.1 Error budget of globular cluster ages in 1990 and 2010.

Input quantity	$\sigma(t)/t$	σ(1990)	$\sigma(t)/t$ %	σ(2010)	$\sigma(t)/t$ %
V_o^{TO}	$0.86\sigma(V_o^{TO})$	$\pm 0^m.10$	9	± 0.03	3
mod	$0.86\sigma(\text{mod})$	$\pm 0^m.25$	22	± 0.10	10
Y	$0.99\sigma(Y)$	± 0.02	2	± 0.02	2
[Fe/H]	$0.3\sigma([\text{Fe/H}])$	± 0.3	9	± 0.10	3

Having abandoned semiempirical methods of distance determinations, such as for example, calibrating the absolute magnitude of RR Lyrae variables because of its extensive use of pulsation and/or evolution theory, efforts concentrated on a more basic approach, relying on the use of trigonometric parallaxes of other distance indicators. The *Hipparcos* mission was instrumental for the subsequent progress in globular cluster age determinations, providing trigonometric parallaxes for local subdwarfs and other metal-poor stars and of local low-mass white dwarfs.

Subdwarfs are main sequence stars with subsolar metallicity, hence they can be used to construct fiducial ZAMS of different metallicities. The distance to individual clusters can then be determined by fitting the cluster ZAMS to the fiducial ZAMS with the same metallicity. Having measured GC distances in this way, the resulting M_V^{TO} is shown in Figure 4.1 as a function of total metallicity [M/H] for a large sample of globular clusters, all with CMDs from HST observations, and extending from very metal-poor to near-solar for the two bulge clusters NGC 6528 and NGC 6553. The procedure followed to construct this figure goes through measuring first the *relative* cluster-to-cluster differences in M_V^{TO}, then measure the absolute value of M_V^{TO} for a cluster with the most accurate input quantities, and then through it calibrate M_V^{TO} for all other clusters. Error bars of M_V^{TO} shown in Figure 4.1 actually refer only to errors in V_o^{TO}, and can be as small as ±0.03 mag for clusters with low reddening and best photometric data, and as large as ±0.15 mag or more for highly obscured clusters. To get the total error in M_V^{TO} the error in distance modulus for the best studied, calibrating cluster (∼ ±0.08 mag) should be added in quadrature.

Before commenting on the results shown in Figure 4.1, it is worth mentioning how progress was achieved with respect to *old* typical errors reported in Table 4.1. HST photometry has led to the reduction of the error in V_o^{TO} from ∼ 0.10 mag to ∼ 0.03 mag in the best studied, low-reddening clusters. The error in absolute

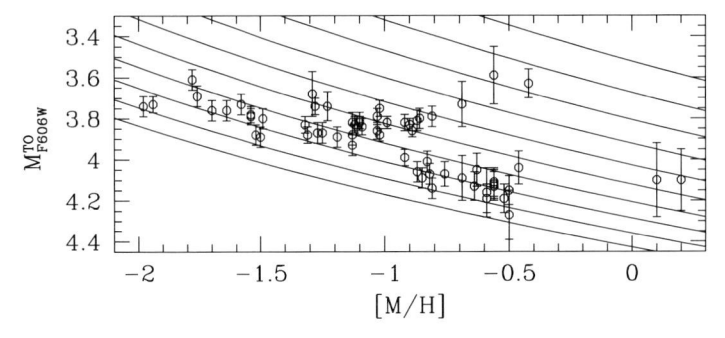

Figure 4.1 The absolute (Vega) magnitude in the HST filter F606W of the main sequence turnoff as a function of the total metallicity, for several ages from 6 Gyr (top) to 15 Gyr (bottom). The absolute magnitude of the turnoff of a set of globular clusters are plotted, together with their estimated relative error bars (adapted from Marin-Franch, A. et al. (2009, *Astrophys. J.*, 694, 1498) with the addition of the two metal-rich clusters NGC 6528 and NGC 6553 for which photometry from Brown, T. et al. (2005, *Astron. J.*, 130, 1693) and Ortolani, S. et al. (1995, *Nature*, 377, 701) was used, lines are fits to stellar evolutionary tracks by Dotter, A. et al. (2007, *Astron. J.*, 134, 376), which assume a helium enrichment $Y = 0.245 + 1.54Z$).

distance modulus has been reduced from ~ 0.25 mag down to $\lesssim 0.1$ mag, primarily thanks to Hipparcos calibration via local subdwarfs. In addition, high-resolution spectroscopy of GC stars on 8–10 m class telescopes brought typical errors in [Fe/H] from ~ 0.3 down to ~ 0.1 dex. All in all, after two decades of effort, the absolute ages of the best studied clusters are now measured with an *internal* error of $\sim 10\%$. But, of course, ages rely on stellar models, and using last generation models from different groups can give age differences on the order of $\sim 10\%$, which can be taken as a *reasonable* estimate of the *systematic* error.

Now, back to Figure 4.1, there are many features worth noticing. Most clusters align with small dispersion along the 13 Gyr line, showing that most clusters, no matter whether in the halo or in the bulge, formed within a short time interval, for which an upper limit of ~ 1 Gyr can be set. A minority of globular clusters appear instead definitely younger than the others, and according to a widespread interpretation they may have formed in dwarf galaxies that were then accreted by the Milky Way. Particularly important are the two most metal-rich clusters in the bulge, which however are also the most heavily obscured of the whole sample, hence with the largest error in V_o^{TO}. One can say that their age is consistent with being the same as that of most other clusters, but one cannot exclude that they may be ~ 1–2 Gyr younger. The limiting factor is here the reddening.

Repeating the same analysis using a different set of theoretical isochrones leads to the same conclusions, with just a 1 Gyr systematic difference, that is, an age of ~ 12 as opposed to 13 Gyr. Thus, one can conclude that most globular clusters in the Milky Way are 12.5 Gyr old, ± 1 Gyr (internal error), ± 1 Gyr (systematic error).

As mentioned above, an alternative distance scale to globular clusters relies on white dwarfs as calibrators. The mass of currently forming white dwarfs in globular clusters is fairly well determined by the condition that the AGB must not extend above the RGB tip (cf. Chapter 1), which implies $M_{WD} = 0.53 \pm 0.02 M_\odot$. Thus, the white dwarf cooling sequence of globular clusters can be fit to white dwarfs of similar mass and known trigonometric parallax to get the distance modulus of the cluster. The limitation of this alternative approach is that it can be attempted only for the nearest clusters, given the faintness of the targets. Nevertheless, in the few cases in which it has been attempted the method gives distances in fair agreement with those derived from the subdwarf method, thus providing an independent consistency check for the derived distances, hence ages.

4.1.2
Globular Clusters with Multiple Populations

The very first, photographic CMD of ω Cen, the most massive globular cluster in the galaxy, showed a wide RGB, much wider than the estimated photometric errors. Spectroscopy soon demonstrated that this was due to a wide metallicity spread, as it was natural to expect given that RGB colors are more sensitive to [Fe/H] than to any other parameter. For the following four decades ω Cen has been considered an exception, as virtually all other clusters (with the possible exception of M22) appeared to be indistinguishable from an SSP. Still, it was a big surprise when even the main sequence of this cluster was resolved into at least three distinct main sequences, and the surprise grew even bigger when it was demonstrated that the bluer main sequence is more metal-rich than the red one. Inescapably, this *color inversion* demands that the blue main sequence stars are highly enriched in helium with respect to the others, with $Y \sim 0.35$, *vis à vis* a primordial (Big Bang) helium of $Y \simeq 0.24$, presumably pertinent to stars on the red main sequence.

Figure 4.2a–c gives an idea of the extreme complexity of the stellar populations of this cluster, with at least five distinct sequences, and possibly others. Similarly complex is the HB morphology of the cluster, which extends to very high temperatures and terminates with an apparent double hook. Figure 4.3 shows another remarkable example of a multipopulation cluster, with three distinct, well separated main sequences, this time all with indistinguishable metallicity, and again the split must be due to different helium abundances.

These two clusters represent extreme cases, but photometric evidence of multiple populations has now been secured for over a dozen clusters, especially among the most massive ones, that is, those over $\sim 10^6 \, M_\odot$. Reconstructing the sequence of events that led to the clusters as we see them today, with their multiple, helium enriched populations, remains a big challenge. However, while posing new puzzles, an enduring one appears to have been solved. Indeed, the complex HB

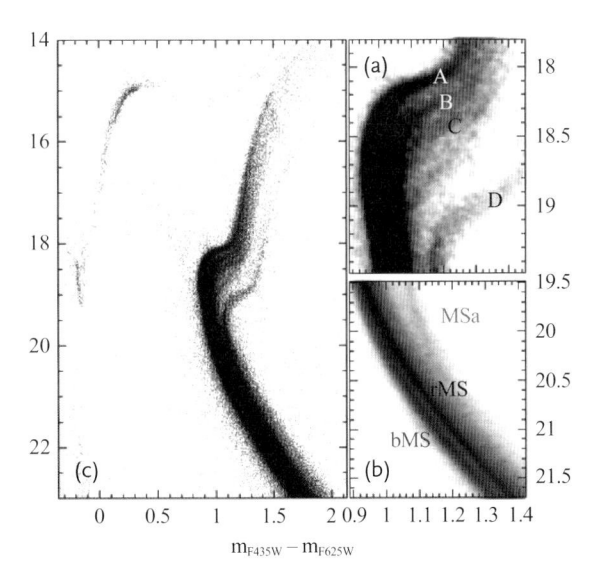

Figure 4.2 An HST/WFC3 color-magnitude diagram of the globular cluster ω Cen, showing the main subpopulations. (a) A blow up of the turnoff plus SGB part of the diagram, with four main components labeled A to D, with a hint of a fifth component between C and D. (b) A blow up of the middle main sequence, with the blue main sequence (bMS) being the helium-rich one, with a metallicity [Fe/H] ~ 0.3 dex higher than the red main sequence (rMS). (c) The large spread of HB stars magnitudes in $0.5 \lesssim m_{F435W} - m_{F625W} \lesssim 1$ is due to most of these stars being RR Lyrae variables caught at random phases (source: Bellini, A. *et al.* (2010, *Astron. J.*, 140, 631)) (Reproduced by permission of the AAS.).

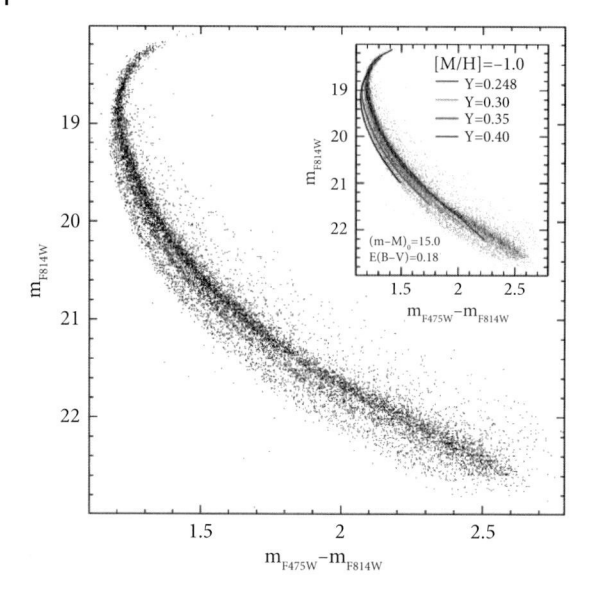

Figure 4.3 The main sequence and turnoff region of the HST/ACS color-magnitude diagram of the globular cluster NGC 2808. The insert shows overplotted isochrones with different helium abundances, from near primordial up to $Y = 0.40$. The latter value appears to be appropriate for stars on the bluest main sequence (source: Piotto, G. *et al.* (2007, *Astrophys. J.*, 661, L53)) (Reproduced by permission of the AAS.). For a color version of this figure please see in the front matter.

morphology of several clusters (in particular with multiple clumps such as in ω Cen and NGC 2808, and others) had escaped our understanding until it became clear that the main driver is indeed a discrete distribution of helium abundances.

Many other, if not all globular clusters show less dramatic but still clear evidence of multiple populations, in particular from sodium and oxygen abundance variations/anticorrelations, and it remains to be established what mix of processes conspired to produce the complex populations of such clusters. Besides successive stellar generations, such processes may include accretion from AGB ejecta, or even star formation growing from the low-mass stellar seeds of the first generation, in the extremely dense stellar environment of the young, massive clusters. Indeed, forming new stars from scratch in a volume already inhabited by over $\sim 10^5$ stars per cubic parsec may be less likely than just grow by accretion on preexisting, low-mass stars. In any event, the discovery of multiple populations in globular clusters has opened a new frontier in stellar population studies, and on objects that were thought to be as simple stellar populations as Nature could possibly produce.

4.2
The Age of the Galactic Bulge

Dating stellar populations in the Galactic bulge provides an indispensable piece of information for the understanding of the formation and evolution of galactic bulges in general. In particular, in no other bulge can we explore spectroscopically the metallicity distribution and reach well below the MS turnoff with deep photometry. Thus, one can reconstruct in much greater detail the star formation history and chemical enrichment of the bulge.

However, age dating bulge stars is complicated by several factors, such as crowding, depth effects, differential reddening, metallicity dispersion, and especially contamination by foreground disk stars. Some of these limitations can be circumvented by studying globular clusters that belong to the bulge, hence straightforward age dating methods can be applied. This is the case of the two globular clusters NGC 6528 and NGC 6553, that physically reside in the bulge, their kinematical properties (radial velocity and proper motion) are consistent with those of the bulge, and their metallicity is well within the metallicity distribution of the bulge. Thus, they have played an important role in the bulge dating efforts.

Since GCs are quite compact objects, their stars greatly outnumber foreground stars over the same area, and therefore disk contamination is just a minor concern. Figure 4.4 shows the metallicity distribution function of stars in Baade's Window (the low-reddening zone of the bulge located at $b = -4°$), and in two other outer fields along the bulge minor axis. The metallicity of the two clusters ([Fe/H] $= -0.2 \pm 0.2$ and -0.1 ± 0.2, respectively for NGC 6553 and NGC 6528) is also indicated, showing that they are indeed representative of the typical metallicity of bulge stars.

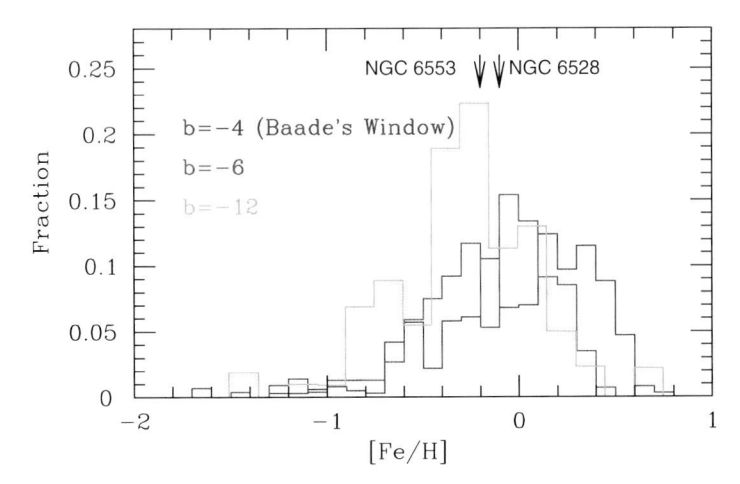

Figure 4.4 The metallicity distribution function of RGB stars in three distinct fields of the Galactic bulge, along its minor axis. The galactic latitude of the fields is indicated. The metallicity [Fe/H] of the two bulge clusters NGC 6528 and NGC 6553 is marked with arrows (data from: Zoccali, M. *et al.* (2008, *Astron. Astrophys.*, 486, 177)). For a color version of this figure please see in the front matter.

4 Age Dating Resolved Stellar Populations

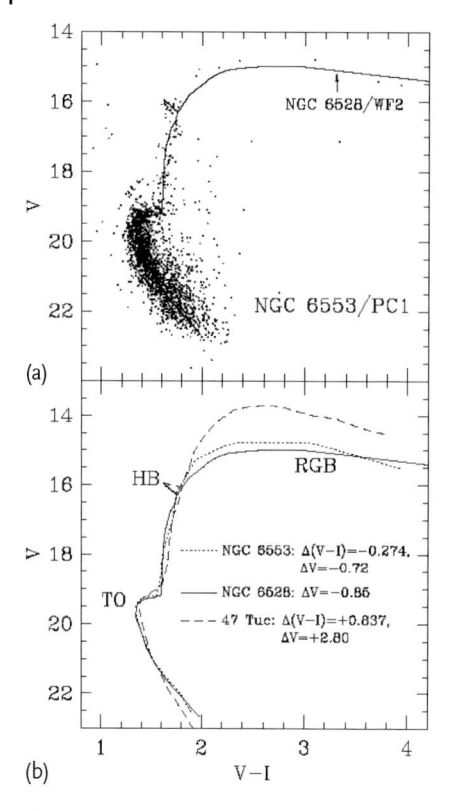

Figure 4.5 (a) The HST/WFPC2 color-magnitude diagram of the bulge globular cluster NGC 6528 with superimposed the mean cluster locus of its near twin cluster NGC 6553. (b) The mean cluster loci of the two bulge clusters NGC 6528 and NGC 6553 and of the inner halo cluster 47 Tuc. Magnitude and color shifts as indicated have been applied in order to bring into coincidence the stubby red HB of the three clusters (source: Ortolani, S. et al. (1995, Nature, 377, 701)).

Figure 4.5a shows the HST/WFPC2 color-magnitude diagram of NGC 6553, with a superimposed ridge line of the color-magnitude diagram of NGC 6528, indicating that the two clusters are almost perfect twins. This is further demonstrated by Figure 4.5b, where the ridge lines of the two clusters are compared to each other, as well as to that of the inner halo GC 47 Tuc ([Fe/H] = −0.7). Vertical and horizontal shifts have been applied to the three loci in such a way to bring into contact their HBs, a procedure that brings into virtually perfect coincidence also their MS turnoffs. This ensures that the age of the two bulge clusters is nearly the same as that of the halo cluster, apart from a small difference that must be induced by the metallicity difference of ~ 0.5 dex, which is difficult to precisely quantify.

Since the age of 47 Tuc is ~ 12 Gyr (cf. Figure 4.1), this comparison is sufficient to demonstrate that the bulge underwent rapid star formation and metal enrichment up to at least near-solar abundance, some 12 Gyr ago.

However, the bulge field population appears to contain stars with supersolar metallicity, well above that of the two clusters, and therefore the bulge may have

4.2 The Age of the Galactic Bulge

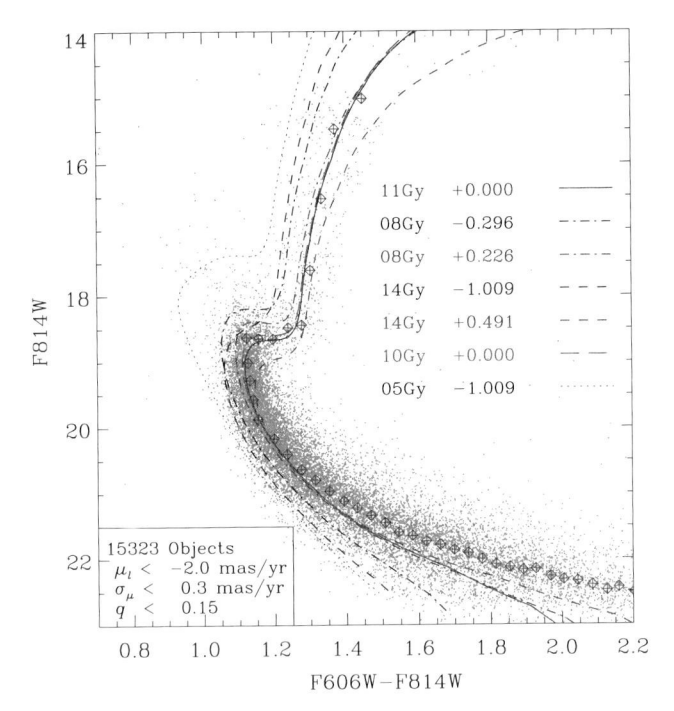

11Gy	+0.000
08Gy	−0.296
08Gy	+0.226
14Gy	−1.009
14Gy	+0.491
10Gy	+0.000
05Gy	−1.009

15323 Objects
$\mu_l < -2.0$ mas/yr
$\sigma_\mu < 0.3$ mas/yr
$q < 0.15$

Figure 4.6 The HST/ACS color-magnitude diagram of a Galactic bulge field with proper motion selected stars to ensure an almost pure bulge membership. Isochrones with different ages and [Fe/H] metallicities as indicated in the insert are overplotted (source: Clarkson, W. et al. (2008, Astrophys. J., 684, 1110)) (Reproduced by permission of the AAS.). For a color version of this figure please see in the front matter.

been forming stars (or accreting them) well after the formation epoch of the two clusters.

Whether the stars with supersolar metallicity are nearly coeval with the two clusters, or have formed much later, can only be assessed by exploring the field population of the bulge itself. Figure 4.6 shows the color-magnitude diagram of a bulge field at $(l, b) = (1.°25, -2.°65)$ including only proper motion selected stars to ensure their bulge membership. In practice, selected stars are in high-velocity, almost radial orbits, which ensures they do not belong to the disk. The average metallicity of bulge stars within $\sim 4°$ from the galactic center is very close to solar metallicity, and the corresponding 11 Gyr old isochrone in Figure 4.6 provides an excellent fit to the data. Therefore, with the bulk of stars in the bulge having formed $\sim 11-12$ Gyr ago, precursors of Milky Way analogs in the verge of forming their bulge are to be found at or beyond redshift ~ 2.

More difficult to ascertain is to which extent the more metal-rich component with [Fe/H] > 0 is younger than the less metal-rich one, basically because of the age/metallicity degeneracy (cf. Chapter 2). Spectroscopic and/or multicolor photometric determinations of the metallicity of individual stars near the turnoff are necessary to break this degeneracy, and therefore proceed to compare isochrones with stars in selected metallicity bins. At the time of writing this manuscript the relevant HST/WFC3 observations have been completed over four bulge fields, and the data are being analyzed. At present, what can be said on the base of Figure 4.6 is that there is no appreciable trace for a genuine intermediate-age population in the bulge (i.e., younger than ~ 8 Gyr). Those few objects between the 8 and 5 Gyr isochrones are in fact likely to be blue stragglers.

4.3
Globular Clusters in the Magellanic Clouds

Different from the Milky Way, the Magellanic Clouds host populous globular clusters of intermediate ages, hence directly documenting the effect of the AGB evolutionary phase on the properties of SSPs. Figure 4.7a,b shows the contribution to the total K-band luminosity of AGB stars brighter than the RGB tip for a sample of clusters in the Large Magellanic Cloud (LMC), as a function of cluster age.

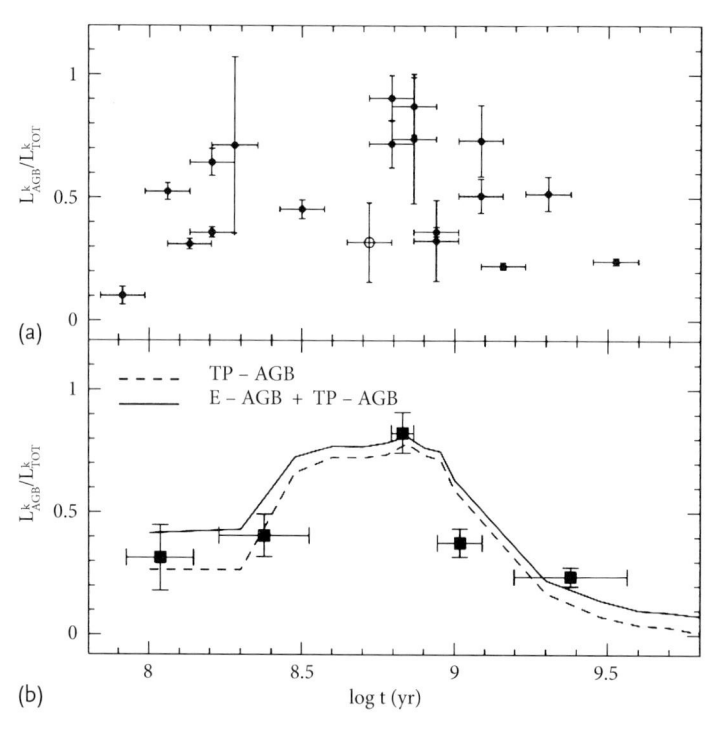

Figure 4.7 Contribution to the total L_K luminosity of AGB stars brighter than $M_K = -6.2$ ($\simeq$ the tip of the RGB in old populations) in a sample of intermediate age clusters members of the LMC. Cluster ages are estimated from their integrated colors, total L_K from aperture photometry, while the contribution of individual bright AGB stars has been summed to derive L_{AGB}^K. Field contamination from the general LMC population has been subtracted. (a) The results for individual clusters; (b) the trend obtained by grouping them into five age bins. Lines in (b) show the same models as in Figure 2.4 (source: Mucciarelli, A. et al. (2006, Astrophys. J., 646, 939)) (Reproduced by permission of the AAS.).

At first sight, individual clusters (Figure 4.7a) do not show a particular trend, but the *K*-band light is often dominated by just 1 or 2 members, making the result prone to small number statistics. To circumvent this limitation, clusters have been grouped into five age bins, the contribution to the total L_K from bright AGB stars computed for the entire bins, and the result is shown in Figure 4.7b. The contribution to the total L_K from bright AGB stars appears to be maximum for an age of ~ 700 Myr, reaching a value of $\sim 80\%$, while in young and old clusters this contribution amounts to $\sim 25\%$.

The dataset used to construct Figure 4.7a,b can be used to estimate the time spent by stars in the portion of the AGB phase at luminosities brighter than the RGB tip, for the ages spanned by these clusters. This allows one to calibrate the corresponding evolutionary tracks, as from Eq. (2.7) one can write:

$$\frac{N_{\text{AGB}}}{(L_K - L_{K,\text{AGB}})_{\text{obs}}} = \frac{b(t)}{(L_K - L_{K,\text{AGB}})_{\text{th}}} \times t_{\text{AGB}} \qquad (4.1)$$

where the suffix "obs" and "th" indicate the observed and theoretical quantities, respectively, and the suffix "AGB" indicates the portion of the AGB at magnitudes

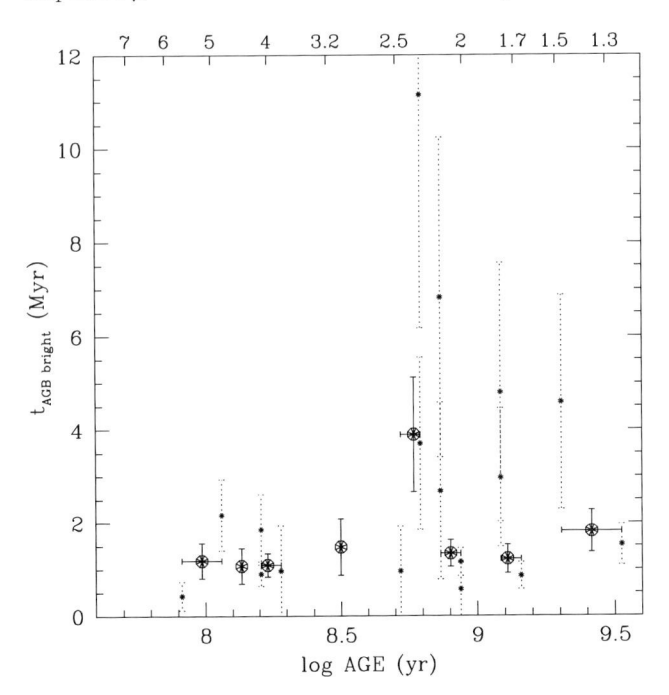

Figure 4.8 Calibration of the AGB lifetime spent at magnitudes brighter than $M_K = -6.2$ obtained from the same cluster sample used in Figure 4.7a,b. The upper axis is labeled with the evolutionary mass on the AGB as derived from Eq. (2.3), but with coefficients suitable to the LMC chemical composition. Small circles show the results for individual clusters, big circles show the results obtained after grouping the clusters into seven age bins. The vertical error bars show the uncertainty related to the Poisson noise on the number of bright AGB stars. The horizontal bars show the bin width.

brighter than $M_K = -6.2$. Notice that, in this equation, the size of the SSP is traced by the K-band luminosity of all its stars except for the bright AGBs, since we aim at calibrating precisely this subphase. The left hand side of Eq. (4.1) can be directly measured for the clusters, while the first factor of the right hand side can be derived from stellar evolution models. The results of this exercise are shown in Figure 4.8 for individual clusters and after grouping them into seven age bins. Again, while the results for individual clusters are very noisy, once binned the clusters show a maximum duration for the bright AGB phase of ~ 4 Myr, for an age of ~ 600 Myr, while this phase appears to last ~ 1.5 Myr at other ages.

While Figures 4.7a,b and 4.8 illustrate the procedure to calibrate the effect of the bright AGB phase on the CMD and integrated light of stellar populations, this calibration is not fully satisfactory. In fact, in this particular sample, ages between 200 and 500 Myr are not well represented, ages older than 3 Gyr are lacking completely, and the binning procedure smears out the features over the age range of the bin. In addition, clusters ages in this dataset have been assigned by means of a calibration of the clusters' integrated colors as a function of age, while turnoff ages would have been desirable. Nevertheless the bright AGB phase clearly has an important impact on the near-IR luminosity and colors of intermediate age stellar populations.

4.4
Stellar Ages of the M31 Spheroid

4.4.1
The Bulge of M31

The next nearest bulge is that of M31, the Andromeda galaxy. At roughly 10 times the distance as Earth is from the Milky Way bulge, crowding is about 100 times worse than it is for the Milky Way bulge, and it is currently impossible to age-date the M31 bulge via the luminosity of the main sequence turnoff. Yet, this offers the opportunity of illustrating one step further on the *age ladder*, and how perverse crowding effects can be.

Early ground-based attempts at resolving the M31 bulge reported the detection of many stars brighter than the RGB tip luminosity. Interpreted as AGB stars, they would have implied an intermediate age for the bulk stellar population of this bulge. At $\sim 2'$ from the center of M31 the K-band surface brightness is $\mu_K \simeq 15/\square''$, and following Eq. (3.1) with mod = 24.4 and $M_{K,\odot} = 3.25$ each resolution element with FWHM = $1''$ samples $\sim 45\,000\,L_\odot$ (bolometric). As illustrated in Chapter 3, this is far too much for doing meaningful stellar photometry in seeing-limited conditions, considering that the RGB tip is at $\sim 2\,000\,L_\odot$.

Figure 4.9a shows how such an M31 bulge field looks in the K-band as seen from the ground with $\sim 1''$ seeing, where several bright objects appear to be *resolved*. If interpreted as bright AGB stars such objects should indeed belong to an intermediate age (i.e., a few gigayears) population. However, the HST/NICMOS2 image in Figure 4.9c shows that such bright objects are resolved into clumps of several in-

4.4 Stellar Ages of the M31 Spheroid

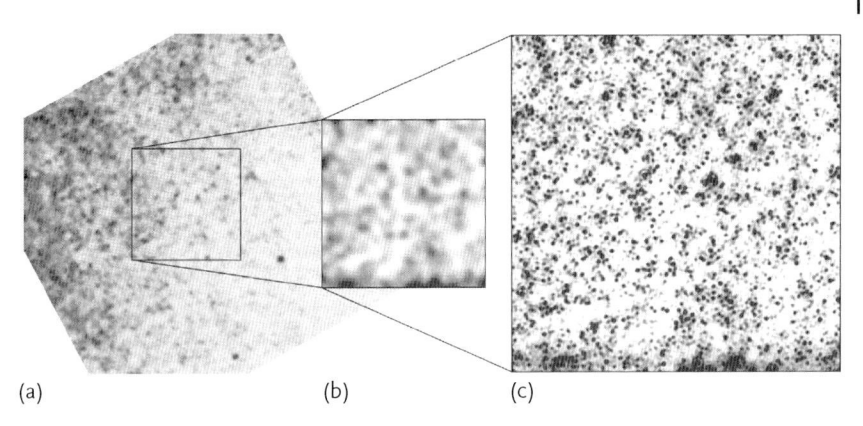

Figure 4.9 (a) The ground-based K-band image of a field in the bulge of M31 with $\mu_K \simeq 15/\square''$, taken with a $\sim 1''$ seeing. (c) The central portion of this field is shown as imaged with the NICMOS2 camera on board of HST, with a PSF with FWHM $= 0''.185$. (b) Zoomed in view is a smoothed version of the NICMOS2 image, so to match the PSF of the ground-based observation (source: Stephens, A. et al. (2003, Astron. J., 125, 2473)) (Reproduced by permission of the AAS.).

dividual stars thanks to the superior resolution of this HST camera (FWHM = $0''.185$). Thus, while the ground-based PSF samples some $45\,000\, L_\odot$ (or $K \simeq 15.25$, $M_K \simeq -9.15$), the NICMOS2 PSF samples only $\sim 1600\, L_\odot$ (or $K \simeq 18.9$, $M_K \simeq -5.5$), allowing meaningful photometry of the brightest RGB stars.

Figure 4.10 shows the K-band luminosity function relative to various M31 bulge fields, ranging in surface brightness from $\mu_K \simeq 15\,\mathrm{mag}/\square''$ down to ~ 16.5. Thus, star counts are affected differently by crowding depending on the surface bright-

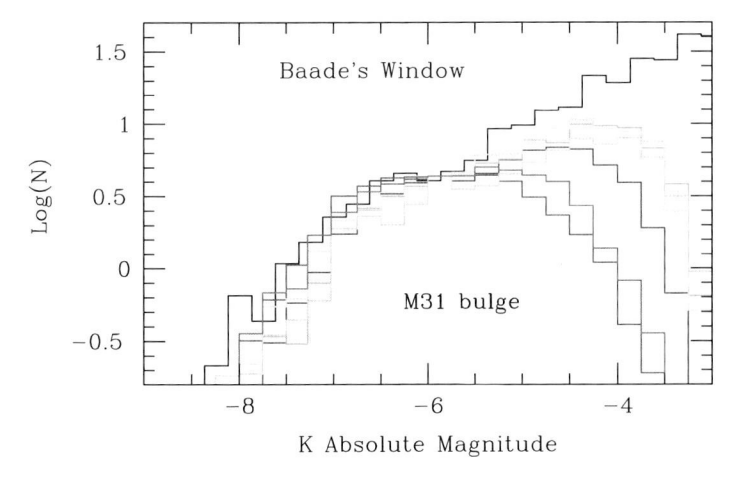

Figure 4.10 The K-band HST/NICMOS2 luminosity function in the bulge of M31 from several fields at different levels of surface brightness. For comparison the luminosity function of the Galactic bulge from 2MASS data is overplotted (adapted from Stephens, A. et al. (2003, Astron. J., 125, 2473)). For a color version of this figure please see in the front matter.

ness: at the bright end unresolved blends cause an artificial extension of the luminosity function, whereas at the faint end incompleteness sets in at a brighter magnitude. In the figure the luminosity function for stars in Baade's Window is also displayed for comparison, showing that where the luminosity function of the M31 bulge is reasonably complete (e. g., for $M_K < -4$ in the fields with the lowest surface brightness) it agrees very well with that of the Galactic bulge. Thus, the near identity of the top end of the luminosity functions of the two bulges can be interpreted in terms of their stellar populations having quite similar ages, hence climbing one step further in the *age ladder*: main sequence photometry in the Galactic bulge has established that the bulge stellar population is very old, older than ~ 10 Gyr, hence the bulk population in the bulge of M31, having a similar luminosity function at the bright end, must have a quite similar age.

Still, this straightforward comparison hides a few important aspects, worth mentioning explicitly. At near-solar metallicity the luminosity of the RGB tip is $M_K \simeq -7$, whereas the luminosity functions in Figure 4.10 extend up by roughly one more magnitude. In the Galactic bulge, where crowding is of no concern at the bright end, the extension of the luminosity function above the RGB tip is due partly to the presence of a few bright AGB stars that are also present in old, metal-rich globular clusters, partly to the depth effect and differential reddening across the field. The two latter effects are not important for the M31 bulge, but crowding is far more severe, and contributes to extend the luminosity function by ~ 0.5 mag, as can be judged by comparing the luminosity functions in Figure 4.10 relative to the fields with highest and lowest surface brightness. Quantitatively, this crowding effect in M31 is comparable to the combined depth and differential reddening effect in the Galactic bulge, and all together conspire to make almost identical the top ends of the two luminosity functions.

4.4.2
The M31 Halo and Giant Stream

The halo of M31 has been the target of much observational effort in recent years. Ground-based wide-field imaging covering many square degrees and reaching a few magnitudes below the RGB tip has mapped the structure of the halo, uncovering a remarkable complexity. The observational study of the M31 halo and its substructures is indeed much easier than for our own galaxy, as all halo stars are at nearly the same distance and concentrated in a relatively small patch of the sky, as opposed to be distributed over all the celestial sphere and at vastly different distances.

As a follow-up of such extensive mapping, ultradeep, pencil-beam imaging of selected areas with HST has reached down to even magnitude ~ 30.5, some 1.5 magnitudes below the main sequence turnoff, and secured accurate photometry for up to $\sim 250\,000$ stars per field. Classical turnoff age dating has then been possible, bringing it to the largest distance so far attempted, with the result of starting to construct the formation and assembly history of the halo of our large neighboring galaxy.

Figure 4.11a–d shows examples of such ultradeep color-magnitude diagrams for three M31 fields along its minor axis, respectively at a projected distance of 11, 21 and 35 kpc from the center, as well as for the most prominent substructure in the M31 halo, known as the *giant stream*. The width of the RGB indicates a wide range of metallicities in each of these fields, from [Fe/H] ~ -1.5 to slightly supersolar. Age dating is then affected by the age-metallicity degeneracy, which can be alleviated using an *ad hoc* procedure based on simulated CMDs. From a set of theoretical isochrones simulated observations are created for an extensive set of age and metallicity combinations, with depth, crowding and photometric accuracy that match those of the actual data. From the resulting set of single age+metallicity CMDs, therefore mimicking SSPs, a large set of CMDs are created for composite populations with different proportions of the various SSPs, that is, with different star formation and chemical enrichment histories. A best fit procedure then sorts

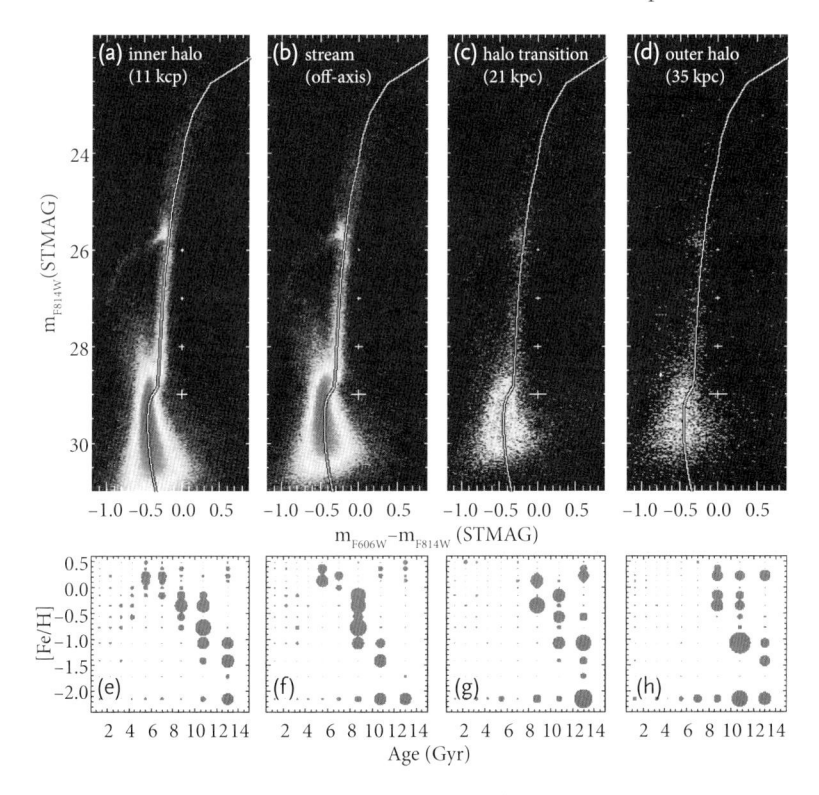

Figure 4.11 (a–d) The HST/ACS color-magnitude diagrams for four different fields in the M31 halo, namely: three fields on the M31 minor axis, at a projected distance of $\sim$ 11, 21 and 35 kpc from the center of the galaxy, plus a field in the M31 giant stream. (e–h) For each field the best fit star formation history obtained using the StarFish code by Harris, J. and Zaritski, D. (2001, *Astrophys. J. Suppl.*, 136, 25); the size of the circles is proportional to the weight of each SSP (age, [Fe/H]) component in the synthetic composite population. The curves in (a–d) show the ridge line of the CMD of 47 Tuc (source: Brown, T. (2009, *ASP Conf. Proc.*, 419, 110). For a color version of this figure please see in the front matter.

out the combination of single age and metallicity populations (SSPs), and their relative proportions, that most closely mimics the observed CMDs. The resulting star formation and chemical enrichment histories are shown in Figure 4.11e–h, with the size of each circle being proportional to the weight of the corresponding SSP in the best fit mix.

At first sight, all four CMDs look quite similar to each other, but at a closer look sizable differences start to appear. The similarity of the CMD and star formation history of the 11 kpc and giant stream fields suggest that debris related to the stream is also present in the former field. On the other hand, the stream itself is seen through the halo, and therefore true halo stars project over the stream. Moving outwards, all the way to the 21 and 31 kpc fields, one can notice the progressive disappearance of the component younger than ~ 10 Gyr, which accounts for $\sim 40\%$ of the population in the other two fields. All in all, the halo of M31 appears to host stars in very extended age and metallicity ranges, quite different in this respect from the halo of the Milky Way where metal-rich, intermediate-age stars are exceedingly rare. Therefore, M31 appears to have experienced a more violent history compared to the Milky Way, with merging with relatively massive galaxies and the associated disk damage having played a more prominent role.

4.5
The Star Formation Histories of Resolved Galaxies

With the advent of panoramic CCD detectors deep observations became feasible, collecting data for very large samples of stars, and extensive CMDs of resolved stellar populations in galaxies of the Local Group started to demand an interpretation. A new field for astronomical research started, focusing on the derivation of the star formation history (SFH) of external galaxies with the synthetic CMD method. At first the method was applied to the intrinsically bright portion of the CMD, thereby sampling young stars and yielding information only on the recent SF episodes. As deeper observations became feasible, digging deeper in the main sequence of nearby galaxies, the *lookback time* sampled by the CMDs increased, thereby opening the possibility of doing *galactic archeology* by deriving the SFH back to a significant fraction of the Hubble time. The advent of HST boosted this field enormously, and currently almost the whole Local Group has been mapped, plus some other galaxies a little further out. The vast majority of studies has concentrated on dwarf galaxies, which have low surface brightness and can be substantially mapped with just a few frames. Although in most cases the old MS turnoffs are still out of reach, important results have been harvested. For example, demonstrating the presence of old stellar populations in all cases in which it was possible to detect them, or showing the existence of stellar population gradients, or of relatively long periods of quasisteady star formation separated by some short hiatus in such activity (*gasping* star formation). Finally, no evidence for appreciable departures from a Salpeter IMF has been found. This section is dedicated to describe some methods for diagnosing the SFH of galaxies with resolved stellar populations, and will do so with the help of a few concrete examples. In most recent years, full coverage of large portions of extended galaxies such as the Magellanic Clouds, M31 and M33 has become possible and is actively being pursued, either from the ground or from space. The SFH diagnostic methods described in this section may help to prepare users to analyze very large databases in a reasonable time, for example generating space resolved maps of the SFH of large nearby galaxies, extending over square degrees, such as those mentioned above.

Before discussing the synthetic CMD method as it is routinely applied, we illustrate the concepts on which it is built, namely the fact that the number of stars counted in a box on the CMD is proportional to the total mass turned into stars in the age range of the stars populating the box. We start by examining SSPs, and then move to composite stellar populations, viewed as collections of SSPs.

4.5.1
The Mass-Specific Production

In Chapter 2 the specific evolutionary flux was introduced and some of its applications have been presented (see also Chapter 3), including the estimate of the number of stars in a certain evolutionary stage, once the sampled *luminosity* of the stellar population is known. However, if the goal is to measure the SFH of a galaxy, one needs to evaluate how much *mass* was turned into stars during given time intervals. Therefore, one needs to use a relation between star numbers and mass of a given (sub)population. For this purpose, we introduce the mass-specific evolutionary flux of an SSP, as

$$B_m(t) = \frac{b(t)}{\mathcal{M}} = B(t)\frac{L_T}{\mathcal{M}} = \frac{1}{\mathcal{M}}\phi(M_{TO})|\dot{M}_{TO}|, \quad (4.2)$$

where $\mathcal{M}$ is the mass in stars at formation of the SSP. The mass-specific evolutionary flux is also given by $B(t)$ times the inverse of the mass-to-light ratio, and is sensitive to age, composition and the IMF as illustrated in Figure 4.12. Unless the IMF is very steep, $B_m(t)$ monotonically decreases with age, as fewer and fewer stars evolve off the MS per unit time, for a given initial mass of the SSP. Its trend with IMF slope depends on the age of the stellar population, but, at ages older than $\sim 100\,\mathrm{Myr}$, an IMF with the Salpeter's slope appears to be particularly productive in terms of stellar flux. Notice that whereas $B(t)$ is almost independent of age, B_m actually decreases by a factor of ~ 40 over one Hubble time, for a Salpeter IMF. Following Eq. (2.7), and using the mass-specific flux, the number of stars in the generic post-MS evolutionary stage j of duration t_j can be written as:

$$N_j = B_m(t)\mathcal{M}t_j = P_j(t)\mathcal{M}, \quad (4.3)$$

where the *(mass) specific production* factor $P_j = B_m(t)t_j$ is introduced. Equation (4.3) allows one to estimate the mass of an SSP, by counting stars in a given stage j, and knowing the production factor from stellar evolution theory and for an assumed IMF. Figure 4.13 illustrates the dependence on age and metallicity of

4 Age Dating Resolved Stellar Populations

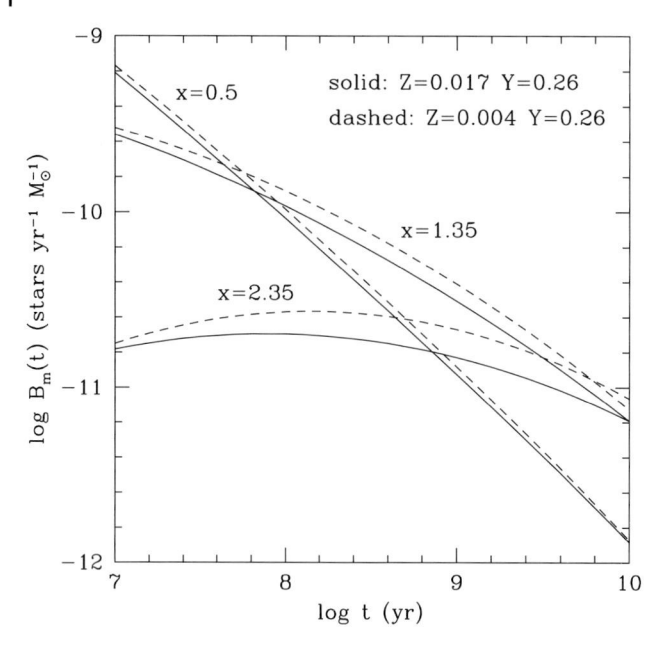

Figure 4.12 Evolutionary flux off the MS per unit mass of the parent stellar population as function of age for three IMFs and two chemical compositions as labeled. The considered IMFs all have the same slope (of $x = 0.3$) between 0.1 and 0.5 $M_\odot$ but differ in the higher mass regime (from 0.5 to 120 $M_\odot$), as the *diet* IMFs considered in Chapter 2. Fitting relations in Chapter 2 for the mass at the turnoff and of the dying star have been used.

the factors P_j for four major post-MS stages, which are identified in Figure 4.14 on three isochrones. The SGB- and RGB-specific productions increase as their parent population grows old, with such an increase becoming quite fast at some intermediate ages. The rapidity of this *transition* and its location in time depend slightly on chemical composition, whereas the HB and E-AGB production factors are rather independent of composition, and both present a maximum around ~ 1 Gyr. This maximum is due to the core mass at helium ignition being minimum at this age (*RGB phase transition*), which translates into a relative maximum of the helium burning lifetime at this age, as already mentioned in Chapter 2. Remarkably, for ages in excess of a few gigayears the mass-specific productions of all four post-MS phases remains almost constant, and a 1000 $M_\odot$ stellar population with a Salpeter-diet IMF has $\simeq$ 11, 5, 1 and 0.1 stars respectively on the SGB, RGB, HB and E-AGB sections of its isochrone, almost irrespective of its age.

Figure 4.15 shows the production factor for all post-MS stages together, resulting from the product of the fluxes shown in Figure 4.12 times the whole post-MS lifetime, from core hydrogen exhaustion to the first thermal pulse, or central carbon ignition, depending on the turnoff mass. The increase of the lifetimes t_j as the stellar population ages largely compensates for the decrease of $B_m(t)$, so that the number of post-MS stars per unit (initial) mass of the parent stellar population tends to increase over a Hubble time, apart from a characteristic peak at $\simeq 1$ Gyr re-

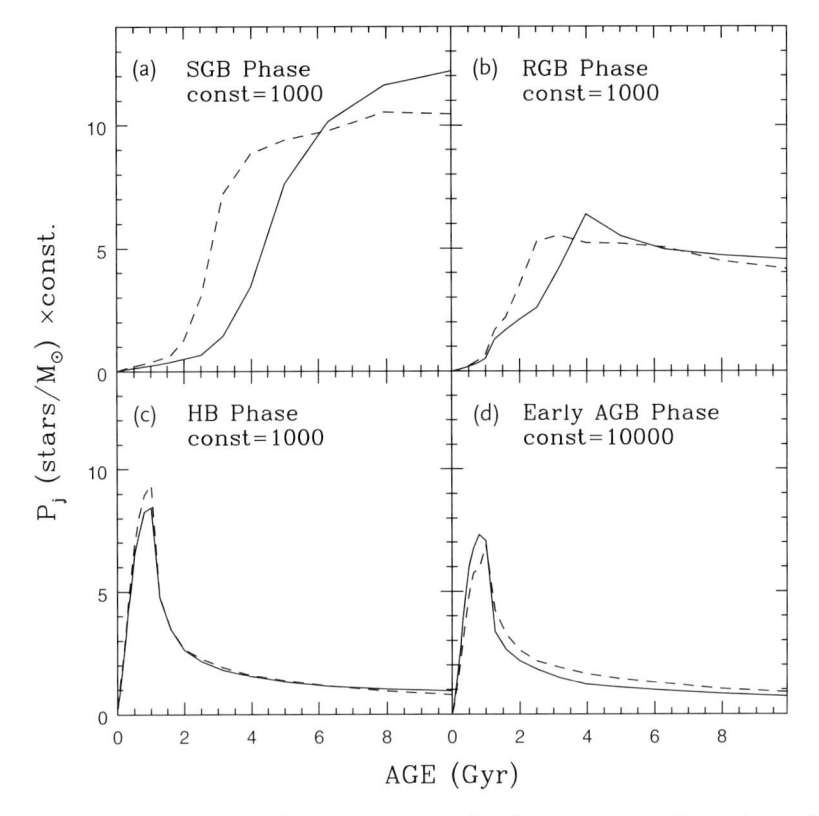

Figure 4.13 The mass-specific production of stars in four major post-MS phases as a function of the age of the parent stellar population. The solid lines refer to solar chemical composition ($Y = 0.26$, $Z = 0.017$); the dashed lines to subsolar metallicity ($Y = 0.26$, $Z = 0.004$). A Salpeter-diet IMF has been adopted. The four phases are identified in Figure 4.14. The production factors refer to a parent stellar population of $10^3\,M_\odot$ for the (a) SGB, (b) RGB and (c) HB phases, to a population of $10^4\,M_\odot$ for the (d) E-AGB phase (the YZVAR (Bertelli, G. et al. 2008, Astron. Astrophys., 484, 815; 2009, Astron. Astrophys., 508, 355) tracks have been used).

lated to the *RGB phase transition*. All in all, the specific production of post-MS stars does not show a big variation over almost the whole Hubble time. For $t \gtrsim 0.5$ Gyr, with a Salpeter-diet IMF and solar composition, it ranges from ~ 4 to 18 stars per 1000 $M_\odot$ of the parent population. A major part of this variation is due to the SGB evolutionary phase, that is, from hydrogen exhaustion near the turnoff to the base of the RGB, whose duration t_{SGB} increases steadily with the age of the population. The durations of the more advanced evolutionary stages (primarily RGB and HB) do not change much with age beyond ~ 1 Gyr, and therefore the corresponding production factors slowly decline with time following the decline of $B_m(t)$. Some dependence on the chemical composition is apparent, although limited to intermediate ages, mostly due to the sensitivity of the SGB lifetimes to the metallicity. The biggest effect is related to the IMF: a flat IMF provides much less post-MS stars at

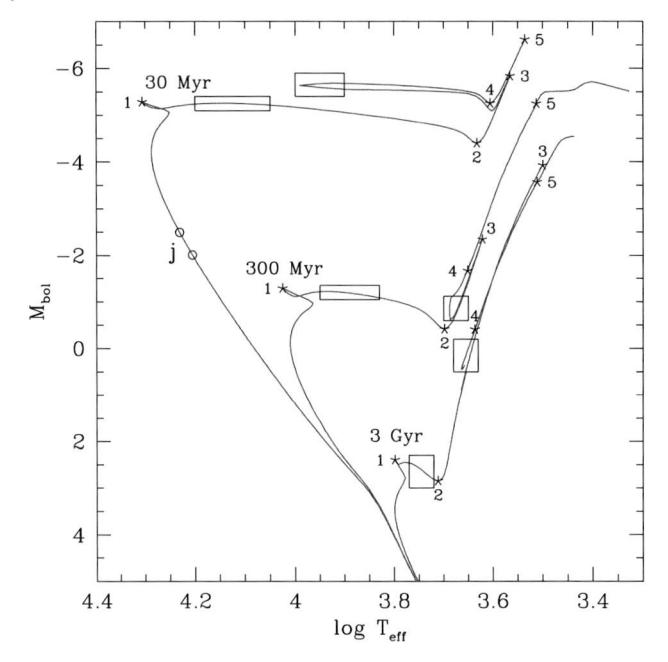

Figure 4.14 Isochrones with solar chemical composition with three different ages as labeled. The numbered asterisks show the partition in post-MS phases adopted for Figure 4.13: from 1 to 2 the SGB phase, from 2 to 3 the RGB phase, from 3 to 4 the HB phase and from 4 to 5 the E-AGB phase. Diagnostic boxes for the determination of the mass of each of three stellar populations with these ages are drawn for illustration (the YZVAR (Bertelli, G. et al. 2008, Astron. Astrophys., 484, 815; 2009, Astron. Astrophys., 508, 355) isochrones have been used).

ages older than a few gigayears, reflecting its lower evolutionary flux per unit mass B_m.

In summary, the initial mass of an SSP can be estimated from the ratio of the number of stars in any given post-MS evolutionary stage over the corresponding specific production, that is,

$$\mathcal{M} = \frac{N_j}{P_j}, \tag{4.4}$$

and possible discrepancies between estimates using different evolutionary stages give an indication of the internal consistency of the result.

The total mass of the SSP can also be estimated by counting stars on a given portion of the main sequence, bounded by two values of the star mass:

$$N_{1,2} = \int_{M_1}^{M_2} \phi(M) dM = \mathcal{M} \frac{\int_{M_1}^{M_2} \phi(M) dM}{\int_{0.1}^{120} M\phi(M) dM}, \tag{4.5}$$

and the ratio of the two integrals can be evaluated independently of the IMF scale factor, once the two bounding masses have been chosen. Again, the total mass $\mathcal{M}$

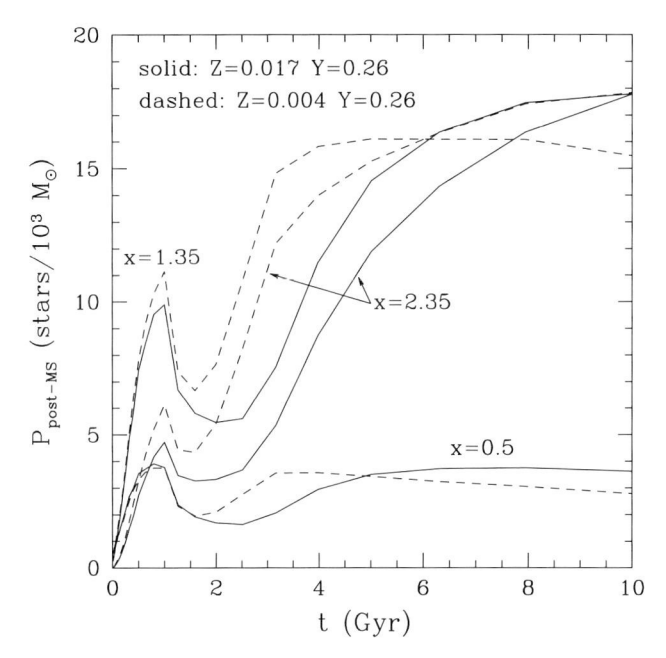

Figure 4.15 Specific production of post-MS stars as function of the age of the parent stellar population for three IMFs and two compositions as labeled (the YZVAR (Bertelli, G. et al. 2008, Astron. Astrophys., 484, 815; 2009, Astron. Astrophys., 508, 355) tracks have been used).

of the SSP is estimated from Eq. (4.5), after counting the number of stars $N_{1,2}$ in the luminosity interval on the main sequence that corresponds to the chosen mass interval.

Turning now to composite populations, let us first consider the example of three bursts of star formation, with ages of 30, 300, and 3000 Myr. The relative HRD will be populated along the three isochrones in Figure 4.14, where boxes which target the three separate star formation episodes can be easily identified, and star counts within these boxes yield the mass turned into stars in each of the three bursts, hence the SFH. However, for a more complex stellar population, formed continuously over a long period of time, a box on the HRD will intercept a range of isochrones, and, in the most general case, we have:

$$N_{\text{box}} = \int_0^{t_{\text{max}}} \psi(t) P_{\text{box}}(t) \, dt \,, \tag{4.6}$$

where $P_{\text{box}}(t)$ is the specific production of stars in the box from the isochrone with age t. If the box samples a narrow age range, P_{box} is different from zero only within this range, and the counts can be readily transformed into the mass of the parent stellar population, given by the integral of the SFR over the sampled age range. Equation (4.6), derived for boxes targeting post-MS phases, can be applied also to bins on the MS by replacing the factors P_{box} with the fraction of stars with mass in the appropriate range, as given in Eq. (4.5). The MS luminosity function (i.e., the counts N_{box} as a function of magnitude for boxes constructed along the MS) results from the convolution of the SFR and the IMF: making a hypothesis on the former allows to recover the latter, or vice versa.

Equation (4.6) can also be written as

$$N_{box} = \Delta \mathcal{M}(t) \langle P_{box} \rangle_{\psi(t)}, \tag{4.7}$$

where $\Delta \mathcal{M}(t)$ is the total mass turned into stars within the age range Δt probed by the box, and $\langle P_{box} \rangle_{\psi(t)}$ is the average of the specific productions of stars in the considered box by the stellar generations sampled, weighted by the star formation rate during the Δt time interval. The quantity $\langle P_{box} \rangle_{\psi(t)}$, totally analogous to P_j discussed for SSPs, is the specific production of stars in the considered box for composite stellar populations.

4.5.2
Decoding the CMD

The decoding of the CMD in terms of SFH is built on the concept that the stellar density across the CMD can be translated into the mass of the parent stellar population by means of Eq. (4.7) applied to *tiles* over the diagram, with the specific productions determined by stellar evolution theory, modulo the IMF. Different tiles probe different age ranges, so that the star formation history is determined as mass transformed into stars within various age bins. The SFR is not a direct output of the method; the star counts in any box depend on the integrated star formation over the age range probed: when this range is wide, the counts divided by the age range will provide an *average* SFR. The time resolution, as well as the intensity of the SFR at the different epochs, depend on the age intervals sampled by the individual tiles. For young stellar populations, whose post-MS star location on the CMD is very sensitive to age, this type of diagnostic is very effective, and it is possible to derive the SFH with high time resolution (e.g., $\sim$ few Myr); conversely, for old stellar populations, it is possible to derive the SFR averaged over longer time intervals (e.g., $\sim$ 1 Gyr or more). Due to the growth of evolutionary lifetimes as the TO mass decreases, the time resolution over the CMD steadily worsens going from young to old ages; this is the reason why the age binning in this kind of application is typically on a logarithmic scale.

Separating isochrones of different ages on the broad band optical CMD is a more difficult task compared to what can be done on the $M_{bol} - \log T_{eff}$ plane. The MS of young isochrones is almost vertical up to the TO, and the blue portion of the blue loops is close to the MS, so that relatively small photometric errors may prevent us separating core helium burning from MS stars. Clearly, the best way to target stars of different ages for the application of Eq. (4.7) would be to consider boxes on the subgiant branches. However, for young ages this portion of the CMD is very sparsely populated, while for old ages it is very faint. For example, the SGB of a 4 Gyr old isochrone is at $M_I \simeq 2$: its stars appear at $I \sim 27$ for galaxies at the edge of the Local Group and at $I \sim 30$ for galaxies in the Centaurus group. On the other hand, all old isochrones climb up to much brighter magnitudes during the later phases, and stars can be easily counted in boxes at, for example, close to the RGB tip, at magnitudes around $M_I \sim -3$. Still, the RGB tip region is crowded with stars spanning a wide age range, so that the temporal resolution of the boxes that one can construct is quite poor. Thus, decoding the CMD of a composite stellar population may not be a trivial affair. A concrete example may help showing what in practice one can get.

Figure 4.16 shows tracks with different TO ages superimposed on the CMD derived for a dwarf galaxy at a distance of ~ 5 Mpc. The observed CMD collects stars with ages up to a Hubble time, albeit in different evolutionary stages: MS and core helium burning for young stars; blue loops and AGB for intermediate age stars, bright portion of the RGB evolution (plus the E-AGB) for stars older than ~ 1 Gyr. The blue edges of the blue loops offer a good location for boxes which target different ages, with a clear trend of decreasing luminosity with increasing age. When a

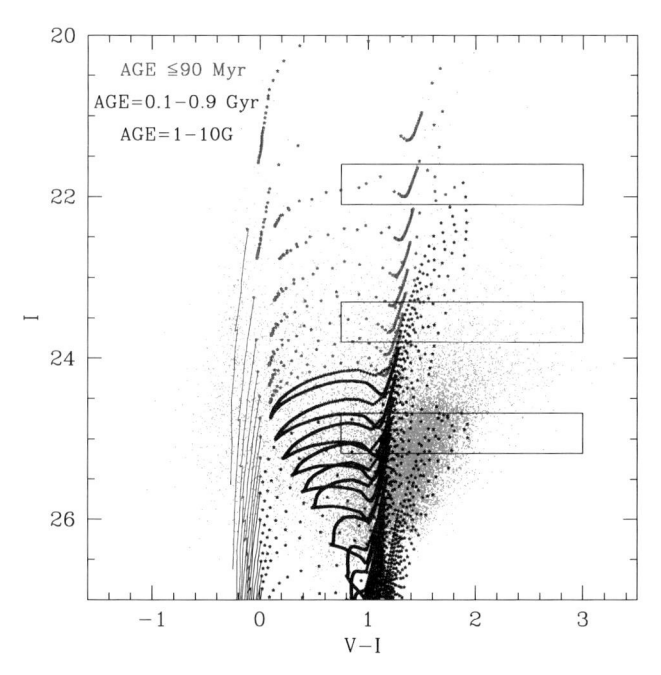

Figure 4.16 Observed CMD of the dwarf galaxy NGC 1705 (small dots) with superimposed stellar evolutionary tracks (with $Z = 0.004$) with various turnoff ages as labelled in the upper left corner, shifted with $(m - M)_0 = 28.54$ and $E(B - V) = 0.045$. The MS portion of the tracks is drawn as a solid line, while in the post-MS portion the density of points is higher where the evolution is slower. The three boxes are shown to illustrate how the CMD traces the SFH: while the brightest box samples only 30 Myr old stars, the others collect a wide range of stellar ages. Although the different age components are in principle separated by color, in practice it is very difficult to tell them apart on the observational CMD (data for the CMD in figure are from Annibali, F. et al. (2003, Astron. J., 126, 275) and the tracks are derived from the Fagotto, F. et al. (1994, Astron. Astrophys. Suppl., 105, 29) database). For a color version of this figure please see in the front matter.

clear sequence of blue helium burning stars appears on the CMD of a galaxy, the SFH at intermediate ages can be nicely recovered from the luminosity function of such stars, modulo the stellar evolutionary tracks which, as discussed in Chapter 1, have some difficulty to accurately predict the extension of the loops. At any rate, very accurate photometry would be needed to clearly sample the blue loop stars: this is not the case for the observational CMD in Figure 4.16 in which the MS and the loops are clearly merged.

A cleaner separation of blue loops from MS stars could be obtained using a passband bluer than the V-band, such as the U-band or even a space-UV band. If this is not available, one may turn to the other side of the CMD. Indeed, the complementary location of young and intermediate age helium burners is on the red side of the diagram, where boxes can be constructed to sample the red portion of this phase. The brightest box drawn on the figure shows one example: red helium burners allow us to reconstruct the SFH with high accuracy, even better than their blue counterparts at very young ages. However, the contamination from intermediate age bright AGB stars soon appears in the luminosity function of the red stars. Virtually any box that we can construct samples a complex stellar population with a wide age range and different specific productions, for example the middle box in Figure 4.16. Proceeding to fainter magnitudes, each bin of the luminosity function of red stars collects objects born over virtually the whole galaxy lifetime. So, the fainter box drawn in the diagram is populated with ~ 0.2 Gyr old core helium burners, plus AGB stars older than ~ 0.3 Gyr, plus RGB stars older than ~ 1 Gyr. The counts in this box yield an integral information on the SFH which is still of great interest, providing a straightforward mean to estimate the total mass in stars of the target galaxy field from stellar counts.

Figure 4.17a,b shows the specific production of SSPs (solid black lines) in two boxes drawn on the M_I–$(V-I)$ CMD as a function of the age of the parent stellar population. Notice that now the specific production is not referred to a particular evolutionary phase, but to the time that stars spend within the target box. Figure 4.17a refers to a box which includes the upper two magnitudes on the RGB (RGB tip box); Figure 4.17b refers to a box sampling the helium burning clump (red clump box). In the RGB tip box, the SSP-specific production of young stellar populations is high, since the box collects young red helium burning stars; as age increases, the specific production drops, since the box samples E-AGB stars, down to a minimum, after which it increases again, when true RGB stars start populating the box. After a local maximum, the SSP-specific production settles on a slowly declining path, reflecting the slow trend with mass of the evolutionary flux as the lifetime on the bright RGB is nearly constant. At all ages, the red clump box samples core helium burning stars, hence the maximum of the SSP specific production at ~ 1 Gyr in Figure 4.17b, corresponding to the RGB phase transition. At ages beyond 2 Gyr a sizable contribution ($\sim 30\%$) from the RGB is also present in this box. It can be noticed that also for the red clump box the SSP-specific production is a slow function of the age beyond ~ 2 Gyr, again driven by the decrease of the evolutionary flux.

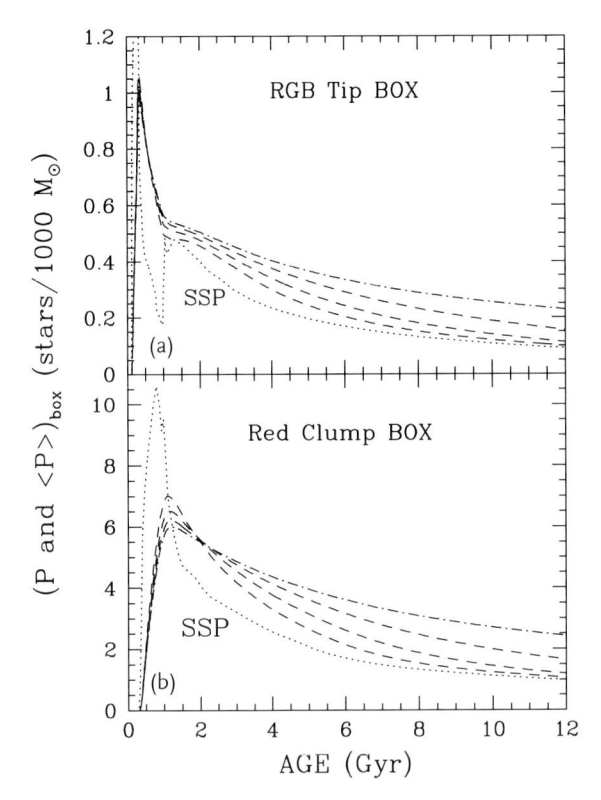

Figure 4.17 Specific production of stars in a box sampling (a) the bright RGB ($-2 \leq M_I \leq -4$; $0.75 \leq (V-I) \leq 2$) and (b) in a box centered on the red clump ($0 \leq M_I \leq -1.75$; $0.75 \leq (V-I) \leq 2$) for SSPs (dotted) and for complex stellar populations. A Salpeter-diet IMF has been adopted. The complex stellar populations are constructed with a constant SFR (dot-dashed), and exponentially decreasing SFR (dashed) with e-folding time of 5, 2 and 1 Gyr. The SSP production factors are plotted as a function of age, while those of composite populations are plotted versus the time elapsed since the beginning of star formation (adapted from: Greggio, L. (2001, ASP Conf. Proc., 274, 444)).

The dashed lines in Figure 4.17a,b show the average specific productions $\langle P_{\text{box}}\rangle_{\psi(t)}$, as they vary in time in a system undergoing continuous star formation according to different laws. In both boxes the average production factors are very high when the system is very young, during the first gigayear of star formation activity; they are instead slowly declining with time beyond a few gigayears. Therefore, for galaxies in which star formation has been proceeding for more that ~ 2 Gyr the specific production $\langle P_{\text{box}}\rangle_{\psi(t)}$ is a weak function of age and SFH, and it can be safely used to convert star counts into the mass of the parent stellar population within a factor of ~ 2, irrespective of the details of the star formation activity.

To summarize, decoding the CMD in terms of SFH is not a trivial task; it is based on the relation between star counts in regions of the CMD and the strength of the star formation episode which originated them, modulo the IMF and stellar evolution. The limitation comes from the fact that in most cases a box on the CMD contains stars with different ages and in different evolutionary phases. However, each star formation episode is sampled in other parts of the diagram, so that redundant information is present on the observed CMD. This is fully exploited in the synthetic CMD method which is discussed later.

So far, we have not mentioned the effect of a metallicity distribution, which further complicates the reconstruction of the SFH. While this may be justified in the case of simple systems like some dwarf galaxies, this is less so for other stellar systems, for which the effect of a metallicity spread must be considered as an additional parameter. In spite of all these problems, the specific productions in post-MS phases do not vary by large factors over a Hubble time, which allows one to conclude that, for example, a complex stellar population has ~ 0.3 stars per 1000 $M_\odot$ in the two upper magnitudes of the RGB, and 3 stars per 1000 $M_\odot$ in the clump, (almost) irrespective of its age and SFH, provided that it is older than ~ 3 Gyr.

4.5.3
The Specific Production Method

The concepts introduced in the previous section allow us to illustrate a simple method to recover some information on the SFH of a galaxy from star counts in diagnostic boxes tailored to target specific age ranges on the CMD. This method does not describe the SFH with high temporal resolution, in favor of an average information which yields a robust estimate of the mass transformed into stars in wide age bins. Figure 4.18 shows the synthetic CMD in near-IR bands constructed with a constant SFR over 12 Gyr and an assumed age-metallicity relation (Z increases with time). The total mass transformed into stars in this simulation is $\sim 3 \times 10^8 \, M_\odot$. The stars on the CMD have been color-coded by their ages, and four boxes on the diagram target specific age ranges: the red supergiant (RSG) box for the youngest component ($\sim$ few 10^7 yr), the blue supergiant box (BSG, around 10^8 yr) and the asymptotic giant branch box (AGB, up to $\sim 10^9$ yr) for intermediate ages, and the red giant branch box (RGB, $\gtrsim 10^9$ yr) for old ages. The age distribution of the stars in these boxes is plotted in Figure 4.19 and shows that there is little age overlap among the four boxes. The simulation allows us to estimate the average specific production in the four boxes, as the ratio between the number of stars in the boxes and the mass turned into stars within the four age bins.

This is a different way to compute the specific productions, but totally equivalent to determining the time spent by tracks within the boxes, deriving P_{box} for each track, and then averaging over the age ranges. The values of $\langle P_{\text{box}} \rangle_{\psi(t)}$ turn out to be $1.26 \times 10^{-4} \, M_\odot^{-1}$ in the RSG box, $2.91 \times 10^{-4} \, M_\odot^{-1}$ in the BSG box, $1.25 \times 10^{-5} \, M_\odot^{-1}$ in the AGB box and $8.54 \times 10^{-5} \, M_\odot^{-1}$ in the RGB box. These specific productions can be used to derive average masses of the parent stellar population in the four age bins from the CMD of any galaxy by applying Eq. (4.7). The resulting SFH will be close to what one could derive with the full synthetic CMD method, to be discussed next. The advantage of the specific production method is that it can be applied very

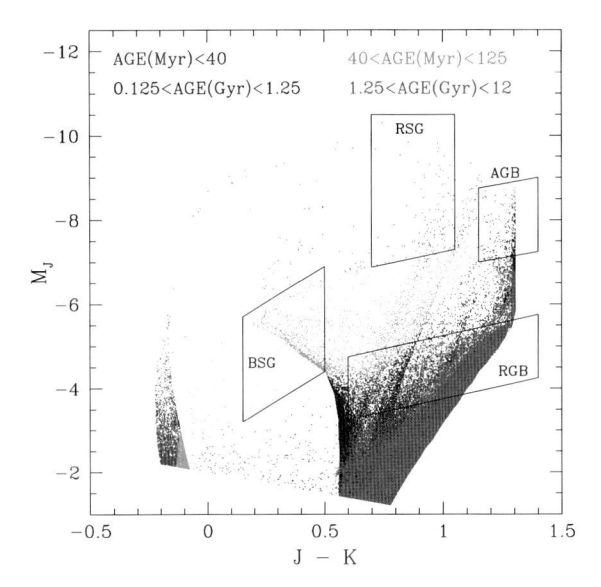

Figure 4.18 Synthetic CMD in the infrared bands produced with a constant SFR over 12 Gyr and an age-metallicity relation in which $Z = 0$ at $t = 0$, $Z = 0.02$ at $t = 7.5$ Gyr and $Z = 0.025$ at $t = 12$ Gyr. The simulation contains 200 000 stars brighter than $M_K = -2$ and corresponds to a total star formation of $2.9 \times 10^8\,M_\odot$, having adopted a straight Salpeter IMF. The four diagnostic boxes target the labeled age ranges, and contain 122 (RSG), 598 (BSG), 340 (AGB) and 22 198 (RGB) stars, respectively. A theoretical modeling of the TP AGB phase is included in the simulation (simulation computed by G.P. Bertelli). For a color version of this figure please see in the front matter.

easily, as soon as an observational CMD is available, so that it allows us to harvest results for many galaxies, or for many portions of a large galaxy, with minimum effort. In addition, the boxes are wide, so that the photometric errors are relatively unimportant, since the bulk of the stars which truly belong to the box are counted.

This is just one example of this method, in which the boxes are constrained by the chosen photometric bands (J and K) and stellar tracks. Where to locate the boxes depends on the specific CMD, which already at first sight may give an idea of the age distribution of the stars in it. The method can also be calibrated empirically, by measuring the average specific production in a particular box from star counts on the CMD of a galaxy for which the SFH is already known.

As mentioned previously, the modeling of the TP-AGB phase is very uncertain. On the other hand, the TP-AGB stars are very bright, especially in the IR, and can be effective tracers of star formation at intermediate ages. TP-AGB stars are characterized by their very red colors, and stand out clearly on the CMD, as can be appreciated in Figure 4.16, where the red stars brighter than the RGB tip at $I \simeq 24.7$ are most likely TP-AGB stars. Unfortunately, as repeatedly mentioned here, lifetimes of stellar models in this phase are uncertain. A calibration procedure similar to that illustrated earlier for the globular clusters in the LMC could be applied to entire galaxies with known age distribution, improving on the statistical

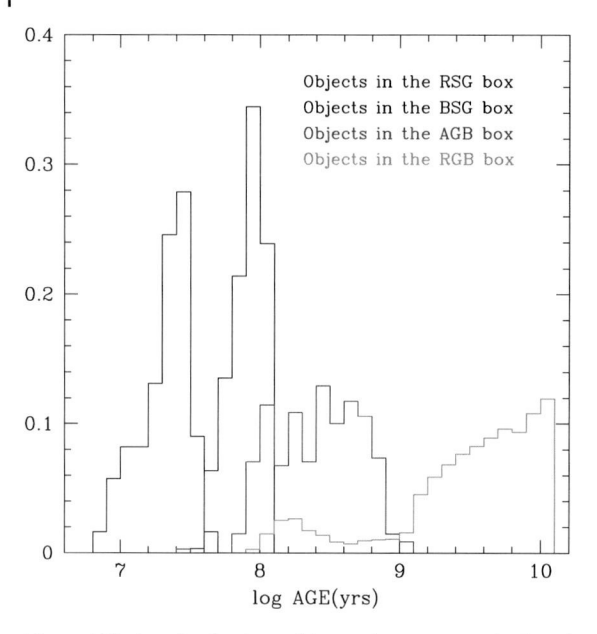

Figure 4.19 Age distributions of the synthetic stars in the four diagnostic boxes in Figure 4.18, normalized to the total number of stars in each box. The color of the lines encodes the box membership. For a color version of this figure please see in the front matter.

significance of the results concerning the evolutionary lifetimes in this phase, and its contribution to the integrated light. This could hopefully bring to an end the enduring controversies regarding the contribution of the AGB stars to the energetics of stellar populations.

4.5.4
The Synthetic CMD Method

Several algorithms have been developed to retrieve the SFH from the distribution of stars on an observed CMD, all basically aiming at minimizing the *distance* between the data and synthetic CMDs constructed from a database of isochrones in which the SFH is a free parameter. In its automatic rendition, partial models of star formation episodes (i.e., stellar generations spanning a small age and metallicity range) are constructed. Each partial model is built with Monte Carlo extractions of stellar mass, age, and metallicity and the corresponding simulated star is placed on the CMD by interpolation on isochrones. Each model is also characterized by a total mass of extracted stars, so that specific productions of stars in the various parts of the CMD are readily determined for all the partial models. We indicate with P_{ij} the specific production of stars in the ith box of the CMD by the jth stellar generation. The total population of the ith box then results from the sum of the contributions of all the stellar generations produced by the simulation:

$$N_i = \sum_j \mathcal{M}_j P_{ij}, \tag{4.8}$$

where $\mathcal{M}_j$ is the total mass of stars formed in the jth stellar generation. The *distance* between the data and models is quantified by the differences between the observed occupation numbers ($N_{i,\text{obs}}$) and their simulated counterparts ($N_{i,\text{sim}}$), and can be evaluated in a variety of ways. For example, using the Poisson likelihood ratio (PLR) as statistical estimator, it can be expressed as (from Dolphin, A.E. (2002, *Mon. Not. R. Astron. Soc.*, 332,91)):

$$-2\ln(\text{PLR}) = 2 \sum_i (N_{i,\text{sim}} - N_{i,\text{obs}}) + N_{i,\text{obs}} \ln \frac{N_{i,\text{obs}}}{N_{i,\text{sim}}}, \tag{4.9}$$

which is the quantity to be minimized in such a way to determine the best combination of the masses $\mathcal{M}_j$ (the free parameters that appear as factors in the $N_{i,\text{sim}}$ values) which accounts for the data. The full set of $\mathcal{M}_j$ values tells how much mass went into stars during each of the corresponding set of (age, metallicity) bins, which is in fact the resulting SFH. The inclusion of photometric errors in the simulation ensures that diffusion in and out of the various boxes due to observational errors are properly taken into account. Additional parameters, which may or may not be adopted *a priori*, include the distance and the reddening, the fraction of binaries, and the IMF. An age-metallicity relation is also sometimes assumed, and other times required as output from the fitting procedure.

The merits of the application of an automatic CMD fitting procedure include its objectivity, the possibility of using as a *black box* a publicly available package, with different researchers treating different datasets in a homogeneous fashion, and a quantitative estimate of the quality of the fit. However, some concerns are also worth mentioning. Namely: (i) the method implicitly assumes that stellar tracks and bolometric corrections are a very good match to reality. In many cases this is not fully justified, for example, the empirical calibration of the TP-AGB phase is still not satisfactory and some of the bolometric corrections of very red/cool stars are still uncertain. (ii) There are unavoidable degeneracies on the CMD, with stars of (very) different ages and metallicities falling on the same CMD box. In this respect, some ways of tiling the CMD may be more adequate than others. However, it is far from trivial to find a *smart* tiling which could work in any situation (see Figure 4.16). For this reason, a uniform tiling of the CMD is typically adopted, with all boxes of the same size. (iii) The statistical estimator tends to give more weight to the more populated regions, but Poisson statistics is only one source of uncertainty affecting the solution, others being the mentioned systematics of the stellar evolution database. Moreover, the most populated regions of the CMD are often the faintest ones, most affected by photometric errors, and it may be advisable not to give too much weight to these regions. For these reasons, the best fit solution from the blind application of the method may not be the truly *best* solution, and the ranking of the models may not reflect their actual reliability. Besides the extensive exploration of the parameter space, the user should make sure to check the solution *by eye* and assess which features of the observed CMD drive a particular fit, and which features exclude other solutions. The CMD analysis and SFH estimates for the four fields in M31 presented earlier in this chapter were performed using this method.

In an alternative approach one can move away from the *single best fit solution* in favor of a *range of solutions* supported by clear observational signatures on the CMD. An astrophysically motivated initial guess simulation is built, adopting reference parameters for the IMF, the distance, the reddening, the metallicity, the age distribution, and so on. Some of these parameters can indeed be gauged by simple superposition of isochrones to the CMD. The comparison of the zeroth-order (or set of zeroth-order) synthetic model to the data suggests variations in the SFH which can improve the fit, and more models are constructed to this end. When a global agreement with the data is reached with respect to some key features, like the luminosity function of blue and red stars, the width of the red giant branch, the population in the red clump, and the like, other SFHs can be tested to define the limits in the parameter space within which plausible models can be accommodated. This approach requires computing many models, each of which can be ranked with a statistical estimator. Different from the *blind* method, the user has some control on the solution, at the expense of objectivity, but allowing a full use of astrophysical knowledge. Sometime the *subjectivity* of the experienced astronomer is more effective than the *objectivity* of the package, but it is more labor intensive. In the following we illustrate an example of the application of this method.

4.5.5
An Example: the Stellar Population in the Halo of the Centaurus A Galaxy

A concrete example can illustrate the procedure and the kind of results that can be achieved when comparing observed and simulated CMDs. Figure 4.20a,b shows the CMD and the I-band luminosity function derived from HST/ACS observations of a field in the outer halo of the giant elliptical galaxy Centaurus A. The width of the RGB clearly shows that a wide metallicity distribution is present in this field, a distribution which can indeed be determined by counting the stars in bins bounded by the overplotted tracks. The effectiveness of this method relies on the fact that the color of the RGB stars is much more sensitive to metallicity than to age. On the luminosity function various typical features are evident: the red clump, populated by core helium burning stars, the AGB bump, corresponding to the slow evolutionary phase at the beginning of the E-AGB, and the RGB tip, whose luminosity allows us to derive an accurate distance to this galaxy. The lack of stars brighter than the tip (the few ones which are present are compatible with being foreground objects) shows that no intermediate or young age populations are present in this field. However, a well populated RGB only sets a lower limit of a few gigayears to the age of the stars. The position of the red clump and of the AGB bump indicate ages over 8–10 Gyr. These positions are quite slow functions of the age and the wide metallicity distribution, plus the photometric errors, smear out the features somewhat.

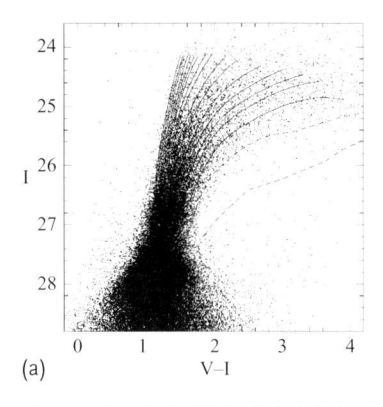

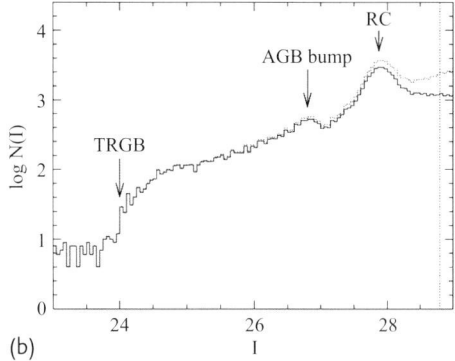

Figure 4.20 (a) VI CMD of a halo field of the giant elliptical galaxy Centaurus A. The overplotted loci are the RGB portions of 1 $M_\odot$ evolutionary tracks with metallicity ranging from [Fe/H] = -2 to $+0.4$. (b) I-band luminosity function of the data shown in (a). The dotted line shows the luminosity function corrected for incompleteness (source: Rejkuba, M. et al. (2005, Astrophys. J., 631, 262)) (Reproduced by permission of the AAS.).

With the aim of getting better age constraints one can extend the analysis to the full stellar distribution on the CMD. Thus, a grid of synthetic CMDs is generated, with ages from 2 to 13 Gyr, a Salpeter IMF, and a metallicity distribution. The latter is determined from the color distribution of the stars on a bright region on the RGB, as mentioned above. For each age, the Monte Carlo simulations go through the following steps: (i) random extraction of a pair (mass, metallicity) according to their distribution functions; (ii) determination of luminosity and effective temperature of the synthetic stars by interpolation within a grid of isochrones of the adopted age; (iii) transformation of the luminosity into V and I magnitudes based on a grid of model atmospheres; and (iv) application of the photometric errors and completeness factors according to the results of artificial star experiments on the actual frames. The random extractions are continued until the region on the CMD between $24.5 \leq I \leq 26$ and $1 \leq V - I \leq 2.3$ is populated with the observed number of stars. The region is selected so that it has a high completeness, small photometric error, and reasonably robust theoretical bolometric corrections. The stellar counts inside this region allow us to derive a robust measurement of the total mass that went into stars in the field.

These *single age* CMDs can be considered as a zeroth-order solution, and compared to the data with a statistical estimator. Figure 4.21a shows the binning adopted to compute the reduced χ^2 of the models to get a quantitative ranking. The grid is designed following several criteria: the fainter boxes are wider to accommodate for the larger photometric errors; strategic boxes are placed on the red clump and AGB bump regions to better constrain these features; wider boxes in the brightest portion of the RGB secure statistically significant population of enclosed stars; the reddest box at the tip of the RGB is very large, because the shape of the theoretical isochrones fails to accurately reproduce the magnitudes and colors of stars in this region (see Figure 4.21b). The merit of each of the models (with

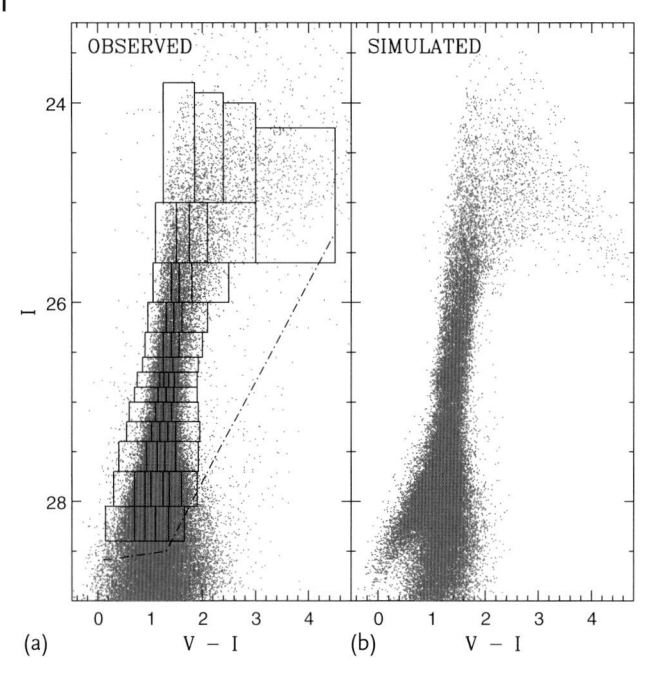

Figure 4.21 (a) The same CMD shown in Figure 4.20a with superimposed boxes used to calculate the χ^2 statistics. The dashed line shows the 50% completeness level. (b) One of the best fitting, single burst simulations, with an age of 11 Gyr. Notice that the upper envelope of the RGB at the redder end is not well reproduced in the synthetic CMD, likely due to inadequate bolometric corrections (adapted from: Rejkuba, M. et al. (2011, Astron. Astrophys., 526, A123)).

the different ages) can be estimated in various ways, using different indicators, such as the fit of the whole CMD, the fit of the V-band and of the I-band luminosity functions, or the number of objects falling in the red clump and in the AGB bump boxes. These different indicators yield somewhat different preferences for the average age of the stars in the field, that is, the minimum χ^2 is reached for different values of the age. In this example, relatively good χ^2 values are obtained for all the indicators for an age of ~ 11 Gyr. In the attempt to improve on this solution, two population models can be computed, contributing to the field with complementary proportions. With more free parameters (the age of the two stellar populations and their relative contributions) the χ^2 obviously improves. In particular the I- and V-band luminosity functions in the region of the red clump are better matched. However, in this example, the χ^2 turned out to be always larger than 1, whereas tests of model-to-model comparison show that the CMD χ^2 statistics does reach a minimum of 1 for the correct age. Adding complexities to the model did not help reducing the χ^2 below that of the two-bursts model, suggesting that the underlying isochrone grid and/or error description is not completely adequate to reproduce the observed dataset. Reassuringly, the total mass in stars of the models

($\sim 1.5 \times 10^7$ $M_\odot$) shows very little variation with age, confirming the arguments made in the previous section.

In summary, the bulk of the stellar population in this Cen A field appears to be old, a conclusion which is mainly driven by the luminosity of the red clump, which becomes too bright and too red if the age is younger than ~ 8 Gyr, and too faint and too blue if the age is older than 12 Gyr. The detailed analysis of the CMD shows that, for a single burst, the best fitting age is 11 Gyr; however, taking into account all the diagnostics mentioned above, the χ^2 statistics formally favors a solution in which some $\sim 80\%$ of the stars are 12 Gyr old, and $\sim 20\%$ are young (~ 2 Gyr old). In the two cases, though, the χ^2 values of the CMD stellar distribution are almost the same, and both greater than 1. Whether the younger component is really present could be tested only by reaching the SGB, some 2 to 3 magnitudes fainter than the red clump, that is, down to $I \simeq 30-31$.

With the known exception of ω Centauri, for many years we have regarded globular clusters as prototypical *simple stellar populations*, arguing that they should be used as calibrators for synthetic stellar populations. Then, in the last few years a kind of revolution has taken place, with exploding evidence that many, if not all, globular clusters actually host two or more distinct stellar populations. Even more surprising, some of such subpopulations appear to be incredibly enriched in helium, which is a formidable challenge to explain. Yet, even if not so simple anymore, globular clusters remain the closest objects to SSPs that Nature provides us to check synthetic populations.

According to a widespread notion, globular clusters may form in merging galaxies, and there is some direct evidence that massive clusters form in nearby mergers. If mergers were a necessary condition for making globular clusters, than their ages could be used to trace the merging history of galaxies. As most globulars for which we measure accurate ages are coeval to ~ 12 Gyr, should one conclude that most of the merging in our galaxy took place such a long time ago? Perhaps not. Maybe globular clusters formed in another, as yet unknown way.

The least implausible scenario for the formation of helium-enriched populations in globular clusters appeal to massive ($M \gtrsim 3.5$ $M_\odot$) AGB stars with hot-bottom burning. Yet, to make as much mass from AGB ejecta as needed to form the helium-enriched populations one needs a stellar population exceeding by at least ~ 10 times the current mass of the cluster. One then appeals to globular clusters as being just the stripped remnants of nucleated dwarfs, torn apart by the galactic tidal field. The picture is getting everything but *simple*.

Globular clusters as well as the bulges of the Milky Way and Andromeda galaxies appear to be at least 10 Gyr old. With current cosmology, 10 Gyr lookback time corresponds to a redshift ~ 2. We could say that much of the spheroids of these two familiar, quite average-sized galaxies *formed at redshift* ~ 2, although this is obviously an oxymoron. But, it is true that at redshift ~ 2 we should be able to see spheroids in formation that will become similar to our own. Perhaps we have already seen them.

The CMD of resolved stellar populations keeps the fossil record of the SFH over the whole Hubble time, and efficient methods for decoding this information are at hand. In spite of the enormous progress in this field, so far only a few nearby galaxies have been analyzed, and results for giant galaxies are limited to the stellar population in tiny fields. Future observing facilities coupling high sensitivity, large field of view and, especially, high resolving power will enable the reconstruction of the SFH of entire giant galaxies up to the Virgo cluster.

Further Reading

Globular Cluster Ages (Absolute or Relative)

Chaboyer, B. et al. (1998) *Astrophys. J.*, **494**, 96.
De Angeli, F. et al. (2005) *Astron. J.*, **130**, 116.
Gratton, R. et al. (1997) *Astrophys. J.*, **491**, 749.
Marin-Franch, A. et al. (2009) *Astrophys. J.*, **694**, 1498.
Renzini, A. (1991), in *Observational Tests of Cosmological Inflation* (ed. T. Shanks et al.), Dordrecht, Kluwer.
Renzini, A. et al. (1996) *Astrophys. J.*, **465**, L23.
Sandage A. (1970) *Astrophys. J.*, **162**, 841.
Stetson, P. et al. (1996) *Publ. Astron. Soc. Pac.*, **108**, 560.
Vandenberg, D. et al. (1996) *Annu. Rev. Astron. Astrophys.*, **34**, 461.
Zoccali, M. et al. (2001) *Astrophys. J.*, **553**, 733.

Multiple Populations in Globular Clusters

Bedin, L.R. et al. (2004) *Astrophys. J.*, **605**, L125.
Carretta, E., et al. (2009) *Astron. Astrophys.*, **508**, 695.
D'Antona, F. and Caloi, V. (2004) *Astrophys. J.*, **611**, 871.
Norris, J.E. (2004) *Astrophys. J.*, **612**, L25.
Piotto, G. et al. (2005) *Astrophys. J.*, **621**, 777.
Piotto, G. et al. (2007) *Astrophys. J.*, **661**, L53.
Piotto, G. et al. (2009) in IAU Symposium No. 258. The Ages of Stars, p. 233.
Renzini, A. (2008), *Mon. Not. R. Astron. Soc.*, **391**, 354.

Age of Galactic Bulge Stars

Brown, T. et al. (2010) *Astrophys. J.*, **725**, L19.
Clarkson, W. et al. (2008) *Astrophys. J.*, **684**, 1110.
Feltzing, S. and Gilmore, G. (2000) *Astron. Astrophys.*, **355**, 949.
Ortolani, S. et al. (1995) *Nature*, **377**, 701.
Zoccali, M. et al. (2003) *Astron. Astrophys.*, **399**, 931.

Metallicity Distribution in Bulges
Milky Way Bulge

Brown, T. et al. (2010) *Astrophys. J.*, **725**, L19.
Fulbright, J.P. et al. (2006) *Astrophys. J.*, **636**, 821.
Fulbright, J.P. et al. (2007) *Astrophys. J.*, **661**, 1152.
Lecureur, A. et al. (2007) *Astron. Astrophys.*, **465**, 799.
McWilliam, A. and Rich, R.M. (1994) *Astrophys. J. Suppl.*, **91**, 749.
Zoccali, M. et al. (2008) *Astron. Astrophys.*, **486**, 177.
Zoccali, M. et al. (2006) *Astron. Astrophys.*, **457**, L1.

M31 Bulge

Davidge, T.J. *et al.* (2005) *Astron. J.*, **129**, 201.
Rich, R.M. and Mould, J. (1991) *Astron. J.*, **101**, 1286.
Stephens, A. *et al.* (2003) *Astron. J.*, **125**, 2473.

Stellar Ages in M31 Halo and Giant Stream

Brown, T. *et al.* (2003) *Astrophys. J.*, **592**, L17.
Brown, T. *et al.* (2004) *Astrophys. J.*, **613**, L125.
Brown, T. *et al.* (2006) *Astrophys. J.*, **652**, 323.
Ferguson, A.M.N. *et al.* (2002) *Astron. J.*, **124**, 1452.
McConnachie, A.W. *et al.* (2009) *Nature*, **461**, 66.
Richardson, J.C. *et al.* (2008) *Astron. J.*, **135**, 1998.

Stellar Populations in the Halo of Centaurus A

Harris, W.E. and Harris, G.L.H. (2002) *Astron. J.*, **123**, 3108.
Rejkuba, M. *et al.* (2005) *Astrophys. J.*, **631**, 262.
Rejkuba, M. *et al.* (2011) *Astron. Astrophys.*, **526**, A123.

Star Formation Histories of Dwarf Galaxies

Grebel, E.K. (1997) *Rev. Mod. Astron.*, **10**, 29.
Mateo, M. (1998) *Annu. Rev. Astron. Astrophys.*, **36**, 435.
Tolstoy, E., *et al.* (2009) *Annu. Rev. Astron. Astrophys.*, **47**, 371.

Synthetic CMD Method

Aparicio, A., *et al.* (1997) *Astron. J.*, **114**, 680.
Cignoni, M. and Tosi, M. (2010) *Adv. Astron.*, 158568.
Dolphin, A.E. (2002) *Mon. Not. R. Astron. Soc.*, **332**, 91.
Harris, J., and Zaritsky, D. (2001) *Astrophys. J. Suppl.*, **136**, 25.
Tolstoy, E. and Saha, A. (1996) *Astrophys. J.*, **462**, 672.
Tosi, M. *et al.* (1991) *Astron. J.*, **102**, 951.
Gallart, C., *et al.* (2005) *Annu. Rev. Astron. Astrophys.*, **43**, 387.

Publicly Available Packages for Simulated CMDs and CMD Fitting

- StarFISH: http://www.noao.edu/staff/jharris/SFH/ (last accessed 2011-06-16)
- IAC-STAR: http://iac-star.iac.es/iac-star/
- SCMD: http://www.astro.yale.edu/demarque/SCMD/cgi-bin/SCMD.cgi (last accessed 2011-06-16)
- BaSTI: http://albione.oa-teramo.inaf.it

5
The Evolutionary Synthesis of Stellar Populations

Chapter 2 deals with the fundamentals of stellar population synthesis, and is restricted to bolometric light. The directly measurable quantities, instead, are fluxes within certain photometric bands (magnitudes), hence the spectral energy distribution (SED), and spectra. How these quantities are synthesized using stellar evolutionary sequences, isochrones and stellar spectra as building blocks is described in this chapter. The case of simple stellar populations (SSP) is presented first, then we shall turn to composite stellar populations.

The synthetic stellar spectra of the type discussed in this chapter have two typical applications: (i) derive galaxy physical properties such as the star formation rate (SFR), mass, age, metallicity, and so on by matching the observed spectrum of an individual galaxy to a set of synthetic spectra, sorting out the best fit case, and (ii) derive the spectra and SEDs of models of evolving galaxies (e.g., the so-called semianalytic models), in order to compare them to observations.

5.1
Simple Stellar Populations

What one needs to do is to write Eq. (2.20) for the generic wavelength λ, so to produce quantities that can be immediately compared to galaxy observables. This is easier than one may think, as the monochromatic flux $\mathcal{F}_\lambda$ of a SSP of composition X_i and age t is given by:

$$\mathcal{F}_\lambda(X_i, t) = \int_{M_{\text{inf}}}^{M_{\text{TO}}} \phi(M_\circ) f_\lambda \, dM_\circ + 9.8 \times 10^{10} b(t) \sum_j \tilde{f}_\lambda(X_i, g_j, T_j) F_j(M_{\text{TO}}) , \quad (5.1)$$

where

$$\tilde{f}_\lambda = \frac{f_\lambda(X_i, g_j, T_j)}{f_{\text{bol}}(X_i, g_j, T_j)} , \quad (5.2)$$

and X_i is the mass fraction of the ith chemical element, $f_\lambda = f_\lambda(X_i, g, T)$ is the flux at wavelength λ of a star with composition X_i, surface gravity g and effective tem-

Stellar Populations, First Edition. Laura Greggio and Alvio Renzini
© 2011 WILEY-VCH Verlag GmbH & Co. KGaA. Published 2011 by WILEY-VCH Verlag GmbH & Co. KGaA.

perature T, f_{bol} is the corresponding bolometric flux, F_j is the fuel consumption during the generic post-MS stage j, and g_j and T_j are the gravity and temperature of stars in phase j. Of course, the sum extends over all the post-MS evolutionary phases.

The integration is performed along the isochrone of age t, and gravity and temperature of stars along the isochrone are functions of composition, mass, and age, that is:

$$g = g\left[X_i, M_\circ(t)\right]$$
$$T = T\left[X_i, M_\circ(t)\right] . \tag{5.3}$$

For the sum over the fuel consumptions g_j and T_j refer to the gravity and temperature of stars of initial mass M_{TO} in its jth post-MS stage, as one can adopt $M_\circ \simeq M_{TO}$ for all post-MS stages (cf. Chapter 2).

Adopting a single slope IMF above $\sim 0.5 M_\odot$, and neglecting the contribution of low-mass stars below this limit, Eq. (5.1) becomes:

$$\mathcal{F}_\lambda(X_i, t) = A \int_{0.5}^{M_{TO}} M_\circ^s f_\lambda(X_i, g, T) d M_\circ +$$
$$+ 9.8 \times 10^{10} A M_{TO}^{-s} |\dot{M}_{TO}| \sum_j \tilde{f}_\lambda(X_i, g_j, T_j) F_j(M_{TO}) , \tag{5.4}$$

where the proper match between the main sequence contribution (the part that is calculated by integration along the isochrone) and that of the post-MS (calculated using the fuel consumption theorem) is ensured by both terms sharing the same scale factor A. Actually, it may be convenient to move the matching point between the two terms from the main sequence turnoff to a more advanced stage, such as for example, the base of the RGB, thus including the SGB phase in the isochrone integration. Formally, suffice to replace M_{TO} in Eq. (5.4) with the mass along the isochrone corresponding to the chosen matching point. This approach is especially effective to properly include in the synthesis the most advanced evolutionary stages, which may not be included in available isochrone sets and are particularly sensitive to mass loss and other assumptions such as the case of the HB and the AGB stages. Thus, this approach allows one to rapidly explore the effect of different temperature distributions (T_j) for HB stars, for example, moving from a red to a blue HB, or adopting different fuel consumptions (F_j) for the TP-AGB stage, compared to those adopted to calculate the isochrones available from databases.

In some realizations of evolutionary population synthesis the calculation is restricted to the first term in Eqs. (5.1) and (5.4), and isochrones are integrated all the way to the end of the TP-AGB. This procedure may give rise to unwanted oscillations due to the discrete ages at which the isochrones are available, if adequate interpolation algorithms are not implemented.

5.2
Spectral Libraries

A critical ingredient in constructing the synthetic spectra of SSPs are, of course, the spectra of the individual stars, that is, the $f_\lambda(X_i, g, T)$ distributions. What is needed are large *libraries of stellar spectra*, extending from the Lyman continuum all the way to at least $\sim 2\,\mu\mathrm{m}$, and covering as widely and uniformly as possible the (X_i, g, T) parameter space. These spectra can come in two flavors: (i) empirical spectra of real stars, and (ii) theoretical spectra generated by model atmospheres. Whatever the choice, it is important to appreciate that there are no perfect libraries. Here are some of their limitations.

5.2.1
Empirical Spectral Libraries

Libraries of excellent signal-to-noise ratio (S/N), flux-calibrated optical spectra exploiting the whole atmospheric window exist, that cover a good fraction of the relevant (X_i, g, T) parameter space. Of course, most of the stars providing these template spectra are in the solar neighborhood, and therefore most of them have near-solar composition. This is not so bad after all, as most stars in the Universe are not far from solar metallicity. Still, spectra of stars with supersolar metallicity are underrepresented in existing libraries, whereas supersolar populations are likely to dominate in the core of giant elliptical galaxies. Indeed, their spectra is noticeably more strong-lined than the integrated spectrum of the Galactic bulge, whose average metallicity is about solar (cf. Figure 4.4).

In most existing stellar libraries and synthetic populations the vector X_i reduces to just one number, the overall mass fraction of metals Z, or the iron abundance [Fe/H]. Yet, at least another composition parameter can play an important role, that is, the abundance of α elements relative to iron, [α/Fe]. In fact, α elements are produced by massive core collapse supernovae, which all explode less than ~ 40 Myr from formation. Thermonuclear, Type Ia supernovae are believed to be major iron producers, but explode with a distribution of delay times from formation that extend from a few dozen million years to the Hubble time (see Chapter 7). Thus, the [α/Fe] ratio provides a diagnostic tool on the duration of star formation in galaxies, especially for those where star formation has ceased long ago. An α-element enhancement ([α/Fe] > 0) is produced if star formation was virtually completed in a time shorter than the average delay time of Type Ia supernovae. Again, empirical spectral libraries include only relatively few spectra of stars with [α/Fe] > 0.

Synthetic populations based on stellar spectra that cover only the optical range, for example, from ~ 3500 to ~ 7500 Å may suffice for some applications, but modeling also the ultraviolet and near-infrared parts of the spectrum is crucial for several of the most important applications of synthetic stellar populations, most notably for the diagnostics of high-redshift galaxies. Instead, rare are the stars for which full spectral coverage is available.

In summary, the major limitations of existing empirical libraries are insufficient coverage at high metallicity, at positive values of the [α/Fe] ratio, and at UV and near-IR wavelengths.

5.2.2
Model Atmosphere Libraries

In principle, there is no domain of the (X_i, g, T) parameter space that cannot be covered by theoretical model atmospheres. In practice, the main limitation come from the complex chemistry characterizing the atmosphere of cool stars, with $T \lesssim 4000$ K, which is dominated by a variety of molecules. Moreover, the atmosphere floats atop a deep convective region, and therefore it is poorly represented by a plane-parallel, stratified model with uniform horizontal temperature. As a result, only few attempts have been made to calculate model atmospheres for stars of spectral type later than the K-type, that is, for M-, S- and C-type stars, and empirical spectra for these stars are typically used in population synthesis models.

Convection and departures from plane-parallel geometry are not unique to these late spectral types. Also G- and K-type stars have outer convective layers, and fully hydrodynamical, 3D model atmospheres have been constructed for them. Compared to 1D model atmospheres, these more sophisticated models yield abundances that generally differ by ~ 0.2 dex, but their calculation is computationally quite heavy, and a full library of 3D model atmospheres is far from being achieved.

So, what should one prefer – empirical spectral libraries or grids of model atmospheres? The answer, in the end, may be just a matter of taste. In the case of empirical libraries we have a set of stellar f_λ, to which we must attach their (X_i, g, T), and this is done via model atmosphere analysis. In the case of model atmosphere libraries, we chose the (X_i, g, T) ourselves, and we need to attach to them their f_λ, which is done by computing the corresponding model atmospheres. Therefore, the two approaches are largely equivalent, as the use of model atmospheres is unavoidable in one way or another, and therefore both eventually suffer from any systematic inadequacy that might affect the model atmospheres. Thus, the common wisdom is to fill with model atmosphere spectra the holes of empirical libraries, and with empirical spectra the holes of theoretical libraries.

5.3
Composite Stellar Populations

Real stellar systems can rarely be represented by SSPs. Even many globular clusters, once considered to be prototypical SSPs, are now recognized to host multiple stellar populations. Galaxies, of course, had a complex star formation history, with the SFR being some function of time $\psi(t)$. Therefore, the emerging flux of a composite stellar population can be written as:

$$F_\lambda(X_i, t) = \int_0^t \mathcal{F}_\lambda(X_i, t - t')\psi(t')dt', \tag{5.5}$$

where $\mathcal{F}_\lambda(t')$ is the spectrum of a SSP as given by Eq. (5.1). Note that this equation assumes the composition X_i to be constant in time, an obvious inconsistency given that star formation is inevitably accompanied by metal production and enrichment of the ISM. In principle, the composition should be a function of time, $X_i(t')$, and Eq. (5.5) should be accompanied by another set of equations describing the chemical evolution of the system. Chemical evolution models have been constructed for a variety of stellar systems, such as galaxies of the various morphological types, and they all rely on assumptions concerning the gas exchanges between the system and its environment. The simplest case, no gas exchange at all, that is, the so-called *closed box model*, is indeed quite simple to calculate, but certainly galaxies do not behave as closed boxes. Indeed, galaxies tend to be continuously fed with fresh gas from the intergalactic medium, while part of the enriched gas is returned to it via galactic winds, and especially so at high redshift.

In theoretical models of evolving galaxies (either semianalytic or hydrodynamical), chemical evolution is relatively easy to incorporate, and therefore the time-dependence of the gas composition can be calculated, hence that of the stars forming at any given time, and the resulting $X_i(t')$ incorporated in Eq. (5.5). In this way the results of evolving galaxy models can be translated into sets of observable quantities, such as spectra, magnitudes, luminosity functions and so on. This corresponds to the second application mentioned at the beginning of this chapter. However, the rates of gas accretion from, and return to, the intergalactic medium are hard to predict from first principles, and are still poorly constrained by the observations, hence assumptions about them – albeit physically motivated – are inevitably quite arbitrary at the present stage.

Concerning the use of synthetic spectra for the diagnostics of observed galaxies, what is needed is a large grid of them (sometimes called templates) among which to chose the one that best fits the data (observed spectrum or SED). In this respect, so-called *self-consistent* models that combine spectral and chemical evolution can be constructed, but ultimately they rely on arbitrary assumptions concerning gas and metal exchanges with the environment. Hence, self-consistency does not ensure that models match reality. In principle, these exchanges could be parameterized in some fashion, but the resulting parameter space would inflate so much, if one wants to have a chance to include cases close to reality, to make very unpractical this approach.

However, one result is quite generic to chemical evolution models, no matter on which assumptions they are constructed: the metallicity tends to reach quite soon near-solar values, and then flattens with time to within a factor of 2 or 3 from solar. These considerations suggest that spectra calculated according to the *inconsistent* Eq. (5.5) can be quite useful after all. The procedure does not pretend to incorporate a credible chemical enrichment history, but assumes instead that the bulk of stars will not be too far from the average metallicity. While avoiding to emphasize this internal inconsistency, spectra constructed in this way are very widely used to estimate galaxy properties such as stellar mass, SFR, reddening, and age, especially for large high-redshift samples. Of course, fitting synthetic spectra to the observed SED of a galaxy can be repeated with two or three different metallicities, hence assigning to the galaxy the (luminosity-weighted) metallicity that gives the best fit. This is a reasonable, legitimate procedure, provided one correctly interprets the resulting metallicity, and one is ready to quote a factor of ~ 2 systematic uncertainty.

Perhaps potentially more troublesome than chemical evolution, is the necessity to assume a star formation history, that is, the function $\psi(t')$ in Eq. (5.5). This is currently done by adopting for $\psi(t')$ simple functional forms, such as for example, constant, exponentially increasing or decreasing, the same with the addition of isolated bursts, or truncated after a certain time, depending on the type of galaxy and its redshift. Concrete examples using this approach are given in the following.

5.4
Evolving Spectra

A series of figures of synthetic spectra are now shown to familiarize the reader with the output of evolutionary population synthesis codes. In particular, the figures are meant to illustrate the age and SFH dependence of the spectra, as well as the main spectral features that appear or disappear as a function of time. With just one exception, all shown spectra refer to solar metallicity, and always refer to the same amount of mass turned into stars.

5.4.1
The Spectral Evolution of a SSP

Figure 5.1 shows the spectral evolution of a solar metallicity SSP as it ages from 1 Myr to 13 Gyr. At $t = 1$ Myr all massive stars are still on the main sequence; the SED is relatively featureless, apart from a prominent Lyman break. By $t = 10$ Myr the most massive stars have already died, whereas others have evolved to the red supergiant (RSG) stage, causing the prominent near-IR hump seen in the figure. In the meantime, the slope of the UV continuum around 1500 Å progressively flattens, as the hottest stars on the main sequence burn out, evolve off the main sequence and die. This trend continues all the way to the oldest population in the figure. By $t = 100$ Myr a prominent Balmer break has developed, along with strong Balmer lines, due to the top of the main sequence now being made of A-type stars. By $t = 300$ Myr the general fading of the spectrum has almost stopped in the near-IR, caused by the appearance of TP-AGB stars that contribute a large fraction of the bolometric and especially near-IR light. Notice also that several metal lines have appeared short of the Balmer break, which indeed progressively weakens, merging with the so-called 4000 Å break due to the growing strength of many metal lines. At $t = 1$ Gyr the TP-AGB contribution is at its maximum, and molecular band

5.4 Evolving Spectra

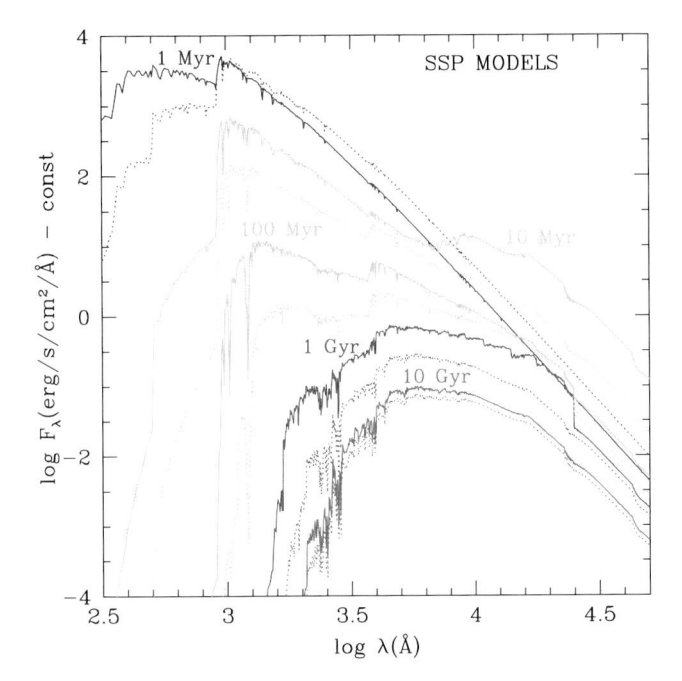

Figure 5.1 The spectral evolution of a solar metallicity SSP from 1 Myr to 13 Gyr. The dotted lines correspond to ages 3 times older than the spectrum preceding them, except for the last one which refers to an age of 13 Gyr. Purely red HBs are assumed at late epochs (figure constructed using M05 database models; Maraston, C. (2005, *Mon. Not. R. Astron. Soc.*, 362, 799)). For a color version of this figure please see in the front matter.

heads are clearly noticeable in the near-IR. At about this age, strong lines of neutral and ionized iron and magnesium have also appeared in the UV around ~ 2800 Å. The final aging between 1 and 10 Gyr and beyond sees a slow down of the rate of fading, whereas the evolution of the features has become more subtle. The molecular bands have disappeared, along with the fuel consumption of TP-AGB stars, whereas the strength of several metallic lines in the optical part of the spectrum has increased, again due mostly to iron and magnesium. The fluxes of the various spectra reflect the actual fading of the population as time goes by. Thus, perhaps the most striking aspect illustrated by the figure is how fast the fading is in the UV, compared to the near-IR, which is a direct consequence of the stellar lifetimes as a function of mass.

The fall of the UV flux with time is due to the hottest stars in the model being those at the main sequence turnoff, and the turnoff gets cooler and cooler as time goes by. This fall can be contrasted if the SSP is to develop a blue HB at late epochs as illustrated in Figure 5.2, which shows the effect of spending the HB fuel with different distributions of effective temperature. Clearly, even a trace of warm/hot HB stars can dramatically affect the UV spectrum of an old stellar population. A flat distribution of HB temperatures also leads to a fairly flat UV spectrum, whereas a bimodal distribution, with half of the HB stars placed at 25 000 K and the other half

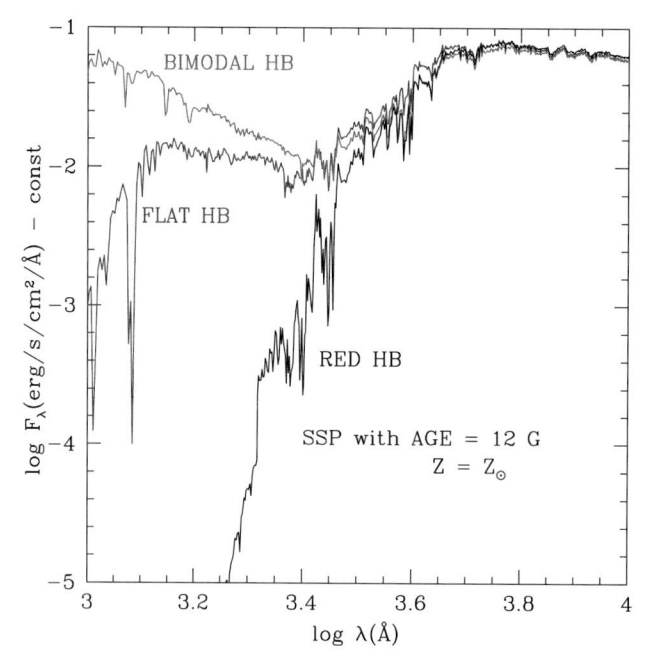

Figure 5.2 The effect of different distributions of stars along the horizontal branch for a 12 Gyr old SSP. With a full red HB the spectrum is intermediate between the 10 and 13 Gyr spectra shown in Figure 5.1. The spectrum labeled "FLAT HB" has been constructed assuming a uniform distribution of HB stars between 4000 and 12 000 K. For the "BIMODAL HB" it is assumed that half of the HB stars burn their fuel at an effective temperature of 4000 K, and the other half at 25 000 K (synthetic spectra kindly provided by C. Maraston).

at 4000 K, generates a prominent *UV rising branch*, which is typical of many giant elliptical galaxies. Actually, to reproduce the level of UV continuum seen in such galaxies only $\sim 10\%$ of HB stars are required to be so hot, whereas the rest should be red. This makes such a bimodal HB the prime candidate for the origin of UV rising branches in elliptical galaxies.

Figure 5.3 shows the effect on the spectrum of a 0.8 Gyr old SSP of including/excluding the contribution of the TP-AGB stars, which peaks near this age. For this half solar metallicity population the major TP-AGB contribution is due to carbon stars, exhibiting strong molecular bands typical of carbon-rich atmospheres (CN, CH, C_2). With increasing metallicity such features tend to disappear, as M-type stars replace carbon stars on the TP-AGB, and prominent TiO bands appear. The model spectrum shown here relies on the calibration of the TP-AGB fuel consumption shown in Figure 2.4.

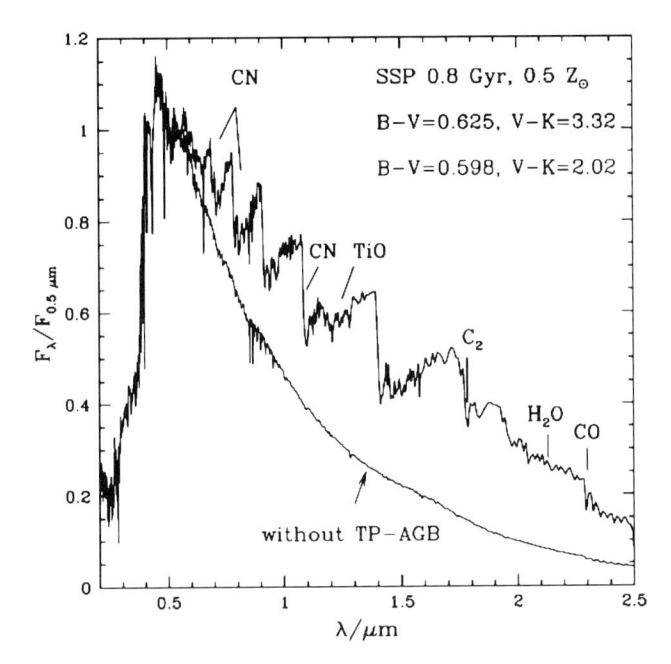

Figure 5.3 Model spectra for 1 Gyr old, $Z = 0.5\,Z_\odot$ SSP with and without the contribution of the TP-AGB phase. The synthetic population with the TP-AGB phase relies on the calibration of the TP-AGB fuel consumption shown in Figure 2.4. A few molecular band heads are indicated (source: Maraston, C. (2005, Mon. Not. R. Astron. Soc., 362, 799)) (Reproduced by permisiion of the Royal Astronomical Society.).

5.4.2
The Spectral Evolution of Composite Stellar Populations

Conveniently simple mathematical laws are usually assumed for the SFH in order to construct sets of template SEDs to be used in best fit procedures. The simplest of them all is a constant SFR, and fairly simple are also SFH laws in which the SFR is assumed to exponentially decline or increase with time (SFR $\propto e^{-t/\tau}$ or SFR $\propto e^{+t/\tau}$, hereafter *direct-τ* and *inverted-τ* models, respectively). Slightly more complicated is a law in which the SFR starts from zero, climbs to a maximum, and then declines exponentially. One such *peaked* law is given by:

$$\text{SFR} \propto \frac{t}{\tau^2} \exp-\left(\frac{t^2}{2\tau^2}\right), \tag{5.6}$$

where τ is a constant. To simulate quenching of star formation any such law can be truncated at any prescribed time, with the resulting composite stellar population evolving passively after truncation.

Although Nature may have conjured to avoid making stars following such a mathematical simplicity, yet each of the mentioned SFH may have some justification. For example, it does not appear that the global SFR in the Galactic disk (and in the disk of other spirals) has changed much over the last several gigayears, and if so assuming a SFR = const. may not be such a bad approximation. Other local galaxies appear to have formed most of their stars in a very remote past, and therefore assuming an age of, say 12 Gyr and an exponentially declining SFR with a $\tau \ll 12$ Gyr may be quite adequate. On the other hand, in other circumstances galaxies may actually grow exponentially or quasi-exponentially for some lapse of time, whereas in others the SFR first increases and then decreases. Therefore, all these possible SFHs may have a certain validity, but one should be prepared to accept that no one simple mathematical law can adequately describe all typical SFH of galaxies. For example, occasionally galaxies may experience a major burst of star formation, then turning passive or return to a more steady SFR. These kinds of events are not difficult to model, for example, suffice to add an SSP contribution at the proper time.

Examples of synthetic spectra generated by the simple laws mentioned above are now showed and intercompared, so as to familiarize with similarities and differences among them.

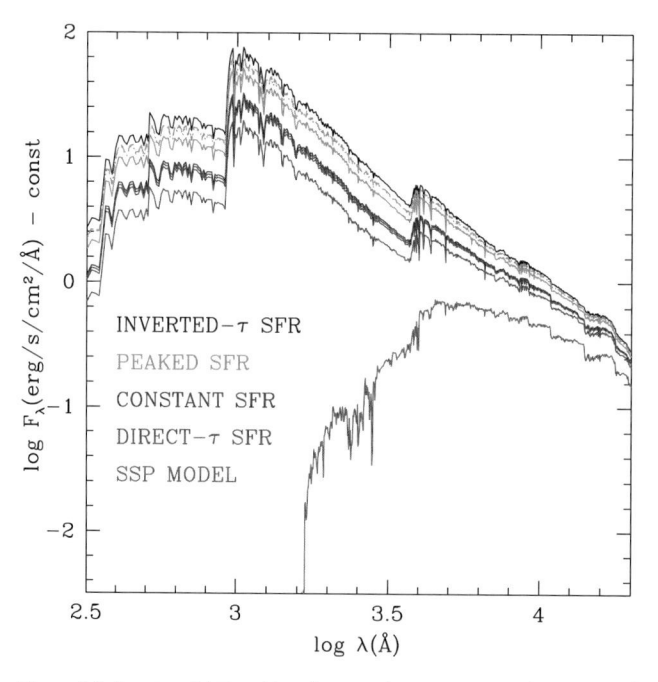

Figure 5.4 Spectra of 1 Gyr old stellar populations with different past star-formation histories (SFH), but normalized to have formed the same mass of stars. Top to bottom the spectra refer to exponentially increasing SFRs with $\tau = 0.5$ Gyr, peaked SFH with $\tau = 1, 4, 7$ Gyr (the last two ones perfectly overlap), SFR = const., and exponentially decreasing SFR, with $\tau = 1, 4, 7$ Gyr (the latter two also overlap). For comparison the spectrum of a 1 Gyr old SSP is also shown (synthetic spectra kindly provided by C. Maraston). For a color version of this figure please see in the front matter.

Figure 5.4 shows the synthetic spectra at $t = 1$ Gyr of stellar populations generated with all four mentioned SFHs, having them acting for 1 Gyr and producing exactly the same mass of stars. For comparison, the spectrum of a 1 Gyr old SSP is also shown. The latter is so different because it represents a stellar population which has evolved passively for 1 Gyr, whereas all other spectra refer to composite populations that are still actively forming stars. They differ rather marginally in shape, whereas their absolute flux in the UV scales with the SFR at 1 Gyr, which is maximum for the inverted-τ model and minimum for the direct-τ model with $\tau = 1$ Gyr. Notice that flux differences among the various spectra decrease towards the near-IR, as all populations have formed the same mass in stars (whereas the current mass differs slightly because of the different mass return). However, near-IR fluxes can still differ by up to a factor ~ 2.

Figure 5.5 shows the spectra of populations evolved with the same star formation laws as those shown in Figure 5.4, now acting for 13 Gyr, with the exception of the inverted-τ models, which are shown in Figure 5.6 for better illustration. Notice the survival of a faint UV rising branch in the peaked SFH model with $\tau = 7$ Gyr, due to a residual, low-level star formation still going on. The direct-τ model with $\tau = 1$

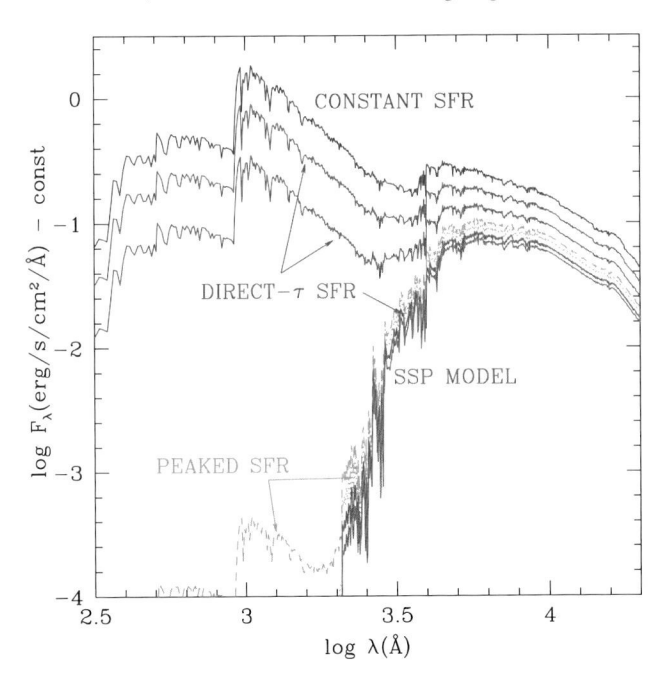

Figure 5.5 The same as Figure 5.4 (without the inverted τ models) but for $t = 13$ Gyr. Displayed top to bottom are the spectra for a population with SFR = const., three spectra for direct-τ models with $\tau = 7, 4,$ and 1 Gyr, then spectra for peaked SFHs with the same τ (largely overlapping). The spectrum of an SSP model for a 13 Gyr old population is also shown. Notice that the scale of the flux coordinate is consistent with that of Figure 5.4 (synthetic spectra kindly provided by C. Maraston). For a color version of this figure please see in the front matter.

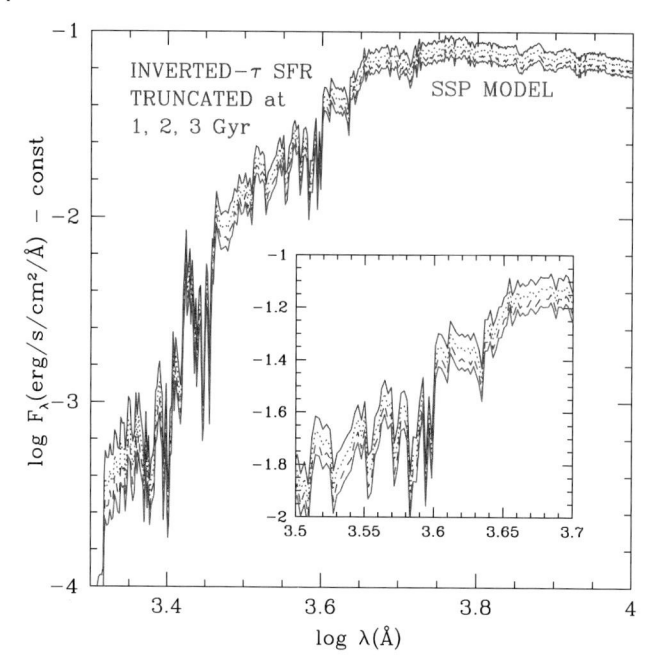

Figure 5.6 The SSP spectrum at an age of 13 Gyr shown in Figure 5.5, now in an expanded scale, is compared to the spectra of $\tau = 0.5$ Gyr inverted-τ models with star formation truncated after 1, 2 and 3 Gyr. The insert shows a blow-up of the spectral region around the Balmer/4000 Å break, to further illustrate the close similarity of all four spectra (synthetic spectra kindly provided by C. Maraston). For a color version of this figure please see in the front matter.

and the peaked models are barely distinguishable in this figure, and are all very close to the SSP distribution. The same holds for the inverted-τ models, for which the SFR is set to zero (truncated) after 1, 2, or 3 Gyr from the beginning of star formation at $t = 0$. These are shown in Figure 5.6 to appreciate the differences on an expanded scale. Notice also the close similarity of the upper three spectra in Figure 5.5, due to still sustained star formation also in models with $\tau = 4$ and 7 Gyr. Conversely, with $\tau = 1$ Gyr the extremely low SFR at $t = 13$ Gyr is barely affecting the spectrum, which illustrates the sensitivity of the spectrum to τ as τ is reduced from 4 to 1 Gyr. Compared to the 1 Gyr old spectra where the Balmer break is very sharp, now the break is smoother as it has merged with the 4000 Å break, with a forest of metal lines having appeared just below 4000 Å, most noticeably in the direct-τ models, which contain a larger proportion of old stars.

Figure 5.6 compares the spectrum of a 13 Gyr old SSP to the spectra of inverted-τ models with $\tau = 0.5$ Gyr of the same age, where star formation has been truncated after 1, 2 and 3 Gyr. Given that the SFR increases exponentially in these models, much of star formation takes place just shortly before truncation. Therefore, in practice the spectrum of the model with SFR truncated at 1 Gyr must be very close to that of a 12 Gyr old SSP, that of the model truncated at 2 Gyr must be very close

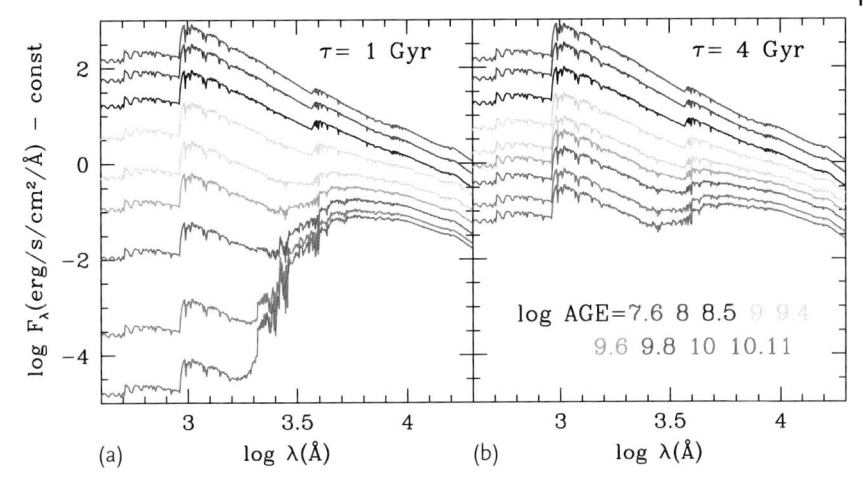

Figure 5.7 The spectral evolution of direct-τ models for $\tau = 1$ Gyr (a) and $\tau = 4$ Gyr (b) for ages between 40 Myr and 13 Gyr (top to bottom) as indicated (synthetic spectra kindly provided by C. Maraston). For a color version of this figure please see in the front matter.

to that of an 11 Gyr old SSP, and so on. In any event, all these spectra are barely distinguishable, apart from their normalization which reflects the slightly higher evolutionary flux $b(t)$ (cf. Chapter 2) of the models with younger average stellar ages. The insert in this figure shows a blow-up of the spectra in the wavelength range around the Balmer/4000 Å break, to illustrate the small sensitivity to age of the spectral features in these old populations.

The spectral evolution of direct-τ models is further illustrated in Figure 5.7. The lowest spectra in the two panels are the same as those in Figure 5.5, referring to

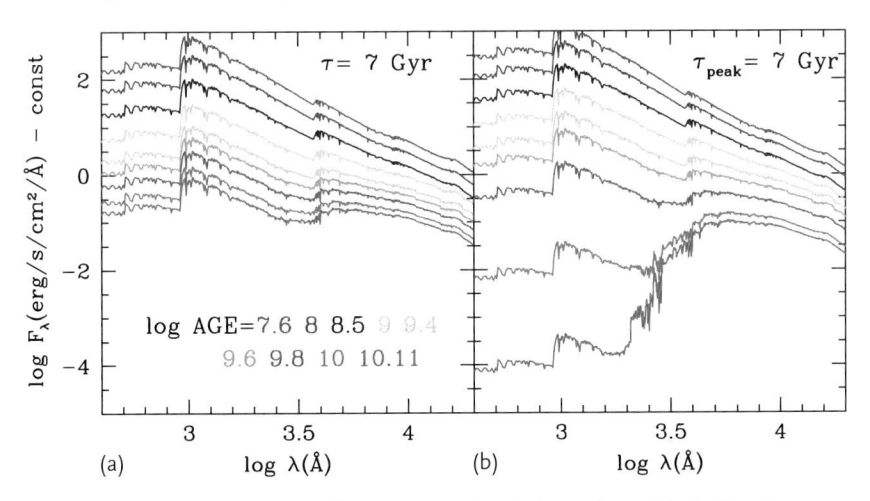

Figure 5.8 The spectral evolution of direct-τ (a) and peaked-SFR (b) models, both with $\tau = 7$ Gyr, is illustrated for ages from 40 Myr to 13 Gyr (synthetic spectra kindly provided by C. Maraston). For a color version of this figure please see in the front matter.

the same age and star formation history. One can appreciate how much faster is the spectral evolution at late times as τ is reduced from 4 to 1 Gyr. Notice that due to the expanded vertical scale a UV rising branch is now appearing in the oldest model with $\tau = 1$ Gyr, due to the corresponding low level of residual star formation, since such models are never totally quenched.

The spectral evolution of direct-τ models is compared to that of peaked-SFR models in Figure 5.8. Notice the faster evolution of the peaked-SFR models (especially at late times), due the square of t/τ in the exponential. Figure 5.9 compares instead direct- and inverted-τ models for ages up to 2.5 Gyr. Direct models have $\tau = 1$ Gyr and inverted ones have $\tau = 0.5$ Gyr. Spectra come in pairs referring to the same age, with the top ones (the youngest) being barely distinguishable. Then the split between the spectra of the two kinds of models grows with age, becoming very wide for the oldest age. Much of the difference between the various spectra is due to different ongoing SFRs, since all models refer to the same amount of mass turned into stars. By construction, direct-τ models give more weight to the oldest stars in the populations, and instead inverted-τ models do it to the youngest ones. For example, the 1 Gyr old direct-τ model is most sensitive to the TP-AGB contribution, and inverted-τ model is less sensitive to it. This can be appreciated in Figure 5.9 by noticing that the molecular band heads in the infrared are much stronger in

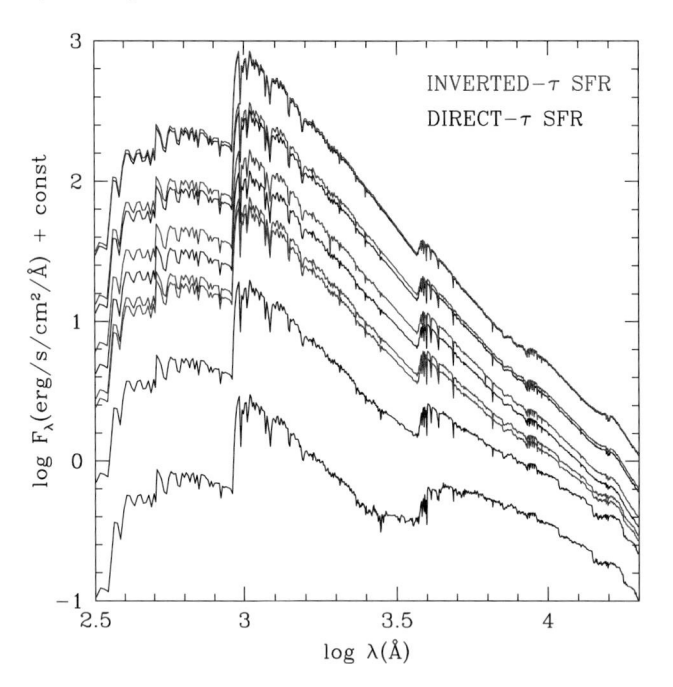

Figure 5.9 A comparison between the spectral evolution of direct- and inverted-τ models, respectively with $\tau = 1$ and 0.5 Gyr, and for ages equal to (40, 100, 250) Myr and 1 and 2.5 Gyr (synthetic spectra kindly provided by C. Maraston). For a color version of this figure please see in the front matter.

5.4 Evolving Spectra

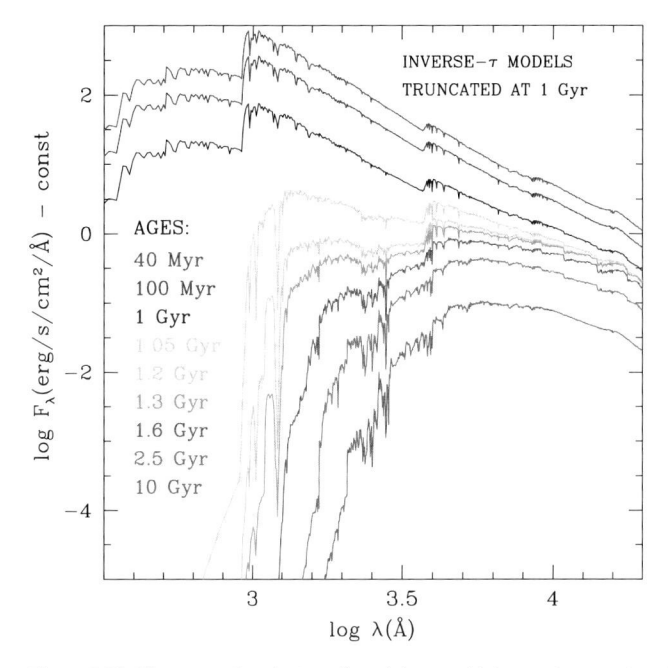

Figure 5.10 The spectral evolution of models with exponentially increasing SFR up to 1 Gyr with $\tau = 0.5$ Gyr, and SFR $= 0$ thereafter. Ages are indicated, and apply top to bottom with increasing age (synthetic spectra kindly provided by C. Maraston). For a color version of this figure please see in the front matter.

the former model than in the latter one. Notice that the spectrum of the inverted-τ models evolve very little after ~ 1 Gyr. This is so because for $t \gg \tau$ the bulk of stars form during the last $\sim \tau$ time interval, regardless of age. Moreover, all models are made with the same total mass in stars, and therefore the evolution of the spectrum saturates not only in shape but also in normalization.

Figure 5.10 shows the spectral evolution of a population with exponentially increasing SFR with $\tau = 0.5$ Gyr, then truncated at $t = 1$ Gyr. Notice how fast the evolution after truncation is, with the spectrum at 1 Gyr (the third from top) fading abruptly in the UV in just 50 Myr. The subsequent passive evolution is displayed, up to an age of 10 Gyr.

It is worth emphasizing again that all displayed spectra refer to models in which exactly the same amount of mass has been turned into stars. From these figures one can appreciate that the near-IR is a far better indicator of the mass of a stellar population than the optical or the UV. Yet, the near-IR alone is not such a good mass tracer either, as stellar populations with the same mass but different star formation histories can have near-IR luminosities differing by a factor of ~ 10 or more.

Those presented here are just a few examples of what can be done using synthetic spectral libraries available *on line*. They are purely meant to stimulate others to do their own experiments, more tailored to specific needs, first without real data, before trying to fit data and models together. One critical issue concerns the choice of the star formation history. Widespread default assumptions, such as an exponentially declining SFR, may be reasonable in some cases, but may dramatically fail in others. One can find many examples in the literature in which such an assumption is made without justification, perhaps just because almost everybody else does the same.

5.4.3
There Are Also Binaries

A SSP is defined as an assembly of single stars, but, as is well known, some 50% of all stars are members of binary systems, and a good fraction of them interact with each other in the course of their evolution. In close binaries, that is, those undergoing mass exchange in the course of their evolution, the fuel consumptions can be radically different from those of single stars. Without entering into the intricate details of binary evolution (cf. Chapter 7 to appreciate some of them), a few comments are in order. Clearly, the fuel consumption theorem can also be applied to binaries, what is more difficult to construct is the *evolutionary flux* for the many kinds of binaries existing in Nature. For this reason, it is not attempted here to generalize the approach presented in Chapter 2 to include binaries, but we restrict the discussion to some qualitative considerations, primarily concerning binaries with low- and intermediate-mass components.

Interacting binaries are commonly classified in three categories, called Case A, B, and C, depending on the evolutionary phase of the primary at the first Roche-lobe contact.

Case A: These binaries experience Roche-lobe contact when both components are still on the main sequence. It is generally understood that Case A binaries just merge to form a single star, with a mass practically equal to the initial mass of the system, M_S. Therefore, in such systems the post-MS fuel consumptions are close to those of a single star of mass M_S, but the corresponding energy release is delayed by a time interval close to the time elapsed before contact. Compared to the case of two single stars with the same masses, the total fuel consumption is decreased by almost a factor of two, because two stars of mass M_1 and M_2 burn more fuel than one star of mass $M_1 + M_2$. This is because the final mass is a rather flat function of the initial mass (cf. Eq. 1.11). However, the SED produced by such binaries would be quite different from the SED of the coeval population of single stars. For example, suppose $M_S = 2\,M_\odot$ and $M_{TO} = 1\,M_\odot$ in the stellar population. The merged product of the binary, a $2\,M_\odot$ star, will appear as a *blue straggler* when on its main sequence, contributing extra blue light, and later will climb to high luminosity on the AGB. contributing extra light in the infrared. The net effect of such binaries is therefore to broaden the SED of the overall population.

Case B: These binaries undergo Roche-lobe contact when the primary has already exhausted hydrogen in its core. If the primary mass is below $\sim 2\,M_\odot$ then the envelope of the primary is either transferred to the companion or lost by the system, helium ignition is avoided, and the primary becomes a helium white dwarf. In practice, its evolution is stopped somewhere on the RGB, and the HB and AGB phases are skipped. The corresponding fuel consumptions vanish, while a few $10^{-3}\,M_\odot$ of hydrogen are burnt by the star at very high effective temperature ($\sim 30\,000 - 80\,000$ K), on its way to becoming a white dwarf. Thus, the system will contribute UV photons, in excess of those that would have been produced by the same two stars if they were not locked in a binary. Potentially more interesting is the subsequent evolution of systems in which the former primary is a white dwarf, and the secondary comes to fill its Roche-lobe and starts spilling material on top of the white dwarf. Depending on the accretion rate, the accreted hydrogen is either burnt steadily, or in thermonuclear runaways (novae of various types), or ejected entirely if the system goes through a *common envelope* phase. The case of steady burning is perhaps more interesting, as far as the SED of stellar populations is concerned. In fact, such accreting and burning white dwarfs can have quite high effective temperatures (up to $\sim 10^6$ K), hence being powerful sources of UV and soft X-ray photons (e.g., supersoft X-ray sources). It is still difficult for binary theory to predict how much hydrogen is burnt in this way, that is, to predict the fuel consumptions of accreting white dwarfs. If it were of the order of some $0.1\,M_\odot$, a binary of this type would produce as much UV radiation as several hundred single stars of the same age during their post-AGB evolution.

Case C: These binaries undergo Roche-lobe contact when the primary is an AGB star. After mass transfer (or common envelope ejection) the primary becomes a carbon-oxygen white dwarf, and part of the AGB fuel consumption is suppressed. For the subsequent evolution of the system some aspects discussed for Case B apply here as well. The binary can be a powerful source of UV and soft X-ray photons, depending on the total hydrogen fuel consumption of the accreting white dwarf. As extensively discussed in Chapter 7, part of these binaries may eventually produce supernovae of Type Ia, as could some of the more massive binaries undergoing Case B contact. If so, there should be a link between the rate of Type Ia supernovae and the UV and soft X-ray luminosity of old stellar populations.

In conclusion, the fuel consumptions in various stages are deeply affected by stars being in a binary system. This includes the suppression of part of the fuel consumption of the cool stages (RGB and AGB), possibly boosting the fuel consumption of very hot stages, depending on the ability of white dwarfs to nuclearly process the materials they are accreting. However, the presence of binaries may have only a minor effect on the SED of stellar populations in the optical and near-IR spectral range.

In the early days of population synthesis the observed spectra of a few dozen stars used to be mixed together in various proportions, trying to see whether or not the combined spectrum was like that of a galaxy. Much progress has been done since then. As extensive sets of stellar evolutionary sequences became available, thus providing the backbone *evolutionary constraints*, full libraries of synthetic spectra were computed and soon appeared freely accessible on the Internet. Nowadays, download these spectral libraries, write a *script* to handle them, and fit them to galaxies is common practice. Users tend to be in a hurry, and since along with the spectra from the database comes the README file, one may be tempted to assume that it contains all one needs to know about the models. Actually, the README file barely informs the user about model ingredients. Yet, no synthetic model is perfect, and on top of it assumed functional shapes of star formation histories are likely to differ from those chosen by galaxies, and therefore all this imprints itself into the results. Thus, sometimes the results are acceptable, and other times less so, and it is difficult to tell which is which. After all, it may be vane to vent a plea for investing time to understand the models first, read the original papers, experiment on one library versus another to sort out their relative systematics, and so on and so on... In the end, progress is very fast anyway, and hopefully the best results will survive longer than others.

Acknowledgments

Spectra for the composite stellar populations shown here have been kindly computed by Claudia Maraston on our request, and the authors are much indebted to her for her efforts.

Further Reading

Synthetic Stellar Populations

This chapter is no substitute for directly accessing the papers describing how the model stellar populations were constructed and spectra calculated. Only in the original papers are all ingredients described. Most widely used synthetic spectral libraries include:

Bruzual, G. and Charlot, S. (2003), *Mon. Not. R. Astron. Soc.*, **344**, 1000B, BC03.

Fioc, M. and Rocca-Volmerange, B. (1997) *Astron. Astrophys.*, **326**, 950, PEGASE.

Le Borgne, D. *et al.* (2004) *Astron. Astrophys.*, **425**, 881, PEGASE-HR.

Maraston, C (2005) *Mon. Not. R. Astron. Soc.*, **362**, 799, M05.

Vazdekis, A. *et al.* (2010) *Mon. Not. R. Astron. Soc.*, **404**, 1639, MILES.

Libraries of Stellar Spectra

Le Borgne, J.-F. *et al.* (2003) *Astron. Astrophys.*, **402**, 433, STELIB.

Sánchez-Blázquez, P. *et al.* (2006) *Mon. Not. R. Astron. Soc.*, **371**, 703, MILES.

Prugniel, Ph. *et al.* (2007) arXiv:astro-ph/0703658, ELODIE 3.1.

Valdes, F. (2004) *Astrophys. J. Suppl.*, **152**, 251, INDO-US.

Synthetic Spectral Libraries via Internet

- Bruzual and Charlot (2003): http://www2.iap.fr/users/charlot/bc2003/ (last accessed 2011-06-16)
- Maraston (2005): http://www-astro.physics.ox.ac.uk/~maraston/Claudias_Stellar_Population_Models.html (last accessed 2011-06-16)
- MILES: http://miles.iac.es (last accessed 2011-06-16)
- PEGASE: http://www.iap.fr/pegase/ (last accessed 2011-06-16)

6
Stellar Population Diagnostics of Galaxies

This chapter is dedicated to illustrate current stellar population diagnostics and procedures to measure fundamental integrated properties of galaxies, such as their star formation rate (SFR), stellar mass ($M_\star$), age and metallicity. In the second part of this chapter the main properties of large samples of galaxies are presented, as derived from such stellar population tools, thus obtaining average trends as a function of mass, environment, and redshift.

6.1
Measuring Star Formation Rates

Star formation involves the birth of massive stars ($M \gtrsim 10\ M_\odot$), which while on the main sequence are hotter than $\sim 20\,000$ K. Star formation is therefore accompanied by the production of an intense ultraviolet radiation field. In turn, hydrogen and other elements in the parent cloud are ionized by such UV radiation, and as these elements recombine strong emission lines start to shine, such as the Balmer and Lyman lines, [OII]$\lambda 3727$, [OIII]$\lambda 5007$, and others. The UV radiation is also absorbed by the dust in the parent star-forming clouds and throughout the galaxy, and the heated dust particles re-emit in the mid- and far-infrared the absorbed energy. Eventually, massive stars explode as supernovae, relativistic electrons are accelerated which emit synchrotron radiation at radio wavelength. Accretion onto the compact remnants of supernovae that exploded in binary systems will also generate X-ray radiation.

Each of these phenomena offers an opportunity to measure the SFR, although we distinguish between primary and secondary SFR indicators. The most direct way to measure the SFR is just counting the number of massive stars, which have formed within a measurable time interval given their short lifetimes. Since in most cases massive stars are not individually resolved, one can alternatively measure the UV luminosity generated by them, which however is partly reduced by dust extinction. Number counts and UV luminosity can be regarded as the *primary* SFR indicators, and to derive SFR values from them one has inevitably to rely on stellar evolutionary sequences for massive stars, which provide lifetimes and UV energy emitted as a function of stellar mass. Once SFRs are measured in this way for a

Stellar Populations, First Edition. Laura Greggio and Alvio Renzini.
© 2011 WILEY-VCH Verlag GmbH & Co. KGaA. Published 2011 by WILEY-VCH Verlag GmbH & Co. KGaA.

suitable number of galaxies, then plots of such SFRs versus the strength of emission lines (e. g., Hα, [OII], and so on), versus the mid- or far-infrared luminosity, and versus the radio and X-ray luminosity provide the empirical calibration of all *secondary* SFR indicators. Note, however, that *all* SFR indicators ultimately rest on evolutionary sequences for massive stars. This section gives the most widely used recipes to estimate the SFR of galaxies, in particular for galaxies at high redshift.

6.1.1
The SFR from the Ultraviolet Continuum

Figure 6.1 shows several UV spectra from 1000 to $\sim$ 2000 Å of composite solar metallicity populations that have turned into stars the same amount of mass, with a Salpeter IMF, and different star formation histories (SFH). The upper four spectra refer to unobscured populations, that is, $E(B-V) = 0$, and their flux level is almost exactly proportional to the actual SFR at the time at which the spectrum is calculated. Note that the continuum slope is independent of the adopted SFH,

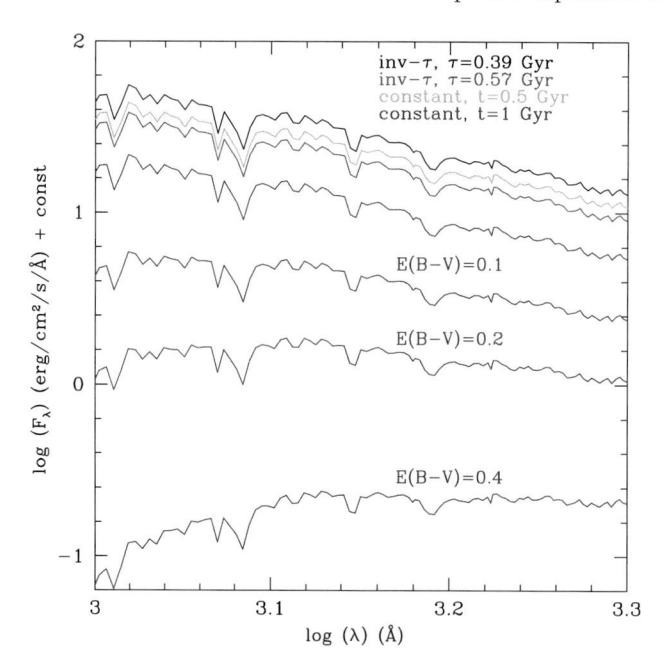

Figure 6.1 The UV spectrum between 1000 and 2000 Å of complex populations with various star formation histories, namely SFR = const. for a duration of 0.5 and 1 Gyr, and for exponentially increasing SFR ($\propto e^{-t/\tau}$) also for a duration of 1 Gyr and two values of τ as indicated. For all four SFHs the same amount of gas is turned into stars, and the spectra refer to the time $t = 1$ Gyr after the beginning of star formation ($t = 0.5$ Gyr for the spectrum in light gray). The upper four spectra are unreddened, whereas the lower three ones correspond to the SFR = const., $t = 1$ Gyr case, with various amounts of reddening, as indicated (Synthetic spectra kindly provided by C. Maraston). For a color version of this figure please see in the front matter.

and would remain so for quite a large variety of other options insofar as galaxies are actively forming stars. This comes from the combination of two effects, namely the slope of the UV continuum in this wavelength range is only weakly dependent on temperature for stars hotter than $\sim 25\,000$ K (with $F_\lambda \propto \sim \lambda^{-3}$) and much of the radiation comes from hot stars at the top end of the IMF. As illustrated by Figure 6.2, this also implies that the slope of the UV continuum is almost independent of the IMF, unless the IMF is much steeper than Salpeter, such as in the case $s = 1 + x = 3.3$. As shown in Figure 6.2, the slope of the UV continuum for the IMF with $s = 2.3$ and no reddening is almost identical to that for the IMF with $s = 1.3$. As discussed in Chapter 8, there are astrophysical reasons that exclude IMFs as flat as $s = 1.3$ or as steep as $s = 3.3$, and therefore the slope of the UV continuum cannot be appreciably different from that relative to the $s = 2.3$ (near-Salpeter) case. In summary, the UV flux at $1000\,\text{Å} \lesssim \lambda \lesssim 2000\,\text{Å}$ of an unobscured, actively star-forming stellar population is proportional to the SFR, and the slope of the UV continuum is almost independent of the previous SFH and of the IMF.

If galaxies were unobscured life would be much simpler than it is. The problem is that stars form out of molecular clouds, which are very dusty; hence, star formation and dust obscuration go hand in hand. Figure 6.1 shows what happens to one

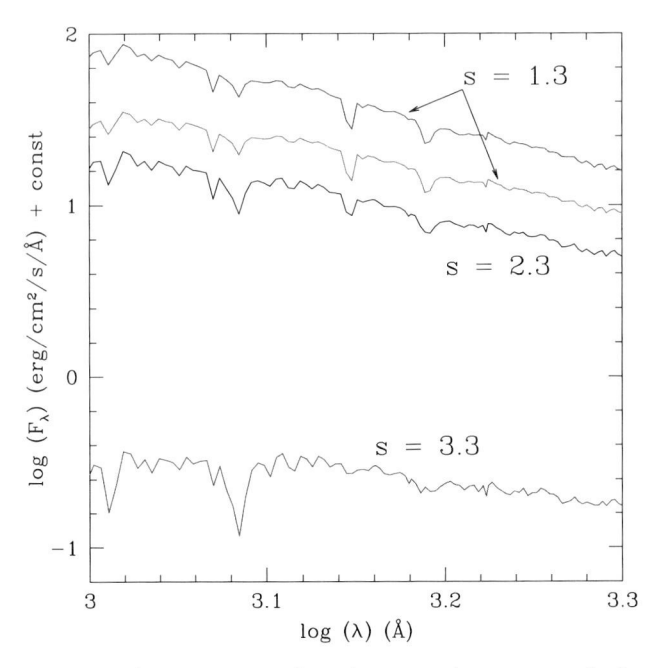

Figure 6.2 The UV spectrum of complex populations obtained with a constant SFR operating for 1 Gyr, for three slopes of the IMF: $s = 2.3$ (as in Figure 6.1), a *flat* IMF with $s = 1.3$ and a *steep* IMF with $s = 3.3$. The case of the flat IMF reddened with $E(B-V) = 0.07$ is also shown as the lower of the two spectra for the $s = 1.3$ case. Note that the SFR is the same in all cases shown here, the relative differences in the UV flux are those implied by the use of different IMFs (Synthetic spectra kindly provided by C. Maraston). For a color version of this figure please see in the front matter.

of the unobscured spectra if a reddening $E(B-V)$ from 0.1 to 0.4 is applied. Extinction as a function of wavelength has been calculated following to the so-called Calzetti Law, according to which the ratio of the extincted ($\mathcal{F}_\lambda$) over the unextincted flux ($\mathcal{F}_\lambda^\circ$) is given by (from Calzetti, D. (2001, *Publ. Astron. Soc. Pac.*, 113, 1449)):

$$\frac{\mathcal{F}_\lambda}{\mathcal{F}_\lambda^\circ} = 10^{-[0.91\,E(B-V)k(\lambda)]}, \tag{6.1}$$

where

$$k(\lambda) = 1.17\left(-2.156 + \frac{1.509}{\lambda} - \frac{0.198}{\lambda^2} + \frac{0.011}{\lambda^3}\right) + 1.78, \tag{6.2}$$

and where $E(B-V)$ is the reddening affecting the stellar population, and the wavelength λ is in μm. Given the strong increase of extinction with decreasing wavelength, the slope and shape of the UV continuum are strongly affected by even a relatively modest amount of reddening, and even more so is the UV flux itself. For example, with $E(B-V) = 0.4$ the extinction is just ~ 1.2 magnitudes in the V band, but it inflates to almost 5 magnitudes at $\lambda \sim 1300\,\text{Å}$.

The dependence of the slope of the UV continuum on reddening offers the opportunity of estimating $E(B-V)$ once the slope is measured, and adopting the appropriate slope for the unextincted continuum. As emphasized above, such slope is not significantly affected by the choice of the IMF. Then, Eq. (6.1) allows us to retrieve the unreddened UV flux and from it the SFR through the relation (from Daddi, E. et al. (2004, *Astrophys. J.*, 617, 746)):

$$\text{SFR}\,(M_\odot\text{yr}^{-1}) \simeq 1.13 \times 10^{-28}\, L_{1500}^\circ\, (\text{erg s}^{-1}\,\text{Hz}^{-1}), \tag{6.3}$$

where L_{1500}° is the luminosity of the galaxy at 1500 Å corrected for extinction. This relation assumes a single slope Salpeter IMF ($s = 2.35$) from 0.1 to 100 $M_\odot$. Adopting instead an IMF that flattens for low-mass stars, the resulting SFR must be reduced accordingly. For example, for a two-slope IMF with $s = 2.35$ for $M > 0.5\,M_\odot$ and $s = 1.35$ for $M < 0.5\,M_\odot$ the numerical coefficient in Eq. (6.3) must be reduced by a factor ~ 0.75, and becomes $\sim 8.5 \times 10^{-29}$. Figure 6.2 also shows that the UV continuum slope obtained with a *flat* IMF ($s = 1.3$) and reddened with $E(B-V) = 0.07$ is almost identical to that of an unreddened population with $s = 2.3$. This difference can be regarded as an upper limit to the uncertainty in the derived reddening due to the uncertainty in the actual slope of the IMF in the high-mass regime.

6.1.2
The SFR from the Far-Infrared Luminosity

The obvious counterpart of dust extinction is dust emission. Therefore the SFR can be estimated also from the energy emitted by dust in the mid- and far-infrared (FIR), that is, in the wavelength range from $\sim 8\,\mu\text{m}$ to $\sim 1\,\text{mm}(L_{\text{FIR}})$. Given the strong wavelength dependence of extinction (cf. Eq. (6.2)), UV photons from young, massive stars provide much of the dust heating and its associated infrared emission. This ensures a fairly direct link between $L_{\rm FIR}$ and the SFR, at least in the optically thick case, when virtually all UV photons are absorbed. The following is a widely used relation based on stellar models, and again holding for a straight Salpeter IMF from 0.1 to 100 $M_\odot$ (from Kennicutt Jr., R.C. (1998, Annu. Rev. Astron. Astrophys., 36, 189):

$$\text{SFR}\,(M_\odot \text{yr}^{-1}) \simeq 4.5 \times 10^{-44}\, L_{\rm FIR}\,(\text{erg s}^{-1})\,, \qquad (6.4)$$

which assumes that all UV photons are absorbed.

The more general case is one in which only part of the UV photons are absorbed by dust, and part escape freely from the galaxy and finally hit our detectors. Thus, the total SFR can be written as

$$\text{SFR}\,(M_\odot\,\text{yr}^{-1}) \simeq 1.13 \times 10^{-28}\, L_{1500} + 4.5 \times 10^{-44}\, L_{\rm FIR}\,, \qquad (6.5)$$

which combines Eqs. (6.3) and (6.4) and where L_{1500} is measured in erg s^{-1} Hz^{-1} and $L_{\rm FIR}$ in erg s^{-1}.

The main uncertainty affecting the FIR diagnostics of star formation comes from the estimate of $L_{\rm FIR}$, that is, the energy emitted in the 8–1000 µm spectral range. The reason is that data rarely exists that covers this whole range, in particular encompassing the peak of the FIR spectral energy distribution at ~ 60–100µm in the rest frame. More often the flux at only one or two wavelengths is available (e.g., only the 24 µm flux from the Spitzer Observatory). Therefore, $L_{\rm FIR}$ is estimated adopting a model SED (that may or may not apply to the particular galaxy) and then matching it to the observed flux. In particular, concerns come from the polycyclic aromatic hydrocarbon (PAH) emission at ~ 8µm, that has a strong dependence on metallicity. Moreover, besides young, SFR-tracing stars, older populations may contribute to the heating of PAH molecules. At the same rest frame wavelengths a contribution by hot dust heated by an AGN is also possible. The situation is now very rapidly improving as the Herschel Observatory has started to provide data all the way to 500 µm, and will further improve when ALMA starts providing high spatial resolution data in the millimeter and submillimeter range.

6.1.3
The SFR from Optical Emission Lines

The UV photons of sufficient energy that are not absorbed by dust can photoionize hydrogen and other elements in the ISM, and upon recombination various emission lines are originated. Clearly, the intensity of such lines is related to the UV flux which in turn is related to the SFR, hence there must be a relation between the SFR and the luminosity of the emission lines. Such a relation can be calibrated empirically, by plotting the line luminosity versus the SFR for a large sample of galaxies, where the SFR is estimated via the UV continuum. Clearly, the emission line luminosity must be corrected for extinction, and emission lines are affected by extinction in two distinct ways: (i) like every other photon, line photons can be directly absorbed by dust, and (ii) dust absorbs the ionizing UV photons, reducing the photoionization rate, hence the recombination rate and so indirectly affecting the emission line luminosity. According to a widely adopted empirical calibration of this effect, the effective reddening to apply to emission lines is related to the reddening affecting the stellar continuum via the relation (from Calzetti, D. (2001, Publ. Astron. Soc. Pac., 113, 1449)):

$$E(B-V)_{\text{lines}} \simeq \frac{E(B-V)_{\text{cont}}}{0.44}. \tag{6.6}$$

The Hα line is the most widely used emission line to estimate SFRs via its empirical calibration (from Brinchmann, J. et al. (2004, Mon. Not. R. Astron. Soc., 351, 1151));

$$\text{SFR}\,(M_\odot \text{yr}^{-1}) \simeq 5 \times 10^{-42}\, L^\circ_{\text{H}\alpha}\,(\text{erg s}^{-1}), \tag{6.7}$$

where $L^\circ_{\text{H}\alpha}$ is the extinction corrected Hα luminosity. The extinction affecting Hα can be evaluated using Eq. (6.6) with $E(B-V)$ derived either from the UV slope, or using emission line ratios, such as for example, Hα/Hβ.

Beyond redshift $\sim$ 0.4–0.5 the Hα line moves out of the optical range and if near-IR spectra are not available SFRs can be estimated from the [OII]λ3727 line. However, the extinction suffered by the [OII] line is higher than for the Hα line, and more difficult to ascertain, which makes [OII]-derived SFRs less reliable. Low-redshift galaxies for which Hα, Hβ and [OII] lines are available can be used to calibrate a SFR-[OII] relation using the SFR values from Hα, that is, Eq. (6.7). The result is (from Kennikutt, R.C. Jr. (1998) adapted by Maier, C. et al. (2009, Astrophys. J., 694, 1099)):

$$\text{SFR}\,(M_\odot\,\text{yr}^{-1}) \simeq 2.36 \times 10^{-41}\, L_{\text{[OII]}} \left(\frac{L_B}{10^{10}\,L_{B,\odot}}\right)^{0.49}, \tag{6.8}$$

where $L_{\text{[OII]}}$ (in erg s^{-1}) is the energy emitted in the [OII] line not corrected for extinction, and the B-band luminosity L_B is in $L_\odot$ units. The last factor is meant to take into account extinction effects, as the $L_{\text{[OII]}}/\text{SFR}(\text{H}\alpha)$ ratio exhibits a correlation with the B-band luminosity, indicating that brighter star-forming galaxies are more extincted. However, the scatter of such correlation is rather large, and therefore [OII] derived SFRs for individual galaxies are quite uncertain. In practical terms, recourse to [OII] is justified when no better SFR indicator is available.

6.1.4
The SFR from the Soft X-ray Luminosity

Star formation results in the production of both single and binary massive stars. In such binary systems the death of the primary star, leaving a black hole or neutron star remnant, prepares the conditions for the subsequent accretion onto the collapsed object of matter from the Roche-lobe overflow of the companion, and the concomitant powerful emission of X-rays. Thus, star formation is accompanied by

X-ray emission from such high-mass X-ray binaries (HMXB), and the X-ray luminosity is indeed found to correlate with the FIR luminosity of star-forming galaxies. Thus, a relation such as Eqs. (6.4) or (6.5) can be used to calibrate an empirical relation between SFR and X-ray luminosity, with the result (from Ranalli, P. et al. (2003, Astron. Astrophys., 399, 39)):

$$\text{SFR}\,(M_\odot\,\text{yr}^{-1}) \simeq 2.2 \times 10^{-40}\,L_{0.5-2\,\text{keV}}\,(\text{erg}\,\text{s}^{-1})\,, \quad (6.9)$$

where $L_{0.5-2\,\text{keV}}$ is the X-ray luminosity in the rest frame 0.5–2 keV energy range.

Similar relations have also been proposed for the hard X-ray luminosity in the 2–20 keV energy range, but here an AGN contribution becomes more likely, which would lead to an overestimate of the actual SFR. Besides from HMXBs and AGNs, X-rays can be also generated by low-mass X-ray binaries (LMXB, that is, a collapsed object plus a low-mass, $\sim 1\,M_\odot$ companion), whose number is not linked to the current SFR. Corrections to subtract the LMXB contribution may be important for galaxies with relatively low SFR. Corrections for removing an AGN contribution may also be appropriate for galaxies for which the X-ray-based SFRs appear to largely exceed those derived from other indicators.

6.1.5
The SFR from the Radio Luminosity

As mentioned earlier in this chapter, shortly after its beginning star formation is accompanied by (core collapse) supernovae, and empirical evidence shows that relativistic electrons are harbored in supernova remnants and emit at radio wavelengths. Thus, the radio power of star-forming galaxies can be empirically calibrated versus the SFR from another indicator. The radio luminosity at 1.4 GHz of star-forming galaxies is indeed found to correlate tightly with their FIR luminosity, for example, at 60 μm, and Eq. (6.4) was used to calibrate the SFR-radio luminosity relation, with the result (from Yun, R.S. et al. (2001, Astrophys. J., 554, 803)):

$$\text{SFR}\,(M_\odot\,\text{yr}^{-1}) \simeq 5.9 \times 10^{-22}\,L_{1.4\,\text{GHz}}\,(\text{erg}\,\text{s}^{-1}\,\text{Hz}^{-1})\,, \quad (6.10)$$

where $L_{1.4\,\text{GHz}}$ is the radio power at 1.4 GHz (rest frame).

The 1.4 GHz luminosity may have other origins besides star formation, most notably AGN activity, hence Eq. (6.10) may greatly overestimate the SFR in such cases. Nevertheless, this relation should hold for the bulk of galaxies, as AGNs are a relatively small fraction of actively star-forming galaxies.

In concluding this section on SFR diagnostics, it is worth noting that no single indicator is perfect, and occasionally all are subject to large errors if all the flux at some wavelength is attributed to ongoing star formation whereas it is not. For example, if a passively evolving galaxy (SFR $\simeq 0$, $E(B-V) \simeq 0$) is mistaken as star-forming and its 1500 Å luminosity is corrected for extinction and entered into Eq. (6.3), one would derive an extremely large reddening and an enormous SFR. As a general rule, it is advisable to use more than one SFR indicator, and first make sure that galaxies are really star forming, and do not harbor an AGN that may affect the result. In summary, all indicators are affected by a systematic uncertainty which typically can be as large as a factor of ∼ 2, and individual galaxies may deviate even more from the mean relation. Moreover, instrumental and database limitations usually force us to use different SFR indicators in different redshift intervals, hence possibly introducing discontinuities or biasing trends. Ideally, it would be better to use the same SFR indicator from $z \sim 0$ all the way to the highest redshifts (e. g., Hα properly corrected for extinction), something that may soon become possible.

Still, all SFR indicators have their roots in evolutionary models for massive stars, and therefore any possible systematic mismatch between such models and reality would have an effect on the derived SFRs, no matter which indicators are used. For example, overshooting from convective cores during the main sequence prolongs the hydrogen burning lifetimes, and more UV photons are produced per unit mass of gas turned into stars. Thus, models with overshooting lead to lower SFRs compared to models without. While early claims called for a large overshooting, comparable to one pressure scale height H_p, more recently small overshooting distances are favored (∼ 0.1–0.25 H_p), leading to SFRs differing by just 10–20% from those obtained using models without overshooting. Massive stars of low metallicity (≲ 0.1 solar) do not experience the post-MS thermal runaway due to increasing metal opacity in the envelope (cf. Chapter 1), and spend the whole helium burning phase at high effective temperatures. Thus, at low metallicities the number of UV photons per unit mass turned into stars is somewhat higher than at solar metallicity, and using Eq. (6.3) would result in a modest (∼ 20–30%) overestimate of the SFR. Finally, massive stars in interacting binaries can also avoid the red-supergiant phase, transfer their envelope to the companion, and become Wolf–Rayet stars. The net result is an increase in the number of UV photons produced by both binary members compared to single stars. Correcting for the effect of binaries, derived SFRs would then be lower, though by an amount that is presently difficult to quantify.

6.2
Measuring the Stellar Mass of Galaxies

Together with the SFR, the stellar mass ($M_\star$) is the other fundamental galaxy quantity that stellar population diagnostics aims to measure. It is worth emphasizing that often different authors may mean different things when referring to the stellar mass of a galaxy, namely: (i) the baryonic mass that went into stars; (ii) the actual mass in stars, equal to the mass that went into stars minus the mass return, equivalent to the mass of living stars and dead remnants; (iii) the mass of only still living stars (not counting the mass in remnants); or even (iv) the mass that went into stars assuming that the mass return is immediately recycled into new stars with 100% efficiency, with or without including the mass in remnants. Clearly, when comparing galaxy stellar masses from different authors one has to pay attention to which specific definition was adopted.

The same holds for the IMF, for which a Salpeter's slope ($s = 2.35$) is almost universally adopted above ∼ 0.5 $M_\odot$, while either maintaining the same slope below 0.5 $M_\odot$, or adopting a flatter slope, for example, $s = 1.35$. Occasionally, to add confusion to an already complex situation, it may happen that one refers to a particular IMF by the name of the author who proposed it, even when the same author has proposed several different IMFs. All these different assumptions and definitions can easily pile up to a factor ~ 2 in stellar masses derived from the same data. Clearly, when comparing stellar masses to dynamical masses the stellar mass that matters is the one including all living stars and the remnants.

The observational data used for measuring stellar masses are made of multi-band photometry, extending as much as possible into the rest-frame near-IR. The blue/UV part of a galaxy spectrum is indeed dominated by its ongoing star formation, and conveys little information on the amount of stellar mass that has accumulated as a result of its previous SFH. For stellar populations older than 0.5–1 Gyr a dominant contribution comes from AGB+RGB stars, which emit most of their luminosity in the near-IR. Therefore, the near-IR is the best tracer of stellar mass.

The procedure to measure stellar masses is straightforward in principle, but in practice may be affected by various sources of systematic errors, among which we shall try to discuss a few. Preliminary to measuring stellar masses is the construction of a great number of *template* synthetic spectra based on different SFHs $\psi(t)$, which are calculated using Eq. (5.5). Then a χ^2 procedure is used to sort among such many SFHs the one that minimizes the discrepancies between the measured and calculated fluxes, that is, the reduced χ^2 of the fit:

$$\chi^2_{\rm red} = \frac{1}{n-\nu-1} \sum_{i=1,n} \frac{[10^{-0.91 E(B-V) k(\lambda_i)} \mathcal{F}_i(t) - F_i]^2}{\sigma(F_i)^2}, \qquad (6.11)$$

where $\mathcal{F}_i(t)$ and F_i are the calculated and observed fluxes in the ith photometric band whose central wavelength is λ_i, $\sigma(F_i)$ is the observational error affecting the ith flux, and the calculated fluxes have been attenuated due to dust extinction by the factor given in Eq. (6.1). The reddening $E(B-V)$ is a free parameter in the overall fit. Thus, the calculated fluxes $\mathcal{F}_i(t)$ are obtained by redshifting the synthetic spectra and convolving them with the set of filters of the observations. This assumes that the redshift is independently known from spectroscopy; if not, redshift becomes an additional free parameter to be determined via the χ^2 minimization. Redshifts computed in this way are dubbed *photometric redshifts* and are indispensable when dealing with extremely large samples of distant galaxies, or for faint galaxies still beyond the current spectroscopic capabilities.

Various types of SFHs are being commonly used to construct the sets of template SEDs, the simplest one being a constant SFR between cosmic time $t_\circ$ (beginning of star formation) and cosmic time t (the cosmic time at which the recorded photons were emitted by the galaxy under study). Quite popular has been an exponentially declining SFR ($\propto e^{-t/\tau}$) between cosmic time $t_\circ$ and t, which however has the built-in assumption that all galaxies are caught at their minimum SFR. Although this may be adequate for galaxies in the local Universe, given the general decline of SFRs over the last $\sim$ 8–10 Gyr, this is hardly a sound assumption for galaxies which instead may be observed while at the peak of their star formation activity.

For this reason, in addition to exponentially decreasing SFRs, an exponentially *increasing* SFR ($\psi(t) \propto e^{+t/\tau}$) is starting to be explored, that is, the inverted-τ models mentioned in the previous chapter.

Quenching of star formation followed by passive evolution (with SFR $\simeq 0$) is possibly the most important event in the life of a galaxy. Thus, no matter what the adopted shape for the SFR as a a function of time, *truncated* SFHs are also considered, that is, those in which the SFR is set to 0 for $t > t_{\text{quench}}$.

Besides the functional shape of the SFH, the free parameters in the fit are: the age of the galaxy $t - t_\circ$, the SFR at some specific time, for example, SFR($t_\circ$), the timescale τ in the case of exponential SFHs, the reddening $E(B-V)$, and t_{quench} in the case of truncated SFHs. Composition, as described for example by [Fe/H] and [α/Fe], could also be added among the free parameters in the fit. In some cases however it is advisable to keep composition fixed while minimizing the χ^2, then compare the results for a few different choices. The most interesting quantities, the current SFR and the stellar mass are then given respectively by $\psi(t)$ and by the integral of the SFR between $t_\circ$ and t, diminished by the mass return $\mathcal{R}(t)$. That is:

$$M_\star = \int_{t_\circ}^{t} \psi(t') dt' - \mathcal{R}(t) , \qquad (6.12)$$

with:

$$\mathcal{R}(t) = \int_{t_\circ}^{t} dt' \int_{M_{\text{TO}}(t'-t_\circ)}^{120} \phi(M) \left[M - M_{\text{f}}(M) \right] \psi \left[t' - t_\circ - \tau(M) \right] dM , \qquad (6.13)$$

where $M_{\text{f}}(M)$ and $\tau(M)$ are the final mass and lifetime of stars of initial mass M (see Chapters 1 and 2). One important aspect of the procedure should be fully appreciated: it seeks the best *compromise* set of free parameters, rather than seeking to optimize the measure of a particular quantity, for example, the current SFR or $M_\star$. This follows from the search of the set of free parameters minimizing the χ^2, while weighing all photometric bands simply by the inverse of their photometric error. However, the current SFR as derived from the SED is most sensitive to the UV part of the spectrum, and the derived $M_\star$ is most sensitive to its near-IR part. Actually, the near-IR flux bears little relation to the current SFR, and the UV flux tells very little about the previous SFH, hence on the accumulated stellar mass. Minimizing the χ^2 does not give the best possible measure of either the current SFR or of the stellar mass, but just the most acceptable *compromise* chosen among the set of template SEDs. Note in particular that this canonical χ^2 procedure gives more weight to those photometric bands that are affected by the smallest errors, not to those which are more directly linked to the physical quantities one seeks to measure (such as the UV flux for the SFR, or the near-IR flux for the stellar mass).

The result of this SED fitting technique can be rather sensitive to the prior, represented by the shape of the adopted SFH. The choice of one particular functional shape should be guided by astrophysical considerations, depending upon the type of galaxy (passive or star forming) and on its redshift.

6.3
Age and Metallicity Diagnostics

The described χ^2 procedure for the SED fit to an individual galaxy picks the SFH and the $E(B-V)$ value that *best fits* the data. Therefore, as such the procedure delivers the *age* $(t-t_0)$ as the time elapsed since the beginning of star formation, and luminosity- or mass-weighted mean ages can be calculated. In the case of truncated SFHs, the procedure delivers also the time $(t-t_{\text{trunc}})$ elapsed since star formation quenching. The procedure is apparently quite robust (it always gives an answer), but delivered ages should be handled with caution, as often they really mean something quite different from what they formally represent. In the following, a specific example can illustrate this problem.

6.3.1
Star-Forming Galaxies

Running the χ^2 procedure on a sample of star-forming $z \sim 2$ galaxies gives ages $(t-t_0)$ spanning from $\sim 10^7$ yr to $\sim 10^9$ yr, irrespective of the adopted shape of the SFH (constant, exponentially increasing or decreasing), with most ages being less than a few 10^8 yr. Since at $z = 2$ the Universe is already $\sim 3\,\text{Gyr}$ old, such short ages imply that these galaxies would have started to form stars just shortly before we happen to observe them. Figure 6.3 shows the spectroscopic redshift

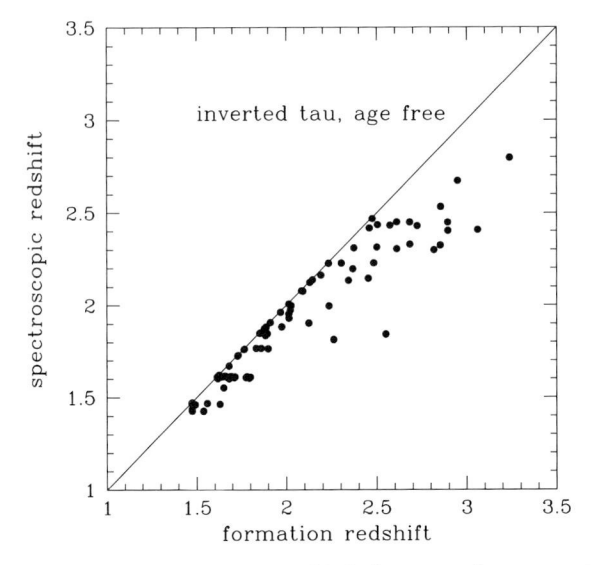

Figure 6.3 The spectroscopic redshift of individual galaxies versus their *formation redshift* as deduced from their age derived from inverted-τ models, with age as a free parameter. The result would have been almost identical using ages derived from models with exponentially decreasing SFR or models with constant SFR (source: Maraston, C. et al. (2010, Mon. Not. R. Astron. Soc., 407, 830)) (Reproduced by permission of the Royal Astronomical Society.).

of each individual galaxy as a function of its formation redshift, the latter being derived by combining the observed redshift and the age ($t - t_o$) of each galaxy as delivered by the fit. Most galaxies appear to have just formed, but we believe that these ages cannot be trusted, and it is worth clarifying why the χ^2 minimization procedure gives such implausibly young ages. It is the consequence of the fact that the very young, massive stars formed during just the last $\lesssim 10^8$ yr dominate the light at most wavelengths in these actively star-forming galaxies, even if they represent a fairly small fraction of the stellar mass. The capability of a very young population to outshine previous stellar generations is illustrated in Figure 6.4. The synthetic spectrum is shown for a composite stellar population, which has formed stars at a constant rate for 1 Gyr, illustrating the contributions of the stars formed during the first half and the second half of this time interval. The young component clearly outshines the old component at all wavelengths, making it difficult to assess the presence and the contribution of the latter one, even if the old component contains half or more of the total stellar mass. Figure 6.4 demonstrates why it is not appropriate to interpret the "age" resulting from these fits as the time elapsed since the beginning of star formation, as it is *mathematically* meant to be in the fitting procedure. It actually is the age of the stars producing the bulk of the light. Moreover, Eq. (6.11) gives more weight to the bands with the smallest photometric errors, which typically are the optical bands ($BVIz$), whereas the near-IR (JHK) and Spitzer/IRAC bands are usually affected by larger photometric errors. Thus, the procedures seeks a better fit in the rest frame UV part of the spectrum (more sensitive to the ongoing SFR), rather than to the rest frame near-IR, which is more sensitive to the total mass, hence to the previous SFH and age. In summary, the best fit parameter t_o should not be read as the time since the beginning of star

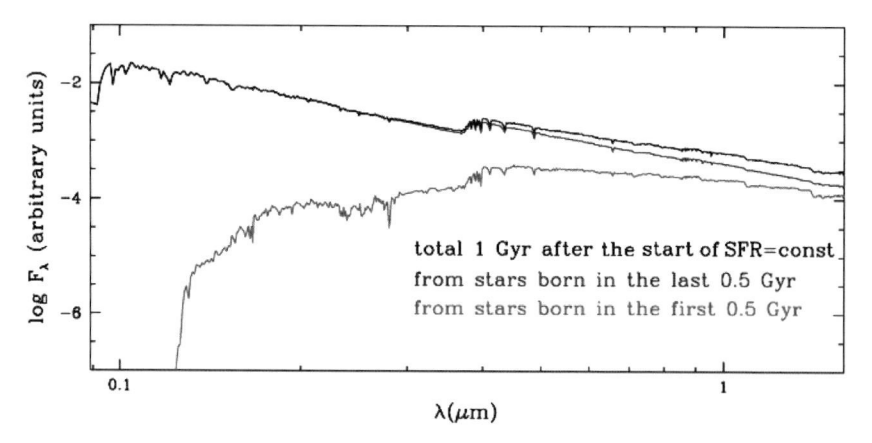

Figure 6.4 The effect of outshining by the youngest fraction of a composite population: the synthetic spectrum is shown for a constant SFR rate over 1 Gyr. The contribution of the stars formed during the first and the second half of this period are shown separately as indicated by the color code, together with the spectrum of the full population (source: C. Maraston *et al.* (2010, *Mon. Not. R. Astron. Soc.*, 407, 830)) (Reproduced by permission of the Royal Astronomical Society.). For a color version of this figure please see in the front matter.

formation, but rather as the time since stars contributing most of the light have formed. This holds especially for the rest frame UV light, for which photometric errors are usually the smallest. Thus, the SED fitting procedure applied to high-redshift, star-forming galaxies is not well suited to sort the most precise stellar mass, or the real age of the galaxy.

6.3.2
Quenched Galaxies

In the case of passively evolving (quenched) galaxies the spectrum or the multiband photometric SED are usually fitted using a set of SSP template spectra, then having age and metallicity as free parameters. The reddening $E(B-V)$ is sometimes included among the free parameters, although a galaxy in which star formation is really quenched is generally expected to be devoid of cold gas and dust. Fits in which large amounts of reddening are required should be looked at with suspicion. Age, metallicity and $E(B-V)$ are all affected by degeneracy with the other two pa-

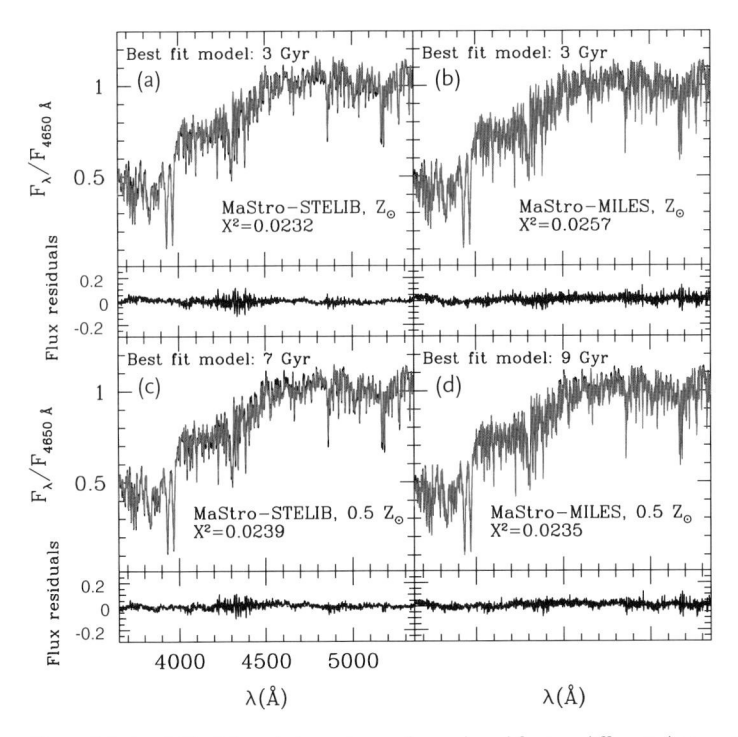

Figure 6.5 (a–d) The integrated spectrum of the open cluster M67 obtained by coadding the spectra of its individual stars weighted by the luminosity function (black lines). Superimposed are the best fit synthetic spectra for the ages and metallicities indicated in each panel, and for two different choices of the stellar spectral library. The corresponding residuals are plotted below each spectrum, and the reduced χ^2 of each fit is indicated (courtesy of C. Maraston). For a color version of this figure please see in the front matter.

rameters, because they all tend to redden the SED, especially in the optical part of the spectrum.

Figure 6.5a–d offers a quite dramatic illustration of the age-metallicity degeneracy, which was briefly touched upon in Chapter 2; it shows the integrated spectrum of the open cluster M67 as synthesized by coadding spectra of individual stars, weighted according to the luminosity function. This integrated spectrum was then fitted with SSP template spectra, seeking the best χ^2. Virtually equally good fits to the data are obtained with an age of 3 Gyr and solar metallicity, or with an age of 7–9 Gyr and half-solar metallicity. For comparison, the metallicity of M67 from individual star spectroscopy is about solar, and its age from the color-magnitude diagram is 3 Gyr.

Thus, this age-metallicity degeneracy can jeopardize attempts at locating in cosmic time the formation epoch of ellipticals and bulges observed in the local Universe, highlighting the need for breaking such degeneracy. One instructive example of how a robust conclusion on ages is achieved with minimal modeling machinery is represented by the use of the tightness of the color-magnitude, or color–σ relations for cluster elliptical galaxies. Figure 6.6 shows the $U - V$ color of elliptical and S0 galaxies in the Virgo and Coma clusters versus their central velocity dispersion σ. Rather than trying to age-date these galaxies one by one, this figure was used by Bower, R.G. et al. (1992, Mon. Not. R. Astron. Soc., 254, 613) to set tight age constraints on all of them at once. Note the remarkable homogeneity of E/S0

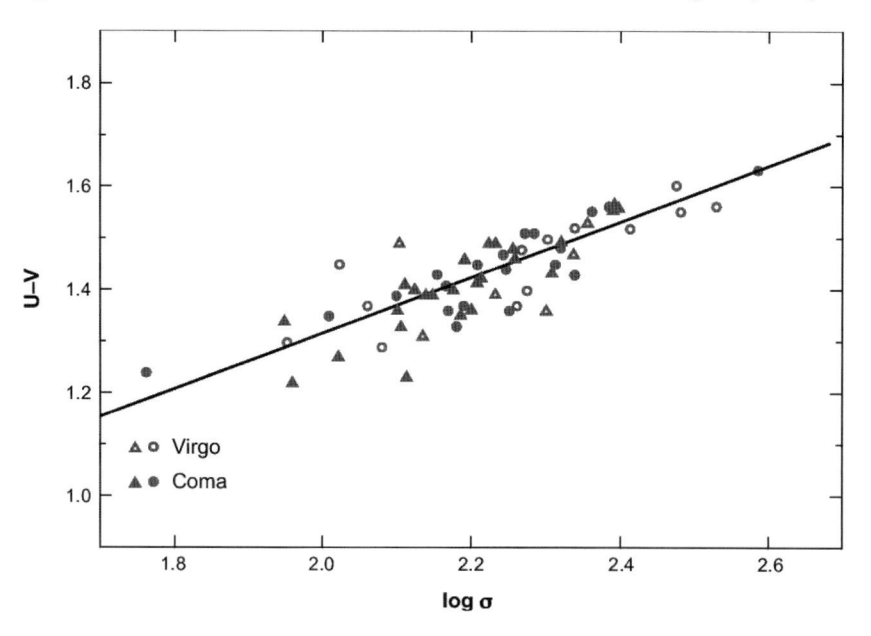

Figure 6.6 The relation between the $(U - V)$ color and the central velocity dispersion (σ) for early-type galaxies in the Virgo (open symbols) and Coma (filled symbols) clusters. Circles represent ellipticals, triangles represent S0 (adapted after Bower, R.G. et al. (1992, Mon. Not. R. Astron. Soc., 254, 613)). For a color version of this figure please see in the front matter.

galaxies in these clusters, with an intrinsic color scatter in the color-σ relation of $\delta(U-V) \lesssim 0.04$ mag. Now, if due entirely to an age dispersion, such color scatter should be equal to the time scatter in formation epochs, times $\partial(U-V)/\partial t$. Expressing δt as a fraction ζ of the age of the Universe $(t_H - t_F)$ when the formation of these galaxies ceased, the upper limit on $\delta(U-V)$ implies:

$$t_H - t_F \lesssim \frac{0.04}{\zeta} \left[\frac{\partial(U-V)}{\partial t}\right]^{-1} \quad \text{(Gyr)}, \tag{6.14}$$

where t_H is the age of the Universe at $z = 0$, and galaxies are assumed to form before a lookback time t_F. The quantity $\zeta(t_H - t_F)$ is the fraction of the available time during which galaxies actually form the bulk of their stars: for $\zeta = 1$ galaxy formation is uniformly distributed between $t \sim 0$ and $t = t_H - t_F$, whereas for $\zeta < 1$ it is more and more synchronized. Using $\partial(U-V)/\partial t \simeq 0.02$ to 0.05 mag/Gyr (for t_F between 13 and 7 Gyr, respectively) as the only ingredient from stellar population models, one derives $t_H - t_F < 2$ Gyr for $\zeta = 1$ and $t_H - t_F < 8$ Gyr for $\zeta = 0.1$, corresponding respectively to formation redshifts $z_F \gtrsim 3.3$ and $\gtrsim 0.8$. A value $\zeta = 0.1$ implies an extreme synchronization, with all Virgo and Coma galaxies forming their stars within less than 1 Gyr time interval, when the Universe was already ~ 8 Gyr old, which seems rather implausible. The inference is that E/S0 galaxies in these clusters formed the bulk of their stars at $z \gtrsim 2$, with later additions providing no more than $\sim 10\%$ of their present luminosity. Making minimal use of stellar population models, this approach provided for the first time a robust demonstration that massive cluster ellipticals are made of very old stars, with the bulk of them having formed more than ~ 10 Gyr ago.

Largely with the purpose of breaking the age-metallicity degeneracy, a set of spectroscopic indices is widely used, known as the Lick/IDS system. Optical spectra of quenched galaxies present a number of absorption features whose strength must depend on the distributions of stellar ages, metallicities and abundance ratios (e.g., $[\alpha/\text{Fe}]$), and therefore may give insight over such distributions. Among these indices, those more promising for breaking the degeneracy are the magnesium-based Mg_2 and $\text{Mg}b$ indices, and the iron $\langle\text{Fe}\rangle$ and $\text{H}\beta$ indices, measuring respectively the strength of MgH + MgI at $\lambda \simeq 5156 - 5197$ Å, the average of two FeI lines such as those at $\lambda \simeq 5270$ and 5335 Å, and of H_β. Given the plethora of iron lines in the spectra of quenched galaxy, other iron lines have been used as well, eventually constructing combination indices such as [MgFe]', defined as:

$$[\text{MgFe}]' = \sqrt{\text{Mg}b(0.72 \times \text{Fe}5270 + 0.28 \times \text{Fe}5335)}, \tag{6.15}$$

and constructed for being (almost) independent of the α-element enhancement $[\alpha/\text{Fe}]$, as shown in Figure 6.7. The advantage of such a plot, and similar ones that can be constructed with other Balmer line strength, and/or other combinations of Mg and Fe indices, is that lines of constant age and lines of constant metallicity intersect at fairly wide angles, therefore offering an opportunity to break the degeneracy.

However, besides formally breaking the degeneracy, plots such as that shown in Figure 6.7 have various drawbacks worth to be aware of, and so do similar plots for

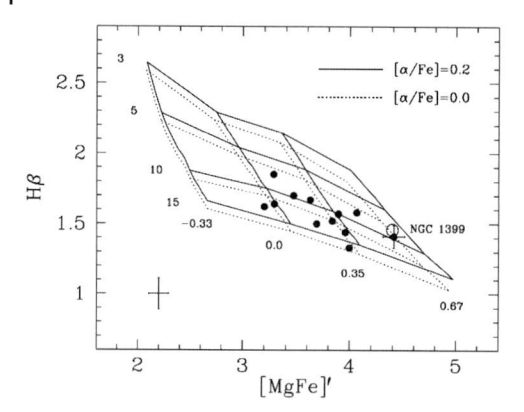

Figure 6.7 The synthetic Hβ – [MgFe]′ grid for SSPs with [α/Fe] = 0.0 and 0.2. Lines are labeled by age, from 3 to 15 Gyr, and by [Fe/H], from −0.33 to +0.67. Filled circles show the central values of the indices for a small sample of elliptical galaxies, including NGC 1399 together with its best fit model (open circle) (source: Thomas, D. et al. (2003, Mon. Not. R. Astron. Soc., 339, 897)) (Reproduced by permission of the Royal Astronomical Society.).

other Lick/IDS indices. One problem is that synthetic models, and the indices computed on them, are never perfect, with theoretical grids being inevitably by some extent shifted, rotated and distorted with respect to reality. A hint to this effect can be seen by comparing grids constructed by different authors, or using different stellar evolution databases, or even comparing ages and metallicities obtained from a same set of synthetic SSPs but using different sets of Lick/IDS indices. Another problem is that real galaxies are not SSPs, but their stars span age and metallicity distributions that can be quite broad. So, which is the physical meaning of ages and metallicities derived by interpolating in SSP grids? Certainly not mass-weighted values, but rather luminosity-weighted ages and metallicities. Another issue concerns the contribution of $\sim$ 7000–10 000 K stars belonging to an extended HB or to the blue straggler component, which are characterized by very strong Hβ absorption, and can bias the result towards younger ages. Last but not least, the Hβ – [MgFe]′ and similar plots are affected by a perverse circulation of the errors which automatically generates an apparent anticorrelation between age and metallicity, even where it does not exist. Indeed, when Hβ is overestimated due to observational errors, then age is underestimated, which in turn would reduce [MgFe]′ below the observed value unless the younger age is balanced by an artificial increase of metallicity. Similarly, if Hβ is underestimated the resulting age will be too old and the metallicity too low.

The main pitfall of the procedure is that the various indices depend on all three population parameters one is seeking to estimate: thus Hβ is primarily sensitive to age, but also to [Fe/H] and [α/Fe], the various Fe indices are sensitive to [Fe/H], but also to age and [Mg/Fe]; and so on. Thus, the resulting errors in age, [Fe/H] and [α/Fe] are all tightly correlated, and one is left with the suspicion that apparent correlations or anticorrelations may be an artifact of the procedure, rather than reflecting the real properties of galaxies.

To circumvent these difficulties, one option is to avoid trusting the results on a galaxy- by galaxy basis, but rather look for patterns in the various index-index plots and compare them to mock galaxy samples generated via simulations that fully incorporate the circulation of the errors. In this case, the real result is not a set of ages and metallicities assigned to individual galaxies, but rather average age and metallicity trends with velocity dispersion, mass and environment, selected for being those that best reproduce the observed patterns in the various index-index plots. This approach requires data for large and representative sets of galaxies. The result of one application of this procedure is shown in Figure 6.8, where lines represent the *average* star formation histories of galaxies in a few mass bins, that is, *not* the time evolution of the SFR within individual galaxies. The SFR(t) of individual galaxies would actually run quite differently, for example secularly increasing until ending abruptly as a result of the sudden star formation quenching. Figure 6.8 can indeed be interpreted as showing a proxy to the distribution function of the quenching times. In any event, the average luminosity-weighted ages, total metal-

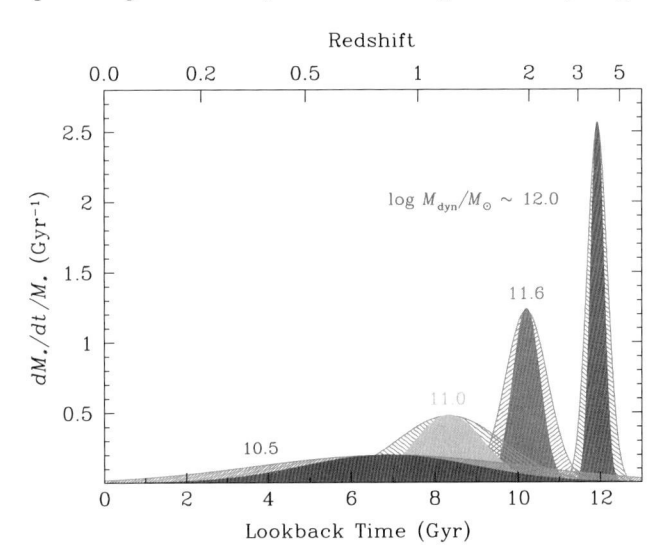

Figure 6.8 The *average* specific SFR as a function of lookback time (and corresponding redshift) for quenched galaxies in the local Universe, as inferred from a large sample of such galaxies and the analysis of the distributions of their Lick/IDS indices. The various stellar masses are indicated by the labels. The gray hatched curves indicate the range of possible variation in the formation timescales that are allowed within the intrinsic scatter of the derived [α/Fe] ratios. Note that these star formation histories are meant to sketch the star formation history averaged over the entire galaxy population (within a given mass bin), in particular reflecting the distribution of their quenching epoch. The star formation histories of individual galaxies are expected to be very different from these averages, for example increasing with time until star formation is suddenly quenched (source: Thomas, D., et al. (2010, *Mon. Not. R. Astron. Soc.*, 404, 1775)) (Reproduced by permission of the Royal Astronomical Society.). For a color version of this figure please see in the front matter.

licities [Z/H] and α-element enhancements derived in this example are:

$$\log(t) = (-0.53 \pm 0.09) + (0.13 \pm 0.01)\log(M_\star),$$
$$[Z/H] = (-2.40 \pm 0.07) + (0.22 \pm 0.01)\log(M_\star),$$
$$[\alpha/Fe] = (-0.95 \pm 0.04) + (0.10 \pm 0.01)\log(M_\star). \quad (6.16)$$

Thus, the most massive ellipticals appear to be the oldest ones, dominated by stars that formed $\sim$ 12 Gyr ago. Then, going to ellipticals of lower mass the luminosity-weighted ages get younger, the duration of the star-formation phase more extended, with quenching taking place at progressively later epochs. The most massive ellipticals are also the most metal-rich (which may appear counterintuitive, with them being the oldest), and are those with the highest α-element enhancement. For the reasons discussed above, the specific ages and durations indicated in Figure 6.8 should be taken with caution, but the qualitative message perhaps

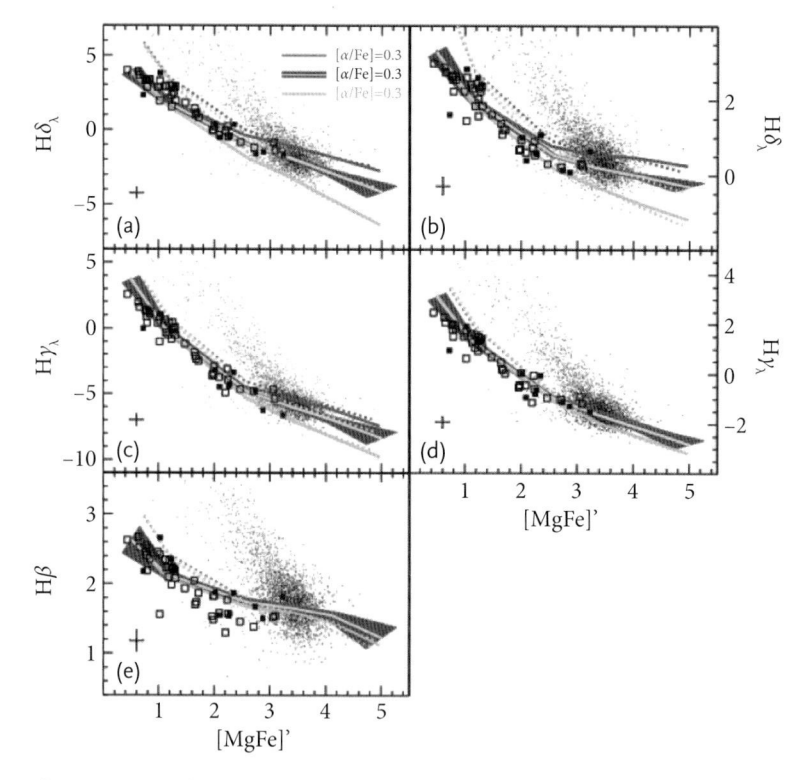

Figure 6.9 (a–e) The Lick/IDS Balmer-line indices for 13 Gyr old synthetic populations with three different values of the α-element enhancement, and total metallicity [Z/H] in the range from -2.25 to 0.67. Note that the [MgFe]' index is fairly insensitive to [α/Fe], cf. Figure 6.7. Data for galactic globular clusters are shown as filled and open squares. The small black points refer to ellipticals galaxies from the SDSS database (source: Thomas, D., et al. (2011, Mon. Not. R. Astron. Soc., 412, 2183)) (Reproduced by permission of the Royal Astronomical Society.). For a color version of this figure please see in the front matter.

counts more than the details: when going to observe galaxies at progressively higher redshifts the first quenched galaxies to disappear must be the less massive ones, and the last the most massive ones – a clear prediction subject to observational test.

Calibration of synthetic stellar populations using the integrated light of globular clusters is always a very useful exercise, as the age and metallicity of the clusters are known with good accuracy from their color-magnitude diagrams and from spectroscopy of individual stars. An example is shown in Figure 6.9a–e, where various Balmer line indices are plotted versus the [MgFe]′ index for 13 Gyr old synthetic populations in a wide metallicity range, and for three different values of the α-element enhancement [α/Fe], as indicated. The plots show fairly good agreement between synthetic indices and those measured for Galactic globular clusters, that is, the synthetic populations predict ages and metallicities in agreement with those of the clusters, the latter being derived from spectroscopy of individual stars and from the color-magnitude diagrams. Several intermediate metallicity clusters appear to have an Hβ index well below the models, which however may be ascribed to an observational effect. Also plotted in Figure 6.9a–e are the indices for a large set of elliptical galaxies with high S/N data: the bulk of these galaxies fall close to the 13 Gyr line, and with metallicities slightly above solar, indicating that the

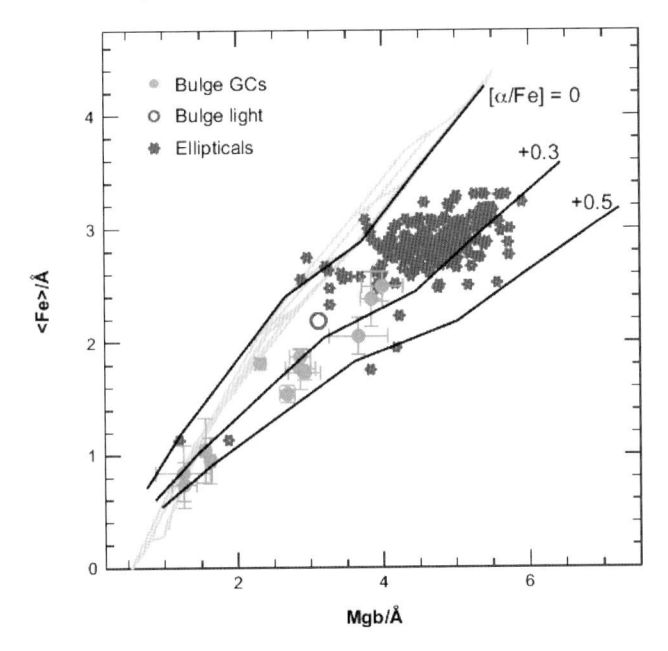

Figure 6.10 The ⟨Fe⟩ index versus the Mgb index for a sample of galactic halo and bulge globular clusters spanning the full range of globular cluster metallicities, the Galactic bulge integrated light in Baade's Window, and for a sample of local elliptical galaxies. Superimposed are synthetic model indices (black lines) for an age of 12 Gyr, [Fe/H] increasing from −2.25 to +0.67, and various α-element enhancements as indicated. The gray grid shows the effect of varying the age from 3 to 15 Gyr (adapted from Maraston, C. *et al.* (2003, *Astron. Astrophys.*, 400, 823)). For a color version of this figure please see in the front matter.

bulk of stars in these galaxies are nearly coeval to the galactic globulars. However, a plume of (less massive) ellipticals clearly departs from the old isochrones, and must contain stellar populations several gigayears younger than the galactic globular clusters. Note that Figures 6.7–6.9a–e use the same synthetic populations, and give consistent results concerning the age of ellipticals. It may seem rather counterintuitive that the galaxies containing the oldest stellar populations also contain the most metal-rich stars. This means that the objects to become massive ellipticals experienced a very fast metal enrichment, early in the history of the Universe.

A fast star-formation timescale of massive elliptical galaxies can also be inferred from their [α/Fe] ratio, for astrophysical reasons that are fully illustrated in the following chapters. Figure 6.10 shows the ⟨Fe⟩ index versus the Mgb index for a set of globular clusters of various metallicities (including in particular galactic bulge globulars of near-solar metallicity), for the integrated light of the bulge in the position of Baade's Window, and for a sample of elliptical galaxies in the local Universe. Overplotted is a set of synthetic indices for three different values of [α/Fe], an age of 12 Gyr and [Fe/H] from -2.25 to $+0.67$. Direct spectroscopy of globular clusters and bulge stars has proved that these objects are α-element enhanced at the level of [α/Fe] $\simeq 0.3$, which offers a consistency check for the synthetic populations shown in the figure, and indicates that most ellipticals must also be α-element enhanced at a similar level.

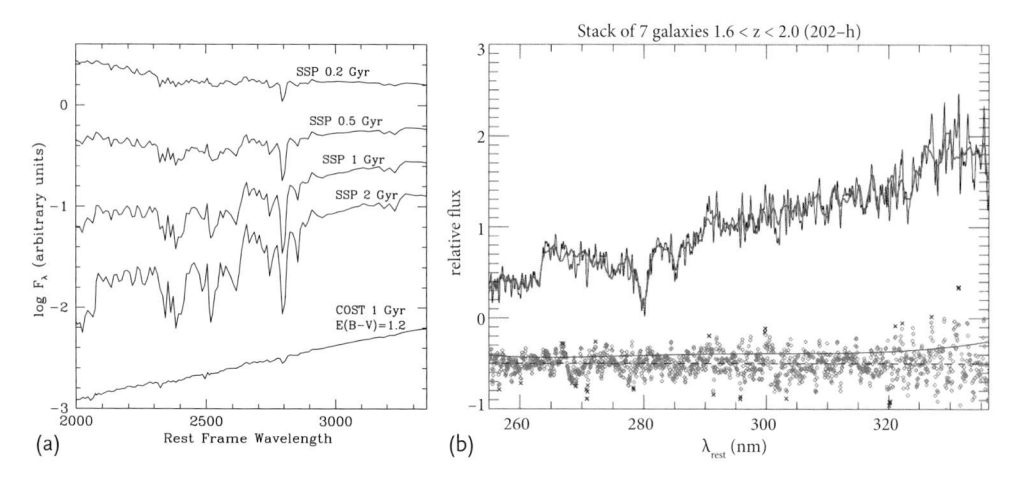

Figure 6.11 (a) The near-UV spectrum of a solar metallicity SSP as it ages from 0.2 to 2 Gyr. Notice the ~ 2 orders of magnitude dimming of the UV flux over this time interval, and that the Mg+Fe feature at $\lambda = 2650$–2850 Å appears only for ages older than ~ 500 Myr. Also shown is the spectrum of a population with SFR $=$ const. and reddened with $E(B-V) = 1.2$. (b) The stacked near-UV spectrum of seven passively evolving galaxies at $1.6 < z < 2$. The black line is the observed spectrum and the gray one is the best fitting SSP template with an age of 1 Gyr and solar metallicity. The points are the residuals (arbitrarily shifted for convenience) and the line is the estimated 1σ noise. Crosses indicated bad pixels that were excluded in the fit ((a) used models from Maraston, C. (2005, Mon. Not. R. Astron. Soc., 362, 799), (b) is from Cappellari, M. et al. (2009, Astrophys. J., 704, L34)) (Reproduced by permission of the AAS.). For a color version of this figure please see in the front matter.

For many years, breaking the age-metallicity degeneracy using synthetic stellar populations and data on local elliptical galaxies has been the prime concern of many, perhaps most, studies in this field. These attempts have achieved only partial success, and probably could not have been otherwise. Of course, the goal was to locate in the history of the Universe the formation epoch of the most massive galaxies, the culmination of the galaxy evolution processes. The advent of large telescopes, both on the ground and in space, has at some point moved the frontier somewhat ahead, revealing the presence of, and giving direct access to, the population of high redshift quenched (elliptical) galaxies. With this giant leap it became possible to map directly the emergence of these galaxies, and their number evolution as a function of mass, environment and cosmic time. Indeed, one may say that the best way of breaking the age-metallicity degeneracy is to travel back in time, 10 Gyr or more, witnessing directly the appearance of this class of galaxies.

In this new context, another magnesium+iron feature, now in the UV at $\lambda \sim$ 2600–2850 Å (rest frame), and due to various FeI, FeII, MgI and MgII lines, proved to be very useful to measure redshifts, ages, velocity dispersions and more of elliptical galaxies at $z \gtrsim 1.4$. In particular, the mere presence of this feature ensures that star formation in the galaxy was indeed quenched since at least ~ 500 Myr, as illustrated by Figure 6.11a,b, thus allowing to estimate the time ($t - t_{\text{quench}}$) elapsed since the quenching of star formation. This is due to the fact that only A-type stars or later exhibit such features, whereas in younger SSPs the feature itself is filled by the strong UV continuum of MS stars more massive than $\sim 3\, M_\odot$. For the same reason, even a low level of star formation is sufficient to wash out the feature, and therefore its mere detection testifies that a galaxy is indeed quenched.

Of course, more difficult to establish is the time interval ($t - t_0$) since the beginning of star formation. In the case of quenched galaxies however, the outshining effect plaguing actively star-forming galaxies is much reduced, and therefore ages derived from the SED fits are more reliable.

6.4
Star-Forming and Quenched Galaxies through Cosmic Times

In the local Universe, as well as all the way to at least redshift ~ 2, galaxies divide into two broad groups in color-magnitude and color-stellar mass diagrams, nicknamed the *red sequence* and the *blue cloud*. Only relatively few galaxies are found to occupy a region with intermediate colors, which is often referred to as the *green valley*. This is illustrated in Figure 6.12a,b, showing the distribution of galaxies on the rest frame $U - B$ color versus stellar mass plane, for large samples of galaxies at various redshift from ~ 0 to ~ 1. The straight line is used to divide the two main groups, so as to study the dependence of the fraction of red-sequence galaxies (in the upper part of the diagram) on stellar mass and environment, and do so as a function of redshift.

Most of the galaxies on the red sequence are morphologically early-type (i.e., ellipticals and S0s), and most of those in the blue cloud are morphologically disk or

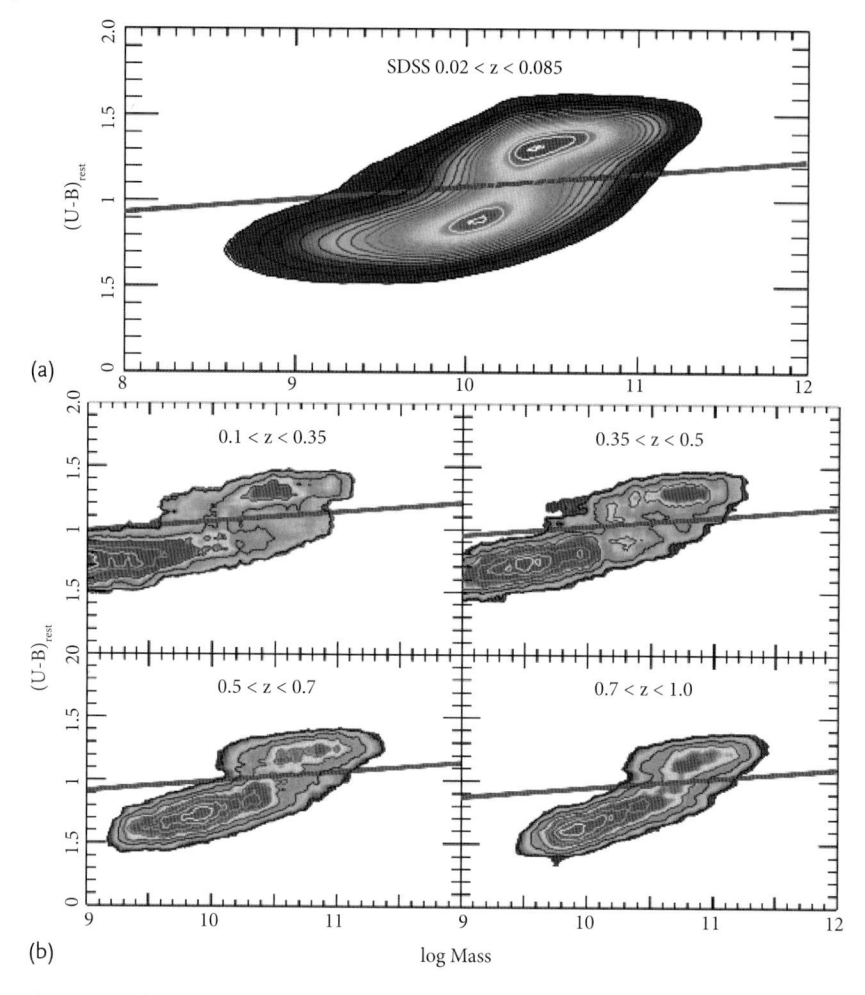

Figure 6.12 The distribution of galaxies in the rest-frame $U - B$ color versus stellar mass plots, for large samples of galaxies at various redshifts. Part (a) refers to the nearby Universe, whereas (b) refer to various redshift slices as indicated. The redshift-dependent line divides the red sequence galaxies from the blue cloud ones (source: Peng, Y. et al. (2010, *Astrophys. J.*, 721, 193)) (Reproduced by permission of the AAS.). For a color version of this figure please see in the front matter.

irregular galaxies. Moreover, most red sequence galaxies show no or little evidence of ongoing star formation, whereas most of blue cloud galaxies are actively star forming. However, exceptions are relatively frequent: there are actively star-forming galaxies that fall on the red sequence because they are highly dust-reddened (including in particular edge-on spirals). There are also morphologically early-type galaxies without ongoing star formation that fall on the blue cloud, quite possibly because star-formation was quenched recently, and they still maintain relatively blue colors on their way to the red sequence. Galaxies in the green valley may well

Table 6.1 Morphology versus color versus spectrum-selected samples (from Renzini, A. (2006, Annu. Rev. Astron. Astrophys., 44, 141), and based on SDSS data elaborated by M. Bernardi (private communication)).

	MOR	COL	SPE
MOR	37 151	70%	81%
COL	58%	44 618	87%
SPE	55%	70%	55 134

be objects caught in their transition from one group to the other, although many appear to host an AGN, and their blue color may be due directly to the AGN, or to some strong emission line falling in the rest frame blue band.

Therefore, being on the red sequence does not ensure that a galaxy is *quenched* (SFR $\simeq 0$), nor that it is morphologically an E/S0 galaxy. Indeed, there are three ways to select candidate quenched galaxies, namely by morphological type (E/S0), by color and by spectrum, and the three criteria are far from being equivalent to each other. The color criterion selects galaxies redder than some limit (such as in Figure 6.12a,b), whereas the spectrum criterion selects galaxies without emission lines and with photospheric absorption lines such as for example, Ca II H and K. Table 6.1 reports that out of 55 000 pure absorption line galaxies only $\sim 55\%$ of them qualify as morphologically early type (E/S0), or that almost $\sim 90\%$ of the color-selected (red) galaxies show a pure absorption line spectrum, but less than 60% among them are also morphologically of E/S0 type. So, one can say that the color selection is likely to pick a quenched galaxy almost 90% of the time, with the rest being highly reddened, star-forming contaminants.

For the local Universe, Figure 6.13a,b shows the contributions to the total stellar mass, separately by red-sequence and by blue-cloud galaxies in several mass bins, along with their contributions to the total number of galaxies. Although red-sequence galaxies represent only 17% of the total number of galaxies in the sample, they contribute $\sim 57\%$ of the total mass. Moreover, $\gtrsim 80\%$ of the stellar mass in red-sequence galaxies belongs to galaxies more massive than $\sim 3 \times 10^{10}\, M_\odot$. By number, dwarfs dominate in the local Universe but they do not contribute much to the total stellar mass in galaxies. Dwarfs are sometimes seen as the "building blocks" of galaxies, but at least in the present Universe not much can be built with them.

6.4.1
The Main Sequence of Star-Forming Galaxies

Having measured SFRs from the Hα diagnostics, that is, from Eq. (6.7) and stellar masses from SED fitting, Figure 6.14a,b shows the SFR $- M_\star$ relation for a large, representative set of local star-forming galaxies, selected for being bluer than the straight line in Figure 6.12b. Notice how tight this relation is, with a dispersion of just ∼ 0.3 dex, and with SFR $\propto M_\star^{-0.9}$, hence the SFR in local, star-forming galaxies scales almost linearly with stellar mass. The specific SFR (sSFR $\equiv$ SFR/$M_\star$) has the dimensions of the inverse of a time, and sSFR^{-1} is the time that the galaxy would have taken to build up its present mass at its present SFR. Galaxies with sSFR^{-1} ≫ of the Hubble time must have experienced a much higher SFR in the past, and those with sSFR^{-1} ≪ of the Hubble time must have been caught in a starburst phase. With the SFR almost linear with $M_\star$, the sSFR turns out to be almost independent of stellar mass, that is, sSFR $\propto M_\star^{-\beta}$, where $\beta = -0.10(\pm 0.1)$.

Figure 6.14a and 6.14b refers respectively to the low- and to the high-density quartiles, in the spatial distribution of galaxies, showing that the SFR–$M_\star$ relation is remarkably independent of local overdensity. This is quite unexpected, as one may think that environment must matter, in one way or another. It does, indeed, as we shall see later, but it does not as far as the SFR-$M_\star$ relation *of star-forming galaxies* is concerned, at least in the local Universe. Evidence exists that this independence of environment persists up to at least $z \sim 1$.

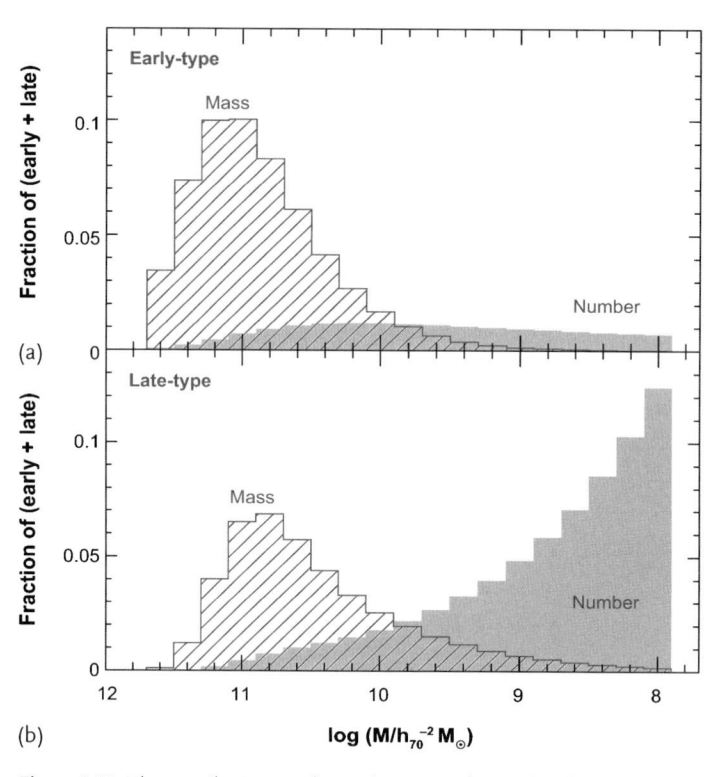

Figure 6.13 The contributions to the total stellar mass and to the number of galaxies by (a) early-type (red sequence) and (b) late-type (blue cloud) galaxies in the local Universe. The relative areas are proportional to the contributions of the early- and late-type galaxies to the total stellar mass and to the number of galaxies. The galaxy database is the same as in Figure 6.12a (figure based on the SDSS data, Baldry I.K. et al. (2004, *Astrophys. J.*, 600, 681)). For a color version of this figure please see in the front matter.

6.4 Star-Forming and Quenched Galaxies through Cosmic Times

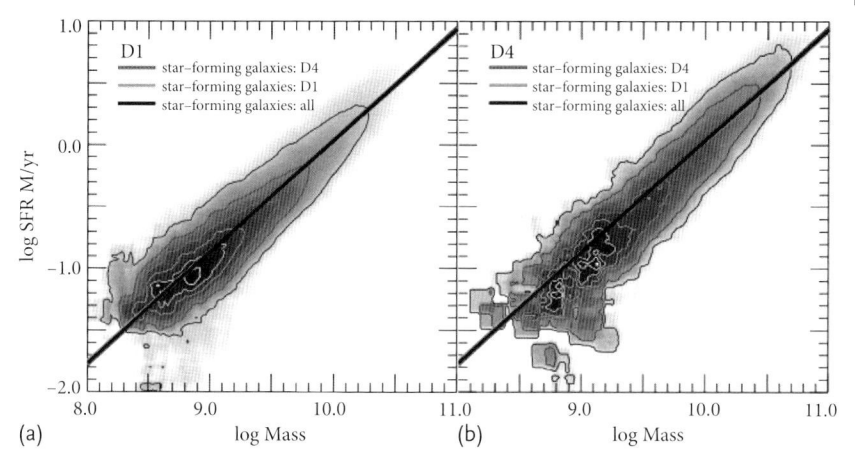

Figure 6.14 The SFR versus stellar mass for a large sample of star-forming (blue cloud) galaxies in the nearby Universe, as derived from the Hα luminosity corrected for extinction. Part (a) refers to galaxies in the low environmental density quartile and (b) to the high-density quartile. Best fit linear relations for the whole galaxy sample as well as for the low- and high-density quartiles are shown in (a) and (b), though they can be barely distinguished, meaning that the SFR–$M_\star$ relation is virtually independent of environment. The galaxy database is the same as Figure 6.12a,b (source: Peng, Y. et al. (2010, Astrophys. J., 721, 193)) (Reproduced by permission of the AAS.). For a color version of this figure please see in the front matter.

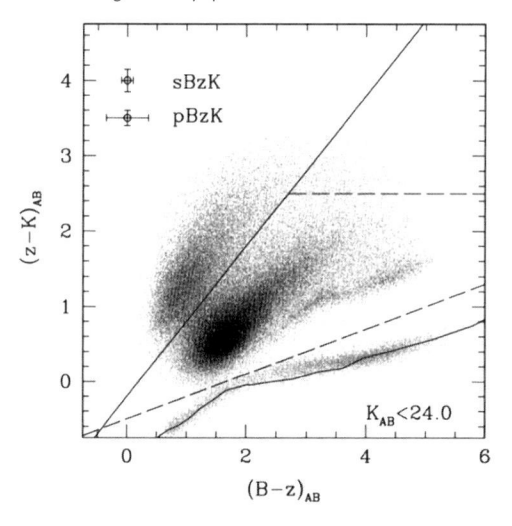

Figure 6.15 The $z - K$ versus $B - z$ plot for a sample of objects down to $K_{AB} \simeq 24$. Candidate star-forming galaxies at $1.4 \lesssim z \lesssim 2.5$ (called sBzKs) lie to the left of the diagonal line, whereas candidate passively evolving (quenched) galaxies in the same redshift interval (called pBzKs) lie within the wedge in the upper-right. Most galaxies in the central part of the plot are at $z \lesssim 1.4$. Stars occupy the sequence in the lower part. Typical error bars for sBzKs and pBzKs are indicated (source: McCracken, H. et al. (2010, Astrophys. J., 708, 202)) (Reproduced by permission of the AAS.). For a color version of this figure please see in the front matter.

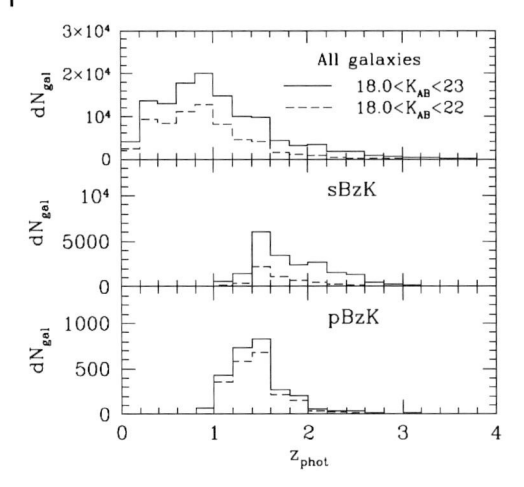

Figure 6.16 The photometric redshifts of all $K_{AB} < 23$ galaxies shown in Figure 6.15 (upper panel), as well as of the star-forming (sBzKs) and passively evolving ones (pBzKs) selected by the BzK criterion for being at $1.4 \lesssim z \lesssim 2.5$ (source: McCracken, H. et al. (2010, Astrophys. J., 708, 202)) (Reproduced by permission of the AAS.).

Moving to higher redshifts, an effective way of selecting galaxies in the redshift range $1.4 \lesssim z \lesssim 2.5$ (both star-forming and quenched) is known as the BzK criterion, being based on the $z - K$ versus $B - z$ two-color plot shown in Figure 6.15. Over 140 000 galaxies are plotted in this figure, whereas over 13 000 stars are all clustering along the line shown in the lower part of Figure 6.15. Candidate star-forming galaxies at $1.4 \lesssim z \lesssim 2.5$ (called sBzKs) are shown in the upper-left part of the plot in Figure 6.15, above the diagonal line drawn to be parallel to the reddening vector. Candidate quenched galaxies in the same redshift range (called pBzKs) occupy instead the wedge-shaped region in the upper-right part of the plot in Figure 6.15, whereas galaxies in other redshift ranges (actually mostly at $z < 1.4$) lie in the central part, where one can still recognize the persistence of the red sequence/blue cloud dichotomy. Figure 6.16 shows the (photometric) redshift distribution for the subsample of $K_{AB} < 23$ galaxies from Figure 6.15, and does so for all of them and separately for sBzKs and for pBzKs, demonstrating the effectiveness of the BzK selection criterion to sort out galaxies in the redshift range $1.4 \lesssim z \lesssim 2.5$. Notice, however, that photometric redshifts of pBzK galaxies may have been somewhat underestimated, because of the dearth of sufficient spectroscopic redshifts for this kind of galaxies to properly train the photometric redshifts.

Figure 6.17 shows the SFR–$M_\star$ relation for a sample of BzK-selected star-forming galaxies, where SFRs are derived from the extinction-corrected UV luminosity, that is, from Eq. (6.3), and stellar masses are based on SED fits. One can notice that the almost linear dependence of SFR on $M_\star$ appears to be preserved at these high redshifts, whereas the SFR at given mass (hence the sSFR) is a factor ~ 20–30 higher than locally. For comparison, also shown are the average SFR–$M_\star$ relation for star-forming galaxies at $z = 0.1$ (the same as in Figure 6.14a,b), and that at

6.4 Star-Forming and Quenched Galaxies through Cosmic Times

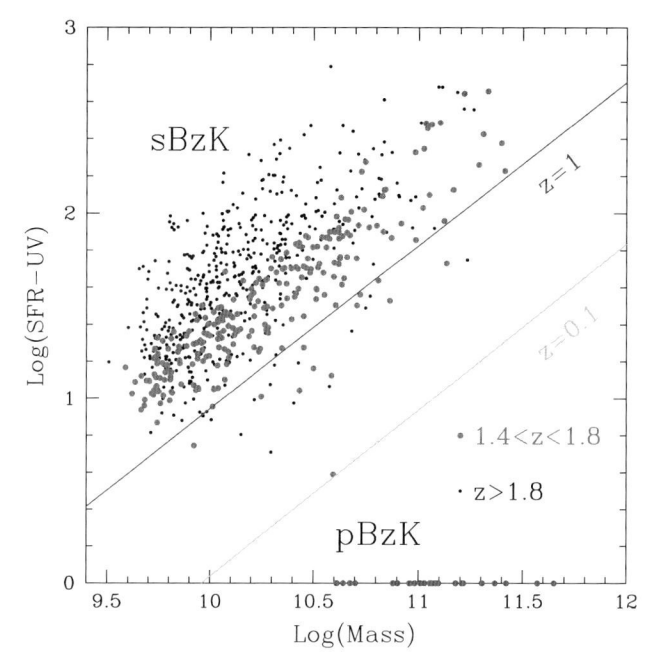

Figure 6.17 The SFR–$M_\star$ relation for a sample of sBzK-selected star-forming galaxies, color-coded by their redshift as indicated. Best fit linear SFR–$M_\star$ relations for star-forming galaxies at $z \sim 1$ and ~ 0.1 are also shown. The pBzK-selected passively evolving (quenched) galaxies at $z > 1.4$ are shown at log SFR = 0 though their SFRs are likely to be substantially lower. The sBzK sample is almost complete above $M_\star \sim 4 \times 10^9 \, M_\odot$ and the pBzK sample above $M_\star \simeq 4 \times 10^{10} \, M_\odot$. Units are $M_\odot \, \text{yr}^{-1}$ and $M_\odot$ (figure used GOODS data from Daddi, E. et al. (2007, Astrophys. J., 670, 156)). For a color version of this figure please see in the front matter.

$z = 1$, where the SFRs were derived from Eq. (6.5). This evolving SFR–$M_\star$ relation for star-forming galaxies, sometimes called the *main sequence of star-forming galaxies*, plays a major role in galaxy evolution, yet substantially different relations can be found in the literature, often shallower than those shown in Figure 6.17. Much of these differences may arise from different criteria used to select star-forming galaxies, in particular if such criteria use observables with a direct dependence on the SFR. In such cases, low-mass galaxies are selected preferentially among those with a higher than average SFR, and therefore the resulting SFR–$M_\star$ relation turns out to be shallower compared to the relation holding for a mass-limited sample (a manifestation of the Malmquist bias). Typical is the case of UV-selected samples of $z \sim 2$ galaxies, where the SFR is virtually independent of mass (i.e., $\beta \sim -1$), hence their sSFR turns out to be inversely proportional to mass. In any event, most recent values reported for β range between ~ -0.4 and ~ -0.1, with the differences between one study and another arising from a combination of differences in the adopted procedures. Besides the selection criteria to cull star-forming galaxies, such differences include the SFR and mass diagnostics, and in particular the star formation histories adopted for the SED fits.

Given its evolutionary relevance, it is worth constructing the SFR–$M_\star$ relation at the various redshifts using as many SFR indicators as possible. For example, the SFR of the sBzKs in Figure 6.15 was measured from the radio fluxes at 1.4 GHz and using Eq. (6.10). Since the vast majority of the objects are not individually detected in the radio, the average radio flux for galaxies in several mass bins was derived by stacking analysis. The result is shown in Figure 6.18a, where the sSFR is plotted as a function of stellar mass and for two redshifts (i.e., $z = 1.6$ and 2.1). The open symbols show the sSFR–$M_\star$ relation as derived from the UV, before extinction correction: clearly, a reddening parameter $E(B-V)$ that systematically increases with stellar mass is required to force the UV-derived SFRs to agree with the radio-derived ones (filled symbols). It is indeed quite reassuring that the same trend of $E(B-V)$ with stellar mass is derived from the slope of the UV continuum.

Figure 6.18b shows the redshift dependence of the sSFR (calculated at $M_\star = 3 \times 10^{10} M_\odot$). The evolution of the main sequence of star-forming galaxies as seen in this figure is captured by the simple relation (from Pannella, M. et al. (2009), Astrophys. J., 698, L116)):

$$\text{SFR} \simeq 270 \left(\frac{M_\star}{10^{11} M_\odot}\right) \left(\frac{t}{3.4 \text{ Gyr}}\right)^{-2.5} \quad (M_\odot \text{ yr}^{-1}), \tag{6.17}$$

which for simplicity neglects the (small) deviation from the linear dependence on $M_\star$. The actual slope of the relations in Figure 6.17 is $1 + \beta \simeq 0.9 \pm 0.1$. An almost equivalent expression derived from the data used for Figure 6.17 is (from Peng, Y. et al. (2010, Astrophys. J., 721, 193)):

$$\text{sSFR} \simeq 2.5 \left(\frac{M_\star}{10^{10} M_\odot}\right)^\beta \left(\frac{t}{3.5 \text{ Gyr}}\right)^{-2.2} \quad (\text{Gyr}^{-1}). \tag{6.18}$$

As mentioned above, the actual value of the slope β of the sSFR–$M_\star$ relation is still relatively uncertain, depending upon the selection criterion used for constructing the galaxy sample, on the specific SFR indicator, and on the procedure to derive the stellar masses. Most recently, data from the Herschel Observatory have started to flow in, SFRs from the far-IR luminosity are being derived using Eqs. (6.4) and (6.5), and an early result is shown in Figure 6.19. By and large the SFRs from the far-IR data appear to be consistent with the SFRs from the radio data shown in Figure 6.18a, although with a slightly steeper sSFR–$M_\star$ relation. This steeper slope may well reflect an intrinsic property of these galaxies, although it might also be the result of a marginal bias introduced by the criterion to select galaxies as star forming. In fact, any such criterion is likely to favor the most actively star-forming objects of the SFR distribution, and especially so towards the low-mass end of the mass function. This may result in an artificial steepening of the sSFR–$M_\star$ relation.

The systematic growth with redshift (lookback time) of the sSFR described by Eqs. (6.17) and (6.18) appears to continue up to $z \simeq 2 - 2.5$, where it comes to a halt, thereafter remaining constant up to the highest redshifts so far explored. This is illustrated in Figure 6.20, with the caveat that data beyond redshift ~ 3 should be regarded as preliminary. This abrupt flattening beyond $z \sim 2$ suggests that sSFR

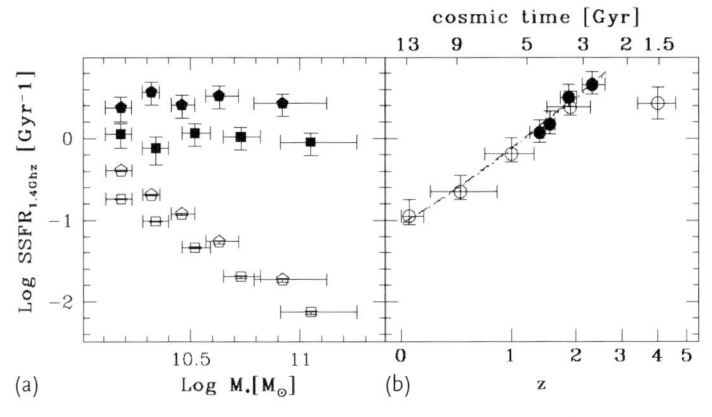

6.4 Star-Forming and Quenched Galaxies through Cosmic Times | 161

Figure 6.18 (a) The specific SFR of sBzK-selected galaxies in five mass bins as a function of stellar mass, as derived from stacking 1.4 GHz data (solid symbols) and from the UV continuum luminosity without reddening correction (open symbols). Pentagons refer to galaxies at $z \sim 2.1$ and squares to galaxies at $z \sim 1.6$. (b) The specific SFR for $M_* = 3 \times 10^{10}\,M_\odot$ galaxies as a function of redshift. Solid symbols refer to SFRs from radio data, whereas open symbols at lower/higher redshifts are from from various other SFR indicators (source: Pannella, M. et al. (2009, Astrophys. J., 698, L116)) (Reproduced by permission of the AAS.).

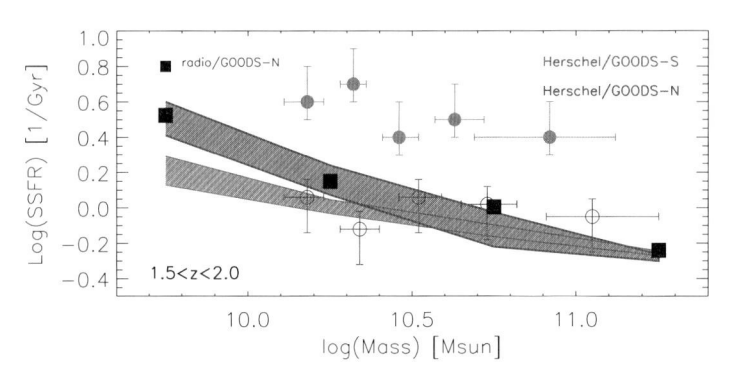

Figure 6.19 The sSFR–M_* relation for a 4.5 μm selected sample down to AB magnitude 23 of $1.5 < z < 2.0$ galaxies in the GOODS fields (shaded areas) with SFRs measured from stacked Herschel data at 100 μm plus models for the far-IR SED. Filled and open circles represent the same data at $\langle z \rangle = 2.1$ and 1.6 shown in Figure 6.18a,b. Filled squares represent SFRs from 1.4 GHz stacked radio data (adapted from Rodighiero, G. et al. (2010, Astron. Astrophys., 518, L25, and private communication)). For a color version of this figure please see in the front matter.

$\sim 2 \times 10^{-9}$ Gyr^{-1} is the maximum rate that galaxies can sustain while on the main sequence, quasi-steady regime. Excess gas feeding would then result in more wind mass loss, rather than more star formation. Figure 6.20 bears a close resemblance to the run with redshift of the global SFR density in the Universe (the *Madau–Lilly* plot), but refers to individual galaxies instead: the convolution of the sSFR(t) with the time dependent (stellar) mass function of star-forming galaxies should in fact give the time (redshift) evolution of the global SFR density.

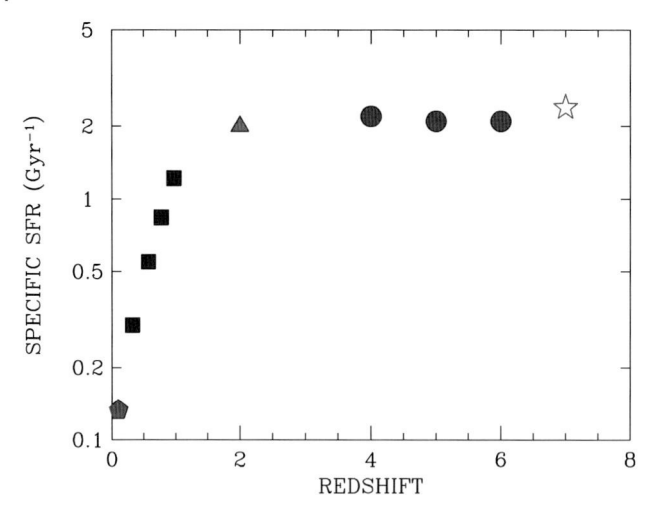

Figure 6.20 The specific SFR for objects on the main sequence of star-forming galaxies as a function of redshift. The sSFR is calculated for $M_\star = 5 \times 10^9 \, M_\odot$. SFRs were measured from the following indicators: Hα (pentagon); a combination of [OII] and 24 µm luminosities (squares); UV luminosity corrected for reddening (triangle); SED global fit for SFR, mass, and so on (circles and star) (adapted from Gonzalez, V. et al. (2010, Astrophys. J., 713, 115), with the addition of the SDSS point at $z = 0.1$ and where squares are from Noeske, K.G. et al. (2007, Astrophys. J., 660, L43), the triangle from Daddi, E. et al. (2007), circles from Stark, D.P. et al. (2009, Astrophys. J., 697, 1493)).

It is worth emphasizing that not all star-forming galaxies lie close to the main sequence: submillimeter galaxies (SMG) are indeed characterized by SFRs ~ 10 or more times higher than those of main sequence galaxies, likely the result of a starburst promoted by the sudden pile up of molecular gas driven by a merging event. Their census is still rather incomplete, partly for the difficulty of getting robust redshifts for them, so their contribution to the global SFR density remains to be well constrained.

In summary, on a SFR–$M_\star$ plot, such as that shown in Figure 6.17, most galaxies belong to two distinct SFR regimes: those actively star-forming on the main sequence, and the quenched galaxies with zero or very low SFR. The exceptions are few, but potentially very interesting. We have just mentioned SMGs, with their exceptionally high SFR. Nearly as rare are galaxies with detected, yet low SFR, much lower than that typical of main sequence galaxies. These are good candidates for being galaxies caught in the act of being quenched, and their rarity indicates that the quenching process is a fast, sudden event rather than a slow decline of the SFR, as for example, parameterized in the τ-models. On the other hand, the tightness of the main sequence demonstrates that star formation in these galaxies proceeds in a quasisteady fashion, as opposed to a sequence of starburst events with large (a factor of 10 or more) excursions in SFR. Indeed, the small dispersion of the SFR for given $M_\star$ demonstrates that these objects cannot have been caught in a special moment of their existence, such as a starburst. Rather, they must sustain such high SFRs for a major fraction of the time interval between $z = 2.5$ and 1.4, that is, for some 1–2 Gyr, much longer than one dynamical time ($\sim 10^8$ yr) typical of starbursts. Even though sustaining SFRs of hundreds of solar masses per year, which are typical of local starburst galaxies, the $z \sim 2$ galaxies shown in Figure 6.17 are not starbursting, they are just forming stars at a very high rate.

6.4.2
The Mass and Environment of Quenched Galaxies

Up to at least redshift ~ 2 quenched galaxies coexist with actively star-forming ones. Thus, one outstanding issue in galaxy evolution is to map the relative number of these two kinds of galaxies, as a preliminary step towards trying to understand the processes that control the transformations from one kind to the other, and in particular so for the quenching processes. A running vertical cut through the panels of Figure 6.12 shows that virtually all low-mass galaxies are blue, hence actively star forming, and that moving the cut to higher masses one recovers an increasing fraction of red (quenched) galaxies. Finally, approaching the top end of the mass distribution, virtually all galaxies are red, and likely to be quenched, as illustrated in Figure 6.21. This trend persists all the way to high redshift, as documented by Figure 6.12 up to $z \sim 1$, and continues even to higher redshifts. The direct evidence is that high mass favors quenching, low mass favors ongoing star formation.

The fraction of red/quenched galaxies is also found to correlate strongly with the overdensity of the environment that galaxies inhabit, where the overdensity is defined as $\delta = (\rho - \bar{\rho})/\bar{\rho}$, ρ and $\bar{\rho}$ being respectively the local and the average density of the Universe at the same cosmic epoch. The density field can be mapped using the 3D distribution of galaxies, and using large samples of galaxies with spectroscopic redshifts. The result is that in regions of high overdensity red/quenched galaxies dominate. On the contrary, in low overdensity regions blue/star-forming galaxies dominate. The trend of the red galaxy fraction with stellar mass and overdensity is illustrated in Figure 6.21 for galaxies in the local Universe: the fraction of red/quenched galaxies increases monotonically with galaxy stellar mass and with local overdensity. Whereas high-mass galaxies are predominantly quenched, no matter where they sit in the density field, low-mass galaxies are quenched only in high-density regions.

Again, the trend of the red/quenched fraction with overdensity is found to persist up to $z \sim 1$, the highest redshift for which the density field has been reasonably well mapped so far. Progress in mapping the density field beyond $z \sim 1$ is rapid, and soon it should be possible to check whether these trends persist even to higher redshifts.

Given this empirical evidence, it is legitimate to introduce the phenomenological concepts of *mass quenching* and *environment quenching*, as clearly both mass and environment matter as far as the quenching of star formation in galaxies is concerned. At this stage we refrain from speculating on the possible physical mechanisms that may be responsible for the two kinds of quenching of star formation. It

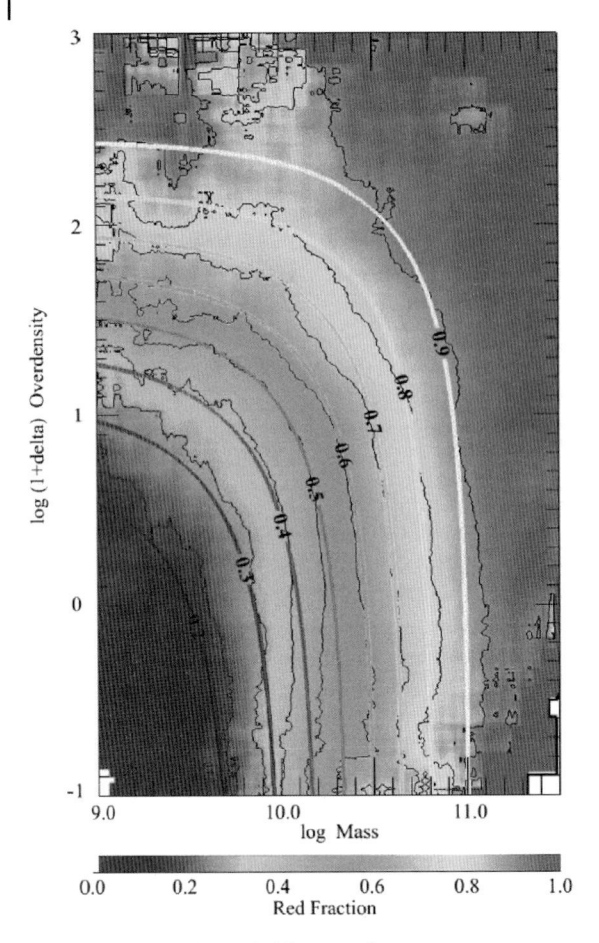

Figure 6.21 The color-coded fraction of red/quenched galaxies in the local Universe as a function of galaxy stellar mass and environment as measured by the local overdensity (source: Peng, Y. et al. (2010, Astrophys. J., 721, 193)) (Reproduced by permission of the AAS.). For a color version of this figure please see in the front matter.

has also been shown that, both locally and at redshift up to ~ 1, the *relative* efficiency of mass quenching is independent of environment, and the *relative* efficiency of environment quenching is independent of mass, thus making the two effects *separable* from each other. Here by *relative* one means with respect to maximum mass (overdensity) for which galaxies in a given environment (of a given mass) are still predominantly star forming.

6.4.3
Mass Functions

Much on the evolution of galaxy populations across cosmic times is captured by the evolution of the galaxy mass functions, that is, the mass function for all galaxies,

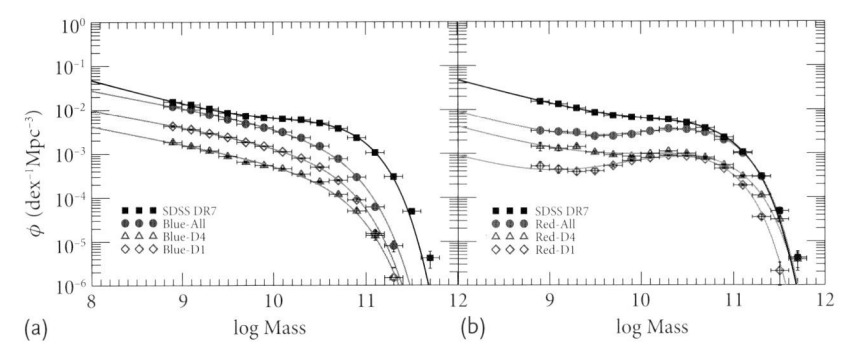

Figure 6.22 The mass functions of galaxies in the local Universe. (a) The global mass function is shown as a black line with solid squares (replicated also in (b)); the other lines show the mass function for the star-forming (blue) galaxies, for the whole such population as well as for the low- and high-density quartiles, respectively D1 and D4. (b) The same for the quenched (red) galaxies. A Salpeter-diet IMF was adopted (source: Peng, Y. et al. (2010, Astrophys. J., 721, 193)) (Reproduced by permission of the AAS.). For a color version of this figure please see in the front matter.

as well as those for the star-forming and for the quenched galaxies separately. Figure 6.22a,b shows all such mass functions for the local Universe, irrespective of the environment, as well as for the high- and low-density quartiles. Note that the global mass function has a characteristic inflection and hump towards high masses, to describe which a simple Schechter mass function:

$$\phi(M_\star)dM_\star = \phi^* \left(\frac{M_\star}{M_\star^*}\right)^\alpha e^{-M_\star/M_\star^*} dM_\star , \quad (6.19)$$

is not sufficient. Indeed, whereas the mass function of the star-forming galaxies is well fit by a single Schechter function, that of the quenched galaxies requires the sum of two Schechter functions, and therefore so does the global mass function. Thus, the global mass function for star-forming galaxies shown in Figure 6.22a,b is well described by Eq. (6.19) with $M_\star^* = 4.7 \times 10^{10} M_\odot$, $\alpha = -1.40$ and $\phi^* = 1.014 \times 10^{-3}$ Mpc^{-3}, whereas the mass function for the red (quenched) galaxies is well fit by the sum of two such functions, both with the same $M_\star^* = 4.7 \times 10^{10} M_\odot$, but one with $\alpha_1 = -0.40$ and $\phi^* = 3.247 \times 10^{-3}$ Mpc^{-3} and the other with $\alpha_2 = -1.40$ and $\phi^* = 0.214 \times 10^{-3}$ Mpc^{-3}. Note that of the two components for the quenched galaxies, one is a scaled down version of the mass function of the star-forming galaxies (a factor ~ 5 lower in normalization), whereas the other one is responsible for the mentioned inflection and hump, and therefore for the high-mass end being dominated by quenched galaxies. Note also that the two low-mass end slopes α_1 and α_2 differ precisely by one unity.

Moving now to high redshifts, Figure 6.23 shows the evolution of the mass function of star-forming galaxies all the way to $z = 1.75$. Notice that $M_\star^*$ remains remarkably constant through the whole redshift interval, as does the slope at the low mass end α. The normalization ϕ^* shows some low-amplitude up and down excursions up to $z \sim 1$ (likely due to cosmic variance in the sample) before starting

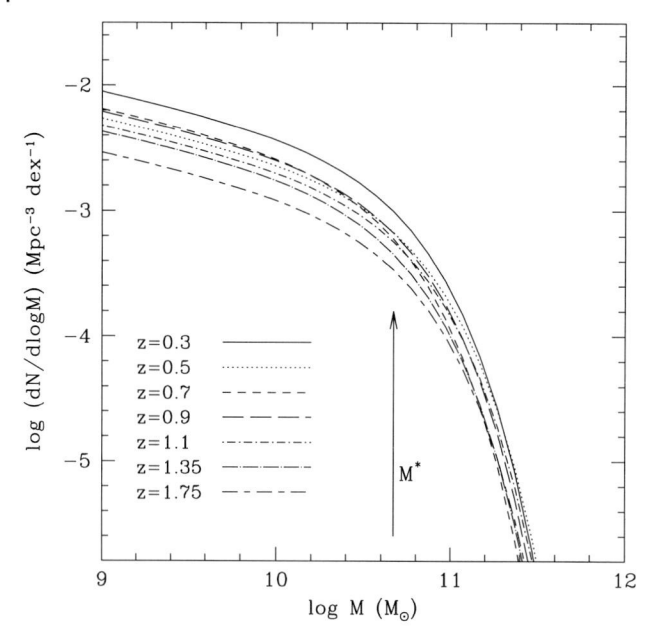

Figure 6.23 The Schechter function fits to the empirical stellar mass function of star-forming galaxies as a function of redshift. The vertical line indicates the value of M_*^*, almost identically the same at all these redshifts. Notice that the secular increase of the mass function is occasionally nonmonotonic, a likely result of a combination of cosmic variance and Poisson noise (figure adapted from Ilbert, O. et al. (2010, Astrophys. J., 709, 644I) using Table 1 in Peng, Y. et al. (2010), with masses shifted down by 0.33 dex to make it consistent with Figure 6.22a,b).

to gradually decline at higher redshifts, with a decrease of a factor ~ 2 between the lowest and the highest redshift bins. Thus, the mass function of star-forming galaxies shows only very modest evolution with redshift.

In contrast with the case of star-forming galaxies, the mass function of quenched (red) galaxies shows very strong evolution with redshift, as illustrated in Figure 6.24a–d for redshifts up to ~ 1. Whereas the number density of the most massive quenched galaxies remains virtually the same up to $z \sim 1$, the number density of lower-mass quenched galaxies rapidly drops with increasing redshifts. As shown in Figure 6.24a–d, incompleteness sets in at higher and higher stellar masses with increasing redshift, which prevents us to unambiguously check to which extent the double Schechter function shape is preserved at high redshifts, and how the two components (and their relative contribution) evolve with redshift. Moreover, quenched galaxies are highly clustered, and therefore cosmic variance tends to affect them more than the star-forming ones.

At still higher redshifts (z from ~ 1 to ~ 2) spectroscopic samples of quenched galaxies become largely insufficient to construct mass functions, and samples with only photometric redshifts may still be affected by biases and contaminations. Although robust mass functions for quenched galaxies are still lacking, all available evidences indicate that the number density of even the most massive quenched

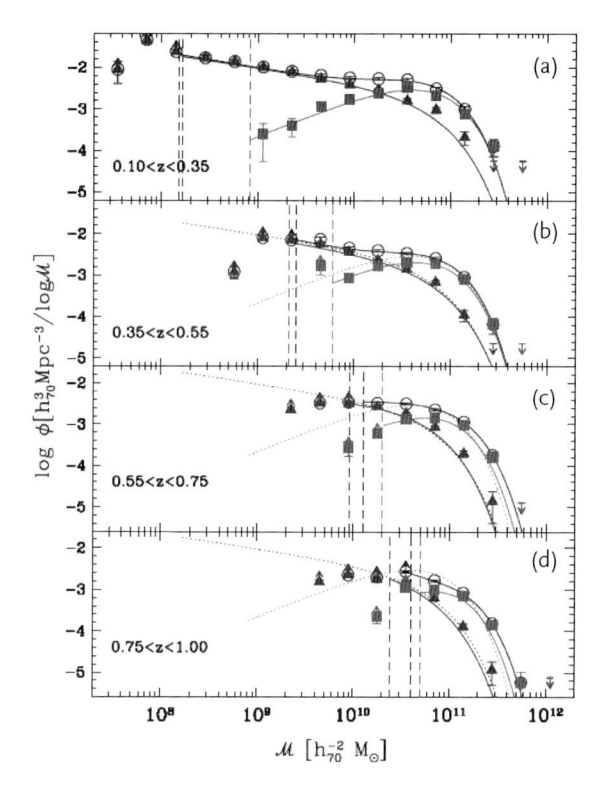

Figure 6.24 (a–d) Galaxy stellar mass function by galaxy types and in various redshift bins. Quenched galaxies and star-forming galaxies are shown as squares and circles, respectively. Continuous lines are Schechter fits. Dotted lines in each panel show the Schechter fits to the first redshift bin. Dashed vertical lines in each redshift bin indicate the limits of the mass completeness (source: Pozzetti, L. *et al.* (2010, *Astron. Astrophys.*, 523, A13)) (Reproduced with permission © ESO.). For a color version of this figure please see in the front matter.

galaxies rapidly drops beyond $z \sim 1$ and almost vanishes by $z \sim 2$. In summary, data on quenched galaxies at high redshifts are still rather incomplete, but there appears to be at least qualitative agreement between the available evidence and the expectations from age dating local ellipticals as illustrated by Figure 6.8: moving to high redshifts the number density of quenched galaxies of low-mass rapidly drops between $z \sim 0$ and ~ 1, whereas that of the massive galaxies starts dropping only beyond $z \sim 1$. This is to say that the epoch of star formation quenching strongly correlates with galaxy mass, the most massive galaxies being the first to be quenched, whereas for lower and lower mass galaxies quenching takes place at progressively later epochs.

Finally, the local correlation of the quenched (red) fraction with galaxy mass and environment (overdensity) together with the double Schechter mass function of quenched galaxies, indicate that at least two independent quenching processes are at work, one most likely related to mass, the other related to the environment, that is, mass quenching and environment quenching.

In this chapter the main global properties of star-forming and quenched galaxies as a function of mass, redshift and environment have been described, as established using purely stellar population tools. In particular, methods for the measurement of the stellar masses and star formation rates of galaxies have been described, their limits highlighted, and some results based on them have been presented. When using large, representative sets of galaxies a few basic, remarkable *simplicities* emerge, namely the tight main sequence of star-forming galaxies, its independence on environment, which implies both an unsuspected uniform behavior of galaxies, and the rapidity of the quenching of star formation leading to the emergence of passively evolving galaxies. The fraction of such quenched galaxies monotonically increases with both mass and local overdensity, and the mass- and environment-quenching processes appear to operate independent of each other. Moreover, the local, *fossil* evidence indicates that quenching must start predominantly first with the most massive galaxies, and then gradually propagate to lower and lower mass galaxies. This set of simplicities, and others, can be combined in a mere phenomenological scheme describing the cosmic evolution of both star-forming and quenched galaxies. This will be the subject of Chapter 9.

Further Reading

Star Formation Rate Calibrations

Kennicutt Jr., R.C. (1998) *Annu. Rev. Astron. Astrophys.*, **36**, 189.

Leitherer, C. et al. (1999) *Astrophys. J. Suppl.*, **123**, 3.

Madau, P. et al. (1998) *Astrophys. J.*, **498**, 106.

Moustakas, J. et al. (2006) *Astrophys. J.*, **642**, 775.

Galaxy Databases Used in this Chapter

COSMOS: Scoville, N. et al. (2007) *Astrophys. J. Suppl.*, **172**, 1.

GOODS: Giavalisco, M. et al. (2004) *Astrophys. J.*, **600**, L93.

SDSS: Abazajian, K.N. et al. (2009) *Astrophys. J. Suppl.*, **182**, 543.

zCOSMOS: Lilly, S.J. et al. (2009) *Astrophys. J. Suppl.*, **184**, 218.

Reviews on High-Redshift Galaxies

Giavalisco, M. (2002) *Annu. Rev. Astron. Astrophys.*, **40**, 579.

McCarthy, P.J. (2004) *Annu. Rev. Astron. Astrophys.*, **42**, 477.

Renzini, A. (2006) *Annu. Rev. Astron. Astrophys.*, **44**, 141.

Fundamental Plane of Elliptical Galaxies

Bender, R. *et al.* (1992) *Astrophys. J.*, **399**, 462.
Bender, R. *et al.* (1993) *Astrophys. J.* **411**, 153.
Djorgovski, S. and Davis, M. (1987) *Astrophys. J.*, **303**, 39.
Dressler, A. *et al.* (1987) *Astrophys. J.*, **313**, 42.

Lick/IDS Indices

Burstein, D. *et al.* (1984) *Astrophys. J.*, **287**, 586.
Greggio, L. (1997) *Mon. Not. R. Astron. Soc.*, **285**, 151.
Kuntschner, H. *et al.* (2001) *Mon. Not. R. Astron. Soc.*, **323**, 615.
Schiavon, R.P. (2007) *Astrophys. J. Suppl.*, **171**, 146.
Thomas, D. *et al.* (2003) *Mon. Not. R. Astron. Soc.*, **339**, 897.
Thomas, D. *et al.* (2005) *Astrophys. J.*, **621**, 673.
Trager, S.C. *et al.* (2000) *Astron. J.*, **120**, 165.
Vazdekis, A. *et al.* (2010) *Mon. Not. R. Astron. Soc.*, **404**, 1639.
Worthey, G. *et al.* (1992) *Astrophys. J.*, **398**, 69.
Worthey, G. (1994) *Astrophys. J. Suppl.*, **95**, 107.

Cosmic Star Formation Rate

Hopkins, A.M. (2004) *Astrophys. J.*, **615**, 209.
Hopkins, A.M. and Beacom, J.F. (2006) *Astrophys. J.*, **651**, 142.
Lilly, S. *et al.* (1996) *Astrophys. J.* **460**, L1.
Madau, P. *et al.* (1996) *Mon. Not. R. Astron. Soc.*, **283**, 1388.

SFR–$M_\star$ Relation

Daddi, E. *et al.* (2007) *Astrophys. J.*, **670**, 156.
Damen, M. *et al.* (2009) *Astrophys. J.*, **705**, 617.
Dunne, L. *et al.* (2009) *Mon. Not. R. Astron. Soc.*, **394**, 3.
Elbaz, D. *et al.* (2007) *Astron. Astrophys.*, **468**, 33.
Erb, D.K. *et al.* (2006) *Astrophys. J.*, **647**, 128.
Gonzalez, V. *et al.* (2010) *Astrophys. J.*, **713**, 115.
Karim, A. *et al.* (2011) *Astrophys. J.*, **730**, 61.
Maier, C. *et al.* (2009) *Astrophys. J.*, **694**, 1099.
McLure, R.J. (2011) arXiv:1102.4881.
Noeske, K.G. *et al.* (2007) *Astrophys. J.*, **660**, L43.
Pannella, M. *et al.* (2009) *Astrophys. J.*, **698**, L116.
Papovich, C. *et al.* (2011) *Mon. Not. R. Astron. Soc.*, **412**, 1123.
Rodighiero, G. *et al.* (2010) *Astron. Astrophys.*, **518**, L25.
Santini, P. *et al.* (2009) *Astron. Astrophys.*, **504**, 751.

Galaxy Mass Function

Baldry, I.K. *et al.* (2004) *Astrophys. J.*, **600**, 681.
Drory, N. *et al.* (2009) *Astrophys. J.*, **707**, 1595.
Fontana, A. *et al.* (2004) *Astron. Astrophys.*, **424**, 23.
Fontana, A. *et al.* (2006) *Astron. Astrophys.*, **459**, 745.
Ilbert, O. *et al.* (2010) *Astrophys. J.*, **709**, 644.
Kajisawa, M. *et al.* (2010) *Astrophys. J.*, **723**, 129.
Marchesini, D. *et al.* (2009) *Astrophys. J.*, **701**, 1765.
Pozzetti, L. *et al.* (2010) *Astron. Astrophys.*, **523**, A13.

Multicolor Selection Criteria for High Redshift Galaxies

Daddi, E. *et al.* (2004) *Astrophys. J.*, **617**, 747.
Steidel, C.C. *et al.* (2004) *Astrophys. J.*, **604**, 534.

7
Supernovae

Supernovae (SNe) are spectacular events in which the luminosity of a star varies dramatically in a short time. From an astrophysical point of view, SNe correspond to the death of stars, giving back to the interstellar medium a great fraction of the original mass transformed into heavy elements by the nuclear reactions. In doing so through energetic explosions they have a dramatic impact on the interstellar medium, eventually powering galactic winds. Therefore in the evolution of stellar populations SN events determine the pace and degree of chemical enrichment of the interstellar and intergalactic media. Modeling of their effect leads to predictions of the chemical trajectories of different elements which, when compared to the observed patterns in stellar systems, allow us to derive information on their star formation timescales. SNe come in two main classes, the traditional distinction being between Type II and Type I according to whether hydrogen lines are detected in the spectra (Type II) or not (Type I). The modern classification is more complicated, with the class of Type I being split into Type Ia, Ib and Ic, according to the presence of specific lines in the early spectra, SiII characterizing Type Ia but absent in Type Ib and Ic events; HeI being prominent for Ib, but less evident in Type Ic event. In the last years dedicated surveys and event follow up yielded detailed data extending from pre-maximum light up to the very late phases. As the dataset grows the SN taxonomy becomes richer and more complicated. However, there still are two main classes of SNe, the distinction being related to the explosion mechanism: core collapse SNe, which include Type II, Type Ib, and Type Ic, and thermonuclear SNe (Type Ia).

Being exceptionally bright, SNe can be detected up to cosmological distances, and can be used as distance indicators. Indeed, the light curves of Type Ia SNe have been instrumental to derive important constraints on cosmological models, having provided the first evidence of cosmic acceleration. Figure 7.1a,b, shows the B-band light curves for a sample of SNIa events, the brightest ones reaching a maximum light peak of $M_B \sim -20$, that is, comparable to the total luminosity of a giant galaxy. There is a substantial dispersion of peak magnitudes, but bright events have a slow decline rate, whereas faint events drop fast, with a reasonably tight correlation between the decline rate and the absolute magnitude at maximum. Thus, measuring the former (a distance independent quantity), the latter can be recovered. Strictly speaking, SNIa are not standard candles, but their light curve can be *standardized*,

Stellar Populations, First Edition. Laura Greggio and Alvio Renzini.
© 2011 WILEY-VCH Verlag GmbH & Co. KGaA. Published 2011 by WILEY-VCH Verlag GmbH & Co. KGaA.

and their use as distance indicator is preserved (see Figure 7.1b). In this chapter we address those features of SN events which are relevant to the evolution of stellar populations, namely their rate of occurrence and their chemical yields. We anticipate the definition of two crucial parameters: the number of SNe per unit mass from one stellar generation, or SN productivity (k_{SN}), and the distribution of the delay times (f_{SN} or DTD), the delay time (t_D) being the age of the progenitor star at explosion. These parameters control the rate at which SNe occur in a stellar population: at an epoch t since the beginning of star formation, the rate of SNe is obtained by summing the contribution of all past stellar generations, each weighted with the star formation rate (ψ) at the appropriate time:

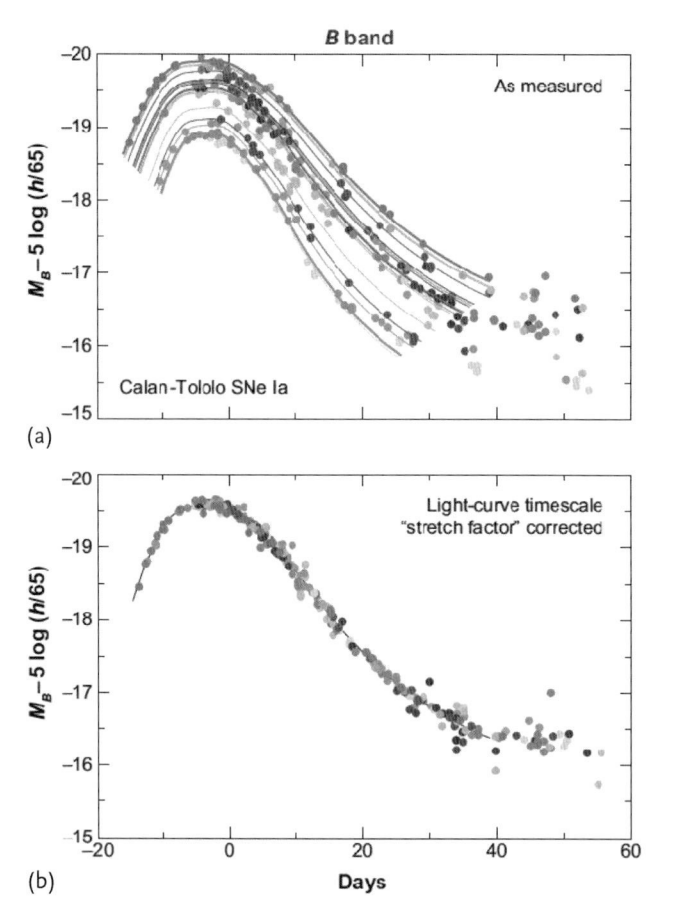

$$\dot{n}_{SN}(t) = \int_0^t \psi(t - t_D) k_{SN} f_{SN}(t_D) dt_D \,. \tag{7.1}$$

Figure 7.1 Light curves of SNIa events from the Calan–Tololo survey (a). (b) Shows the light curves of the same events as in (a), after applying the *stretch* correction (source: Frieman, J.A. et al. (2008, Annu. Rev. Astron. Astrophys., 46, 385)) (Reproduced with permission of Annual Reviews, Inc.). For a color version of this figure please see in the front matter.

Thus the productivity and the distribution of the delay times (DTD) control the pace at which the SNe products, as a metabolism of stellar populations, are released to the interstellar medium. For an instantaneous burst in which $\mathcal{M}$ solar masses of stars are formed, Eq. (7.1) becomes:

$$\dot{n}_{SN}(t) = \mathcal{M} k_{SN} f_{SN}(t_D = t), \qquad (7.2)$$

which shows that the SN rate following an instantaneous burst of star formation is proportional to the DTD through the product of the total mass of formed stars and the SN productivity.

7.1
Observed SN Rates

The measurement of SN rates in galaxies is not trivial, since the efficiency with which events are detected needs to be carefully evaluated. In a supernova survey, the number of detections in a given galaxy is

$$n_{SN} = R_{SN} \Delta t_{obs} \epsilon_{SN}, \qquad (7.3)$$

where R_{SN} is the SN rate in the specific galaxy, Δt_{obs} is the duration of the survey, and ϵ_{SN} is the detection probability, which depends on the SN light curve, the sampling strategy, extinction and other effects. The product $CT = \Delta t_{obs} \epsilon_{SN}$ is the time lapse within which the SN can be detected in the specific galaxy, that is, the *control time*. Since SN events are fairly rare (e.g., 1 every 10–100 yr per galaxy), one needs to monitor a large number of galaxies in order to collect a statistically significant number of events, especially if one wants to derive the rates of different kinds of SNe, and in different kinds of galaxies. For a galaxy sample the total number of events is obtained by summing the contribution of all individual galaxies:

$$n_{SN,TOT} = \sum_{g}(R_{SN} CT)_g \qquad (7.4)$$

where the subscript g emphasizes that both the rate and the control time of a given SN kind can be different in different galaxies. In order to evaluate R_{SN} one has to specify how the rate scales with the galaxy properties. Traditionally it has been adopted that the SN rate is proportional to the B-band luminosity of the parent galaxy, which leads to

$$n_{SN,TOT} = \frac{R_{SN}}{L_B} \sum_{g}(L_B CT)_g \qquad (7.5)$$

so that the rate is derived as

$$\frac{R_{SN}}{L_B} = \frac{n_{SN,TOT}}{\sum_{g}(L_B CT)_g}, \qquad (7.6)$$

Table 7.1 Observed SN rates. Rates in SNuK are normalized to the L_K luminosity of the parent galaxy; rates in SNuM are normalized to the mass of the parent galaxy. The latter is derived by adopting a relation between the parent galaxy $B - K$ color and its M/L_K ratio based on theoretical models; therefore they are not purely observational quantities. Errors refer to 1σ uncertainty (source: Mannucci, F., et al. (2005, Astron. Astrophys., 433, 807)).

Type	Ia	Ib/c	II	Ia	Ib/c	II
	Rate in SNuK			Rate in SNuM		
E/S0	$0.035^{+0.013}_{-0.011}$	< 0.007	< 0.1	$0.044^{+0.016}_{-0.014}$	< 0.009	< 0.013
S0a/b	$0.046^{+0.019}_{-0.017}$	$0.026^{+0.019}_{-0.013}$	$0.088^{+0.043}_{-0.039}$	$0.065^{+0.027}_{-0.025}$	$0.036^{+0.026}_{-0.018}$	$0.12^{+0.059}_{-0.054}$
Sbc/d	$0.088^{+0.035}_{-0.032}$	$0.067^{+0.041}_{-0.032}$	$0.40^{+0.17}_{-0.16}$	$0.17^{+0.068}_{-0.063}$	$0.12^{+0.074}_{-0.059}$	$0.74^{+0.31}_{-0.30}$
Irr	$0.33^{+0.18}_{-0.13}$	$0.21^{+0.26}_{-0.14}$	$0.70^{+0.57}_{-0.43}$	$0.77^{+0.42}_{-0.31}$	$0.54^{+0.66}_{-0.38}$	$1.7^{+1.4}_{-1.0}$

and measured in supernova units (SNu), corresponding to events per century per $10^{10}\, L_{B,\odot}$. Alternatively one may assume that the SN rate is proportional to the K-band luminosity, or to the galaxy mass. Correspondingly, the SN rates can be measured as:

$$\frac{R_{SN}}{L_K} = \frac{n_{SN,TOT}}{\sum_g (L_K C T)_g}, \quad (7.7)$$

that is, the rate in SNuK (events per century per $10^{10}\, L_{K,\odot}$), or as

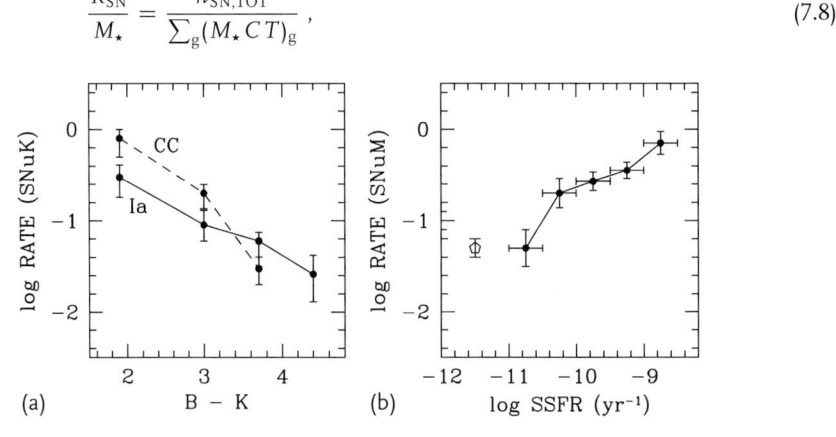

$$\frac{R_{SN}}{M_\star} = \frac{n_{SN,TOT}}{\sum_g (M_\star C T)_g}, \quad (7.8)$$

Figure 7.2 (a) SN rates per unit L_K as function of the $B - K$ color of the parent galaxy. (b) SNIa rates per unit mass as function of the sSFR. In this case the current SFR is an average over the last 0.5 Gyr. Notice that some modeling is necessary to derive both the rate per unit mass and the sSFR. The isolated pentagon, plotted at an arbitrary sSFR, shows the rate determined for the passive galaxies, that is, galaxies for which the recent SFR is below the detection threshold of the method used to evaluate it (data for (a) from Mannucci et al. (2005, Astron. Astrophys., 433, 807); for (b) from Pritchet, C.J. et al. (2008, Astrophys. J., 683, L25)).

that is the rate in SNuM (events per century per 10^{10} $M_\odot$). Table 7.1 shows the rates in SNuK and in SNuM as determined from a database constructed over a ~ 20 yr monitoring of a sample of $\sim 10\,000$ (nearby) galaxies. For all SN types, the rates per unit L_K appear to increase going from early- to late-type galaxies. This trend indicates a higher efficiency of SN production in younger stellar populations. There are different ways to quantify this effect, as shown in Figure 7.2a,b. Figure 7.2a shows the SN rates per unit L_K (for both CC and Ia events) as a function of the parent galaxy $B - K$ color; Figure 7.2b shows the SNIa rate per unit mass as a function of the specific star formation rate (sSFR, that is, the ratio between the recent SFR and the total stellar mass of the parent galaxy). Bluer $B - K$ colors and/or higher sSFR characterize younger stellar populations, which appear to have larger SN rates per unit luminosity and/or mass. The correlations are naturally explained in the framework of stellar evolution theory, as we show next.

7.2
Core Collapse SNe

Core collapse (CC) SNe exhibit a large variety of light curves and spectra. As already mentioned, this class includes Type Ib and Ic (no hydrogen lines in the early spectra) and Type II SNe, which are distinguished in four classes (SNIIP, SNIIL, SNIIn and SNIIb), according to the shape of the light curve and spectral characteristics. There's no doubt that (most) CC SNe correspond to the death of massive stars which evolve all the way up to the formation of a massive iron core, whose collapse is triggered by the photodissociation of the iron nuclei. The collapse of the core and its *bounce* at the neutron star formation are responsible for initiating the explosion of the outer layers, which include a nuclearly processed mantle and any hydrogen-rich envelope surviving the previous evolution. The successful ejection of these layers is secured by the absorption of a tiny fraction (~ 0.01) of the energy flux of neutrinos being released by the collapsed core. Other mechanisms (e. g., electron capture in a collapsing ONeMg core, or pair instability SNe in very massive objects) may be responsible for a few events, respectively at the lower and upper mass range of SNII progenitors. A CC SN leaves a compact remnant, which may be a neutron star or a black hole depending on whether the remnant mass is lower or higher than about 2 $M_\odot$.

As discussed in Section 1.2.2, during their quasi-static evolution, massive stars suffer different degrees of mass loss, so that the pre-SN configuration ranges from a WR star (with no hydrogen envelope) to a red supergiant, with a massive hydrogen envelope. The plateaux (i. e., a phase at constant luminosity) in the light curve of SNIIP traces the presence of a massive hydrogen-rich envelope at the time of explosion; at the other extreme, SNIb and Ic arise from bare cores explosions; the other SNII kinds likely originate from intermediate configurations. It is worth noting that the occurrence of SNIIn, showing narrow hydrogen emission lines due to interaction between the ejecta and the circumstellar material, strongly supports the notion that some CC SNe suffer a high mass loss prior to explosion. The relative fraction of Type II to Type I b/c SNe (empirically estimated) is 0.75/0.25. Such a high fraction of SNIb/c events requires a relatively low lower limit to the mass of single stars WR progenitors: for example assuming a Salpeter IMF and CC SNe progenitors with mass between 8 and 120 $M_\odot$, stars more massive than $\sim 21\ M_\odot$ should become WR stars. This requires a very high mass loss rate for single stars; alternatively, a sizeable fraction of SNIb/c could originate in close binary systems, where the envelope of the SN progenitor is removed during Roche-lobe overflow. At any rate, Ib/c events are closely associated with star-forming regions, likely more than SNII; this suggests that their progenitors are more massive than those of Type II.

7.2.1
Theoretical Rates

Since CC supernovae are the end-product of the evolution of massive stars, the distribution of the delay times can be easily derived given the relation between the mass of the progenitor and its lifetime. The number of events with delay time between t_D and $t_D + dt_D$ is equal to the number of stars with mass between M_D and $M_D + dM_D$, such that the evolutionary lifetime of M_D equals t_D:

$$\dot{n}_{CC}(t_D)|dt_D| = \phi(M_D)|dM_D|. \tag{7.9}$$

Indicating with $M_{CC,n}$ and $M_{CC,x}$ respectively the minimum and maximum stellar mass which give rise to a CC SN event, Eq. (7.9) yields:

$$f_{CC}(t_D) = \frac{\phi(M_D)|\dot{M}_D|}{\int_{M_{CC,n}}^{M_{CC,x}} \phi(M) dM}, \tag{7.10}$$

which shows that the DTD for CC SNe is proportional to the death rate of stars with appropriate mass. Since by definition

$$k_{CC} = \frac{\int_{M_{CC,n}}^{M_{CC,x}} \phi(M) dM}{\int_{M_i}^{M_s} \phi(M) M dM}, \tag{7.11}$$

where the integral at the denominator extends over the total mass range of the stellar population, Eq. (7.1) becomes:

$$\dot{n}_{CC}(t) = \int_0^t \psi(t - t_D) \phi(M_D) |\dot{M}_D| dt_D, \tag{7.12}$$

with $\phi(M)$ normalized to the total mass of the stellar population. The evolutionary lifetimes of CC SN progenitors are short (less than ~ 40 Myr), so that one can approximate the SFR term in Eq. (7.12) with the current rate $\psi_0 = \psi(t)$. Applying this approximation, and changing the integration variable from delay time to mass of the dying star we get

$$\dot{n}_{CC}(t) = \psi_0 \int_{M_D(t)}^{M_{CC,x}} \phi(M_D) dM_D, \tag{7.13}$$

which shows that the CC SN rate is proportional to the ongoing SFR, through a factor which increases with time as the mass of the dying star decreases. Figure 7.3a shows the temporal behavior of the rate of CC SNe per unit SFR for a stellar population formed at a constant rate, having adopted that the progenitors have masses between 8 and 120 $M_\odot$. Different (diet) IMFs are explored, all with a slope $(1 + x) = 1.3$ between 0.1 and 0.5 $M_\odot$, but with different slopes $(1 + x = 2.35, 2$ and $2.6)$ in the range between 0.5 and 120 $M_\odot$. The adopted relation between the progenitor mass and its lifetime results from a fit to YZVAR models smoothly merged with Limongi, M. and Chieffi, A. (2006, Astrophys. J., 647, 483) models, all for solar composition, and is:

$$\log M_D = \begin{cases} 0.9455(\log t_D)^2 - 14.081 \log t_D + 53.4894 & \text{for } \log t_D \leq 7.17 \\ -0.5229 \log t_D + 4.8838 & \text{for } \log t_D \geq 7.17 \end{cases}, \quad (7.14)$$

where mass is in solar units and delay time in years.

Soon after the start of star formation, the most massive progenitors start to explode; as time progresses the rate increases since stars with lower mass contribute to the death rate, up to a maximum, when even the least massive progenitor start to explode. After this epoch the CC rate remains constant, all possible progenitor masses being included in Eq. (7.13). The rate depends on the IMF: flatter IMFs (in the relevant mass range) provide a higher CC SNe rate; for a Salpeter-diet IMF, in a galaxy forming stars at a constant rate of 1 $M_\odot$/yr, after the first $\sim$ 40 Myr from the beginning of SF one gets 1 CC event per century. For a universal IMF we thus expect that the CC SN rate is basically proportional to the current SFR in all galax-

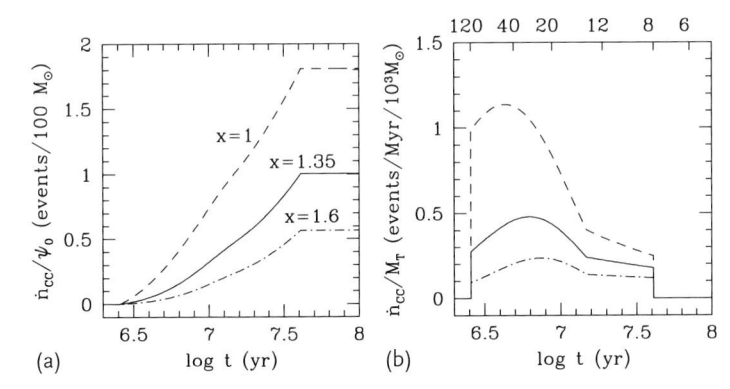

Figure 7.3 Time evolution of the rate of CC supernovae for a constant SFR (a) and for an instantaneous burst (b). The rate in (a) is normalized to the value of the SFR; the rate in (b) is normalized to the total stellar mass (in $M_\odot$) formed in the burst. The different linetypes refer to different slopes of the IMF ($\phi(M) \propto M^{-(1+x)}$) in stars more massive than 0.5 $M_\odot$, as labeled. The mass of CC SNe progenitors is adopted to range from 8 to 120 $M_\odot$. The upper axis in (b) is labeled with the value of the mass, in $M_\odot$, of the star exploding at the corresponding epoch in the lower axis.

ies which have been forming stars since ~ 40 Myr ago. In this case, the correlation shown in Figure 7.2a,b reflects the higher specific star formation rate in the bluer galaxies, as long as L_K traces the total galaxy stellar mass.

Figure 7.3b shows the time evolution of the CC rate past an instantaneous burst of SF, as given by

$$\dot{n}_{CC}(t) = \mathcal{M}\phi(M_D)|\dot{M}_D|, \tag{7.15}$$

i.e., the rate is equal to the death rate of the SSP, and $\mathcal{M}$ is the mass of the starburst. As before, the first events occur at a delay time equal to the lifetime of the most massive CC progenitors; then the rate is determined by the interplay between the derivative of the mass of the dying star, which decreases with time, and the IMF term, which provides more and more stars as the stellar population grows older. The result is an early peak in the death rate, which occurs a little later for steeper IMFs. At a delay time equal to the lifetime of the least massive progenitor the rate goes to zero. Also in this case flatter IMFs correspond to higher rates; for a Salpeter slope, a burst forming 10^7 $M_\odot$ of stars produces between 0.5 and 0.17 events century, depending on its age.

The productivity of CC SNe, depends not only on the IMF slope, but also on the mass range of the progenitors. Figure 7.4 illustrates this dependence as a function of the IMF slope. For each slope the upper and lower curves are obtained with

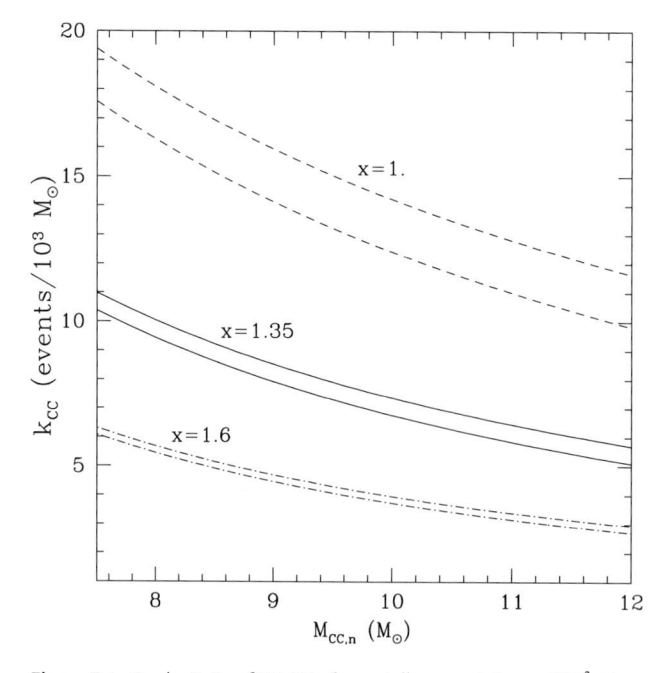

Figure 7.4 Productivity of CC SNe from stellar populations of 10^3 $M_\odot$ as a function of the lower mass limit of the progenitors. The linetype encodes the IMF slope as in Figure 7.3a,b; uppermost (lowermost) lines refer to $M_{CC,x} = 120$ (50) $M_\odot$.

$M_{CC,x} = 120$ and $50\,M_\odot$ respectively, showing that the productivity is quite insensitive to the upper mass limit for $M_{CC,x} \gtrsim 50\,M_\odot$, unless the IMF is particularly flat. Thus, if the most massive stars, undergoing core collapse do not produce a detectable SNII event, for example because of sizable fall back of the envelope, the CC SNe statistics are barely affected. More important is the dependence on the lower cut off mass, the productivity decreasing by a factor of $\simeq 0.7$ if $M_{CC,n}$ goes from 8 to $10\,M_\odot$ for a Salpeter diet IMF. As discussed in Chapter 1, the fate of stars in this mass range is uncertain because if mass loss is strong enough, they end their evolution as ONeMg WDs without reaching the Chandrasekhar mass, thereby avoiding core collapse. Given the sensitivity of the CC SNe productivity on $M_{CC,n}$, indications on this limit can be derived from empirical k_{CC} values. For a stellar population in which SF has been active over the last 40 Myr, Eq. (7.13) can be written as $\dot{n}_{cc} = \psi_0 \times k_{CC}$; thus the productivity can be estimated as the ratio between the rate of CC events and the recent SFR. As discussed in Chapter 6, observational values of SFR are constrained by the light produced by massive stars; the ratio $\dot{n}_{CC}/\psi_0$ traces the fraction of them which end up as CC SNe, sensitive to $M_{CC,n}$. The ratio between the cosmic CC SN rate and the cosmic SFR derived with a diet Salpeter IMF suggests $k_{CC} \sim 5 \times 10^{-3}\,M_\odot^{-1}$, which, for the same IMF, indicates $M_{CC,n} \sim 12\,M_\odot$. However, the detection of individual progenitors in pre-explosion images suggests $M_{CC,n} = 8 \pm 1\,M_\odot$. On the other hand the uncertainties on both the cosmic SFR and the cosmic CC rate could easily account for a discrepancy of a factor of 2 on k_{CC}. Current work on the comparison of the CC rate and the current SFR in a sample of nearby galaxies is consistent with a lower limit of $\sim 8\,M_\odot$ for the SNII progenitors.

7.2.2
Nucleosynthetic Yields

Massive stars provide most of the metal enrichment in the Universe, although some specific elements are mostly manufactured by SNIa or intermediate-mass stars. The metal yield from massive stars results from the sum of two contributions: the products of the nucleosynthetic burning during the quasi-static evolution and of that from explosive nucleosynthesis at the time of the SN event. The former contribution can be released to the interstellar medium already during the star's evolution, through stellar winds; if not, it is ejected at explosion. The latter results from the outward propagation of the shock wave generated by the core *bounce*, heating up the star's layers thereby promoting further nuclear processing. The explosive nucleosynthesis is confined to the inner portions of the star.

Due to intrinsic difficulties to describe the physical processes which occur when a massive star dies (most notably the interaction between neutrinos and the dense mantle) there are still no self-consistent hydrodynamical models which follow the evolution through core collapse and ensuing explosion. The explosive nucleosynthesis is then computed by assuming some energy deposition in a deep layer inside the star, and following the evolution with a sophisticated hydrodynamical code which includes a very large reaction network. The way in which the energy deposition is simulated, its magnitude and its location inside the star are all parameters which have an impact on the results. In general, the energy is released just inside the Fe core, but not all the mass out of this location will be able to escape in the explosion, as the innermost material falls back and remains locked in the remnant. The *mass cut*, that is, the mass coordinate which separates the remnant from the ejecta, is a fundamental parameter for the nucleosynthetic production from core collapse supernovae, and it is highly dependent on the modeling.

The difficulty of predicting the outcome of the core collapse of a massive star, in particular whether the stellar mantle and envelope are ejected or fall back, and where precisely the mass cut is located, can be grasped by considering a few numbers. The binding energy of the newly formed neutron star is $\sim 10^{53}$ erg, which are released within ~ 1 s in the form of neutrinos. Part of these neutrinos are absorbed in the inner part of the mantle, and this energy and momentum deposition causes the ejection. The observed energy release of the SN, that is, the kinetic energy of the ejecta, is $\sim 10^{51}$ erg, and therefore the absorption of just $\sim 1\%$ of the neutrinos is sufficient to power the supernova. For a proper theoretical prediction, the hydrodynamical calculation should be able to determine the energy transfer from the neutrino flux to the collapsing matter with an accuracy of a fraction of a percent, which may explain why some calculations fail to produce a supernova, whereas other succeed. Therefore, the yield of elements produced close to the *mass cut* are particularly uncertain, ^{56}Fe being one of them. The yield of lighter elements located in some outer zones is also relatively model dependent, being sensitive to the development of various semiconvective regions and convective shells during the presupernova evolution.

On top of this, mass loss systematically affects the yield from massive stars in various ways: (i) by directly subtracting fuel to nuclear burning, (ii) by reducing the mass of the inner core where more advanced processing occurs, and (iii) by affecting the mass of the remnant, where the products of the most advanced burning sink. The effect of mass loss during the WR stage is particularly important: not only does it alter the mass of the CO core, but also the final C/O ratio, and thus the ensuing nucleosynthesis. Other processes are relevant to the nucleosynthetic yields, like overshooting and rotation, which result in larger helium cores at given initial stellar mass, mimicking the behavior of a star with a larger initial mass.

To summarize, the yields from massive stars are subject to uncertainties that we shall try to quantify, and depend on the presupernova evolution, as well as on the physics at explosion. They depend on the initial metallicity of the star as well, through its effects on the stellar structure and on the mass loss rate. The theoretical uncertainty in the nucleosynthetic output of CC supernovae can be gauged by comparing the predictions between different sets of models in the literature. We choose here to illustrate the results of three sets of solar metallicity models (labeled as WW, LC and KUNTO), restraining to those elements which are mostly used in stellar population studies: O, Mg, Si and Fe (Figure 7.5a,b). The WW and LC sets are computed with similar prescriptions for the explosion, that is, by requiring a given amount of kinetic energy at infinity (of $\sim 10^{51}$ ergs for LC and the WW models shown as dotted lines). The two sets differ for the presupernova evolution, since LC models take mass loss into account throughout the whole stellar life, while WW

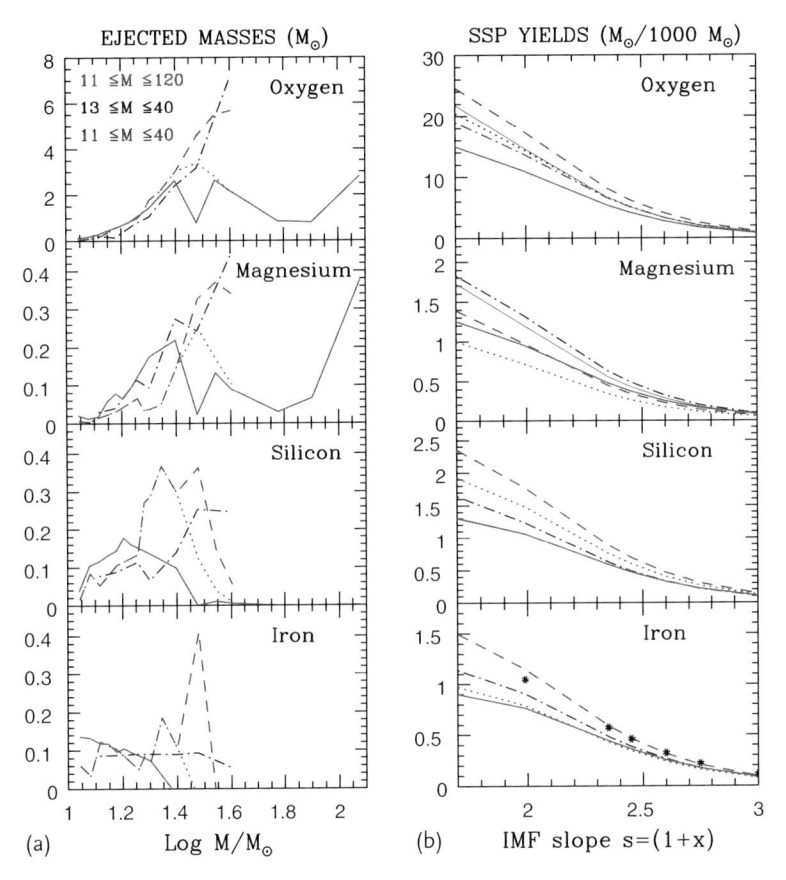

Figure 7.5 Stellar (a) and SSP (b) yields from CC SNe from four sets of solar composition literature models: LC (solid), KUNTO (dot-dashed), and WW (dotted and dashed, the latter showing the effect of enhancing the explosion energy for stars with $M \geq 30\,M_\odot$) (the acronyms refer to the references given at the end of this caption). Notice that KUNTO models range from 13 to 40 $M_\odot$; WW models from 11 to 40 $M_\odot$, while LC models range from 11 to 120 $M_\odot$. The ejected masses and the SSP yields refer to the newly synthesized material, that is, the original amount of the element present in the ejecta has been subtracted. The SSP yields, computed for a family of *diet* IMFs, adopt an upper limit of 40 $M_\odot$ (thick lines) or 120 $M_\odot$ (thin line, relative to LC models). Black dots in (b) for iron show a semiempirical estimate of the Fe yield, assuming that CC progenitors range from 8 to 120 $M_\odot$, and that each event provides 0.057 $M_\odot$ of Fe, which is the average yield of the events plotted in Figure 7.6 (stellar yields are from Limongi, M., and Chieffi, A. (2006, *Astrophys. J.*, 647, 483, LC); Kobayashi, C., et al. (2006, *Astrophys. J.*, 653,1145, KUNTO); Woosley, S.E., and Weaver, T.A. (1995, *Astrophys. J. Suppl.* 101, 181, WW)). For a color version of this figure please see in the front matter.

models do not. KUNTO models also take mass loss into account, but the explosion is modeled forcing that each supernova ejects 0.07 $M_\odot$ of ^{56}Fe, virtually imposing the *mass cut*.

At given initial stellar mass, sizable variations exist among the model predictions which can only in part be ascribed to these different prescriptions. Other physical effects taking place during the pre-SN evolution are important to determine the outcome. In this context it is instructive to mention the reason for the minimum in oxygen and magnesium ejected mass at 30 $M_\odot$ in the LC dataset: for this particular model, the carbon burning convective shell extinguishes before the collapse begins, promoting a core contraction which produces a particularly compact core. As a result, the fall back for this model is more pronounced than for the adjacent masses, resulting in a local minimum of the yields. The occurrence of convective shells and their partial overlap throughout the evolution is very sensitive to the input physics and the way in which convection is modeled; at the same time these phenomena have important effects on the final results. Therefore, the nonuniform behavior of the ejected masses, as well as the differences between the various authors, are not surprising.

In the economy of the chemical evolution of a stellar population what matters are the SSP yields, as all these oddities are washed out when convolving the stellar yields with the IMF. These are shown in Figure 7.5b as functions of the slope of the IMF, as they result from the integration of the ejected masses shown in Figure 7.5a over the IMF, normalized to a total mass of 1000 $M_\odot$ of stars with mass between 0.1 and 120 $M_\odot$. No extrapolation of the ejected masses has been implemented to compute the SSP yields, which then lack the contribution from CC SNe less massive than 11 $M_\odot$, for the LC and WW sets, and 13 $M_\odot$ for the KUNTO set. This contribution is likely small because these stars provide little amounts of newly synthesized material. All the thick lines in Figure 7.5b, lack the contribution from stars with mass between 40 and 120 $M_\odot$, while the thin line shows the yields from the whole LC dataset, which extends up to 120 $M_\odot$.

For steep IMFs, the model predictions are very close to each other, reflecting the agreement of ejected masses at the low-mass end of CC SNe progenitors. For a flatter IMF the yields are more sensitive to the contributions from massive CC SNe, where the four sets predict very different ejected masses, and thereby SSP yields. The upper mass cut off turns out to be relevant for the oxygen and magnesium yields, especially for flat IMFs. All in all, however, the SSP yields do not appear dramatically different among the various sets, and over the explored parameter space. For a Salpeter diet IMF, the differences are at most $\sim$(40, 48, 41 and 32)%, respectively, for the SSP yields of oxygen, magnesium, silicon and iron.

As mentioned previously, the theoretical yield of iron is very uncertain: most of the iron originates from radioactive ^{56}Ni, which decays to ^{56}Co (with a half-life of 6.1 days), which in turn decays to ^{56}Fe (with a half-life of 78.7 days), and ^{56}Ni is synthesized in layers very close to the *mass cut*. On the other hand the mass of ^{56}Ni ejected in a CC SN event can be measured empirically from the light curve at late phases, which are powered by the radioactive decay of ^{56}Co. The mass of the progenitor star can also be evaluated, for example by detecting it on pre-explosion images and comparing its position on a color-magnitude diagram to evolutionary tracks. Alternatively, the ejected mass can be derived by comparing the evolution of the light curve and spectral features to corresponding models; this allows an estimate of the progenitor mass, having made a hypothesis on the remnant mass and the mass loss suffered during the presupernova evolution. At present, though, the available datasets are insufficient to meaningfully compare observations to the model predictions shown in Figure 7.5a. Nevertheless we can draw some information by considering the distribution of measured ^{56}Ni ejected masses. Figure 7.6 shows such distribution for a collection of data on SNIIP in the literature with no particular selection criteria applied. It appears that most events provide a small amount of ejected nickel; the famous SN 1987A occurred in the LMC (and observed extremely well) released 0.075 $M_\odot$ of nickel and would be a rather efficient Fe producer. One event (SN 1992am) released a particularly high nickel mass; some other events with a very high ejected nickel mass can be found in the literature, but usually regard Type Ib or Ic SNe, which come from high-mass progenitors. The general impression is that most CC SNe come from stars at the lower boundary of the mass range, and provide little Fe to the interstellar medium; some very productive events do occur, but they seem rare. All in all, the best current estimate of the Fe yield from CC SNe can be derived by adopting the average ^{56}Ni ejected mass per event, multiplied by the productivity of CC SNe of a stellar population. This semiempirical evaluation is shown in the last plot in Figure 7.5b as black dots, which reassuringly happen to be very close to the theoretical yields.

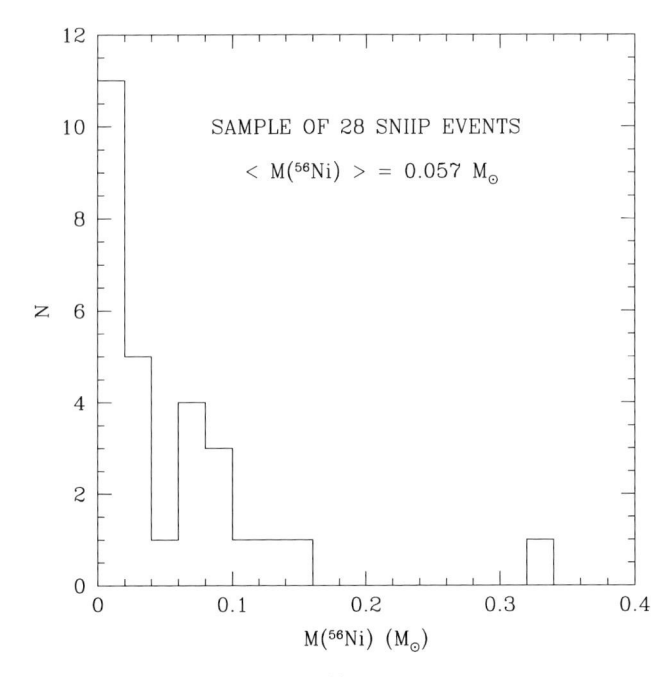

Figure 7.6 Distribution of ejected ^{56}Ni masses in a sample of Type IIP SN events. The masses have been measured from the light curve in the nebular phases (ejected ^{56}Ni masses are from Hamuy, M. (2003, *Astrophys. J.*, 582, 905) and Zampieri, L. (2008, *AIPC*, 924, p. 358); for the five events in common the average between the two measurements has been adopted).

7.3
Thermonuclear Supernovae

Ignition of nuclear fuel under highly degenerate conditions leads to a thermonuclear runaway, because the gas does not expand in response to the local energy generation. The local temperature then increases, further promoting nuclear burning, which may lead to explosion. Type Ia SNe are believed to arise from this kind of explosion, as stellar evolution naturally provides ideal candidates for these events, that is, WDs which manage to reach ignition conditions because their mass grows due to accretion from a companion. The most relevant observational facts which support this model are:

- Hydrogen lines are not detected in the SNIa spectra, excluding progenitors which retain their hydrogen-rich envelope until explosion, that is, red supergiants, blue supergiants and AGB stars.
- SNIa occur in all galaxy types, including ellipticals, and are not exclusively associated with young stellar populations. This implies that the delay time of their progenitor varies within a wide range. WDs are the evolutionary end point of stars with $M \lesssim 8\,M_\odot$, whose evolutionary lifetimes range from ~ 40 Myr up to a Hubble time.
- Light curves and spectra of SNIa are remarkably similar, especially compared to other SN kinds. WDs have a characteristic structure which easily accounts for the uniformity of SNIa events.
- The light curve of SNIa is well explained as powered by the radioactive decay of ^{56}Ni to ^{56}Co and then of ^{56}Co to ^{56}Fe. Ignition of helium or of carbon under highly degenerate conditions leads to the production of ^{56}Ni in high quantity, when the nucleosynthesis proceeds to full incineration and iron group elements are the ultimate burning *ashes*.

Three kinds of WDs can be produced by stellar evolution: oxygen-neon-magnesium (ONeMg) WDs ($8\,M_\odot \lesssim M \lesssim 11\,M_\odot$), carbon-oxygen (CO) WDs ($M \lesssim 8\,M_\odot$), and helium WDs ($M \lesssim 2\,M_\odot$), if mass loss removes the hydrogen-rich envelope of a low-mass star while climbing the RGB. While the mass of a ONeMg WD grows towards ignition conditions, electron capture reactions on magnesium and neon subtract pressure support and the WD collapses (accretion induced collapse (AIC)), until oxygen deflagrates. The result is an electron capture SN, which leaves a neutron star remnant. The light curves and spectra of this kind of explosion may account for some peculiar SNIa, but not for the bulk of the events, because of their low energetic and ^{56}Ni output. In addition, the narrow mass range producing ONeMg WD makes these candidates statistically unfavored. CO and helium WDs instead are much more abundant, and account for the energetics of typical SNIa events. Ignition of carbon in a CO WD can take place when the mass reaches the Chandrasekhar limit, of $\sim 1.4\,M_\odot$. If the WD is made of 50% carbon and 50% oxygen, and if the total mass is incinerated to ^{56}Ni, the energy produced in the explosion is $\sim 2.2\times 10^{51}$ ergs, largely exceeding the binding energy of the WD ($\sim 5.1\times 10^{50}$ ergs).

Since the radiative losses are negligible, $\sim 1.7 \times 10^{51}$ ergs are available to accelerate the ejected material, fully adequate to explain the observed velocities of the ejecta ($\sim 11\,000$ km/s). A similar amount of nuclear energy is produced by the full incineration to nickel of a 0.8 $M_\odot$ helium WD. However, the explosion of a helium WD is unlikely in nature, because when a sufficiently massive ($\sim$ few 0.1 $M_\odot$) helium layer has accumulated on top of the WD, a helium shell flash occurs which rapidly quenches. A succession of shell flashes follows, progressively deeper inside the star, until helium is ignited in the center. Similar to what happens at the helium flash in single stars, the WD gently expands, degeneracy is lifted and the helium WD is rejuvenated as a core helium burning, helium star. Therefore, the accepted paradigm for SNIa precursors consists in CO WDs, which manage to explode either because the mass of the CO core reaches the Chandrasekhar limit (Chandra explosions), or because the accreted helium layer ignites in a detonation runaway, which causes the implosion of the CO core until the core itself detonates (double detonation). In the latter case explosion occurs before the total mass of the WD has reached the Chandrasekhar limit (sub-Chandra explosions).

The explosion of a Chandra CO WD accounts very well for the light curves, nucleosynthetic yield, and spectra of the bulk of SNIa events, for which reason it is considered the most successful model. Sub-Chandra explosions perform worse, especially because helium detonation provides ^{56}Ni in the outer layers, so that high velocity lines from iron group elements should be observed, which is not the case. However, from an evolutionary point of view, compared to Chandra events, sub-Chandra explosions should be more easily realized, and it is embarrassing that these events have no obvious observational counterpart yet.

7.3.1
Evolutionary Scenarios for SNIa Progenitors

Figure 7.7 illustrates the standard evolutionary scenarios for SNIa progenitors. It starts with a close binary system in which the primary is an intermediate-mass star, so that the first Roche lobe overflow leaves a CO WD. Low-mass stars ($M \lesssim 2\,M_\odot$) do produce CO WDs, but in a close binary they can do so only if the first mass exchange takes place after the helium flash. Since the stellar radii at helium flash are large (hundreds of solar radii), and since the distribution of the original separations A_0 scales as $1/A_0$, in most cases low-mass primaries in close binary systems will fill their Roche lobe (RL) while on the first red giant branch, thereby producing helium, rather than CO, WDs. Therefore, reasonable lower and upper limits to the primary mass (M_1) in SN Ia progenitor systems are 2 and 8 $M_\odot$. When the secondary fills its RL there are two possibilities: either the CO WD accretes and burns the hydrogen-rich material remaining compact inside its RL, or a common envelope (CE) forms around the two stars. In the first case, the WD grows in mass and when conditions are met, a SN Ia explosion occurs; in the second case, due to the action of a frictional drag force, the CE is eventually expelled, while the binary system shrinks. If the secondary is more massive than $\sim 2\,M_\odot$, the outcome of the CE is a CO WD + helium star system. Further evolution causes the helium

EVOLUTIONARY PATHS

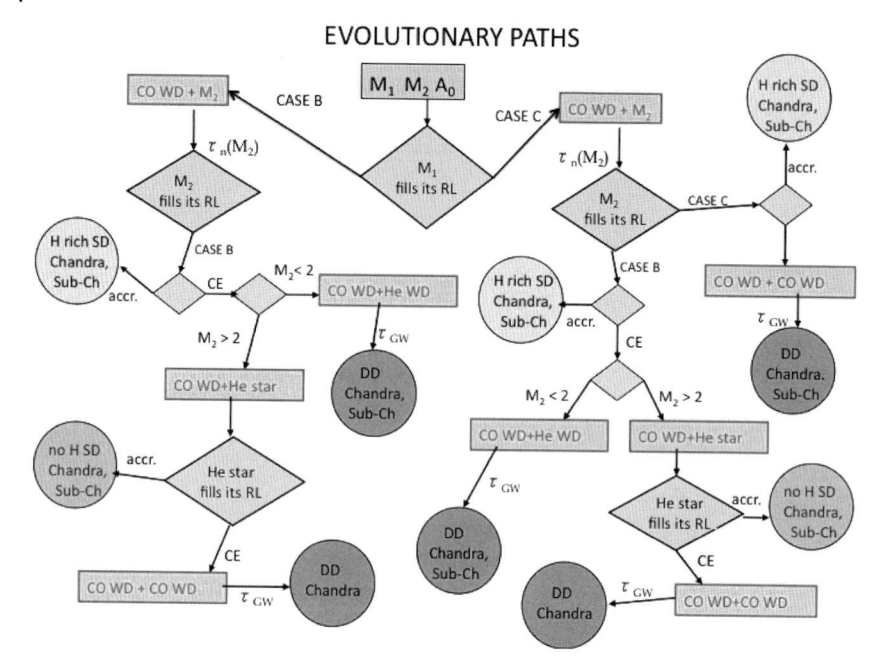

Figure 7.7 Evolutionary paths for close binaries with intermediate-mass components, leading to SNIa explosions. The RL overflow can occur when the expanding star is in the shell hydrogen burning phase (Case B), or in the double shell burning phase (Case C). See text for more details. For a color version of this figure please see in the front matter.

star to expand and fill its RL, pouring helium-rich material on the companion, and the CO WD has again the possibility of accreting and growing up to explosion conditions. Both these paths to the SNIa event are single degenerate channels (SD, that is, the CO WD accretes from a nondegenerate companion), but in the first one (hydrogen-rich SD channel) some hydrogen is expected in the vicinity of the SNIa which may show up in the spectrum, while in the second one (helium star channel) no hydrogen-rich material should be around. If the CO WD does not accrete the donated helium, a close CO+CO WD system may be produced. Angular momentum losses via gravitational wave radiation will eventually lead the two WDs to merge, and, if the total mass exceeds the Chandrasekhar limit, explosion may arise. This is the double degenerate channel (DD, that is, the CO WD accretes from a degenerate companion) to SNIa, as provided by intermediate mass binaries. A close CO + CO WD system is also the outcome of a CE phase occurring when the secondary star fills its Roche lobe during its AGB phase.

Notice that a DD system may also be formed when the secondary is a low-mass star, which fills its RL when its helium core is degenerate. In this case, the CE phase leaves a close binary with one CO and one helium WD, but, typically, the total mass of the DD system does not reach the Chandrasekhar limit. Therefore this channel provides mainly sub-Chandra events; while potentially important for the SNIa, especially at late delay times, this path cannot account for the bulk of SNIa events, for the reasons mentioned above.

In Figure 7.7 it is assumed that the RL overflow of the secondary is either a Case B (the donor is in the shell hydrogen burning phase) or a Case C (the donor is climbing the AGB), but there is also the possibility that the secondary fills its RL while still on the MS (Case A RL overflow). This channel, which may provide SNIa events, has been neglected in Figure 7.7 to avoid overcrowding the diagram. The outcome of a Case A RL overflow can be a nova, if the accretion rate is lower than $\sim 10^{-7}$ $M_\odot$/yr, a SNIa, if the accretion rate is larger than this and the CO WD manages to accrete and burn enough material to reach the Chandrasekhar limit, or a CO WD + helium star system, if the explosion is avoided. Thus, the possible products of a Case A RL overflow are included in Figure 7.7; however, the delay time for the SNIa in this channel can be shorter than the time it takes for the secondary to evolve off the MS ($\tau_n(M_2)$). This needs to be taken into account when discussing the DTD.

In summary, there are two main channels through which a CO WD can evolve to explosive conditions, the SD and the DD, and pros and cons can be found for both channels. From a theoretical point of view, the realization of explosion conditions of the H-rich SD channel requires that the accretion rate is in a narrow range around $\sim 10^{-7}$ $M_\odot$/yr, to avoid nova explosions (for lower accretion rates), or the formation of a CE (for higher accretion rates). Interestingly, in the latter case a strong radiative wind from the WD may develop, with the result of stabilizing the mass transfer, thus allowing the CO WD mass to grow. Part of the donated material, though, is lost in the wind, so that only binaries with relatively massive components can secure a Chandra explosion. On the other hand, according to theoretical models the merging of two CO WDs could result into an AIC, rather than a SNIa explosion.

From an observational point of view, the detection of circumstellar material for some events (e. g., SN 2006X) favors the single degenerate model, while the lack of it in most events is more easily understood within the double degenerate model. A few events have been reported to require a super-Chandrasekhar WD mass: these are more easily explained as DD mergers. Meanwhile, single degenerate systems containing a WD with mass close to the Chandrasekhar limit have been discovered, which are very good candidate SNIa precursors (e. g., RSOphi). Statistics of potential progenitors for the two models is also not conclusive. On the one hand there are many types of interacting binaries containing a WD plus a nondegenerate star, like cataclysmic binaries, symbiotic systems and supersoft X-ray sources. On the other hand, the inventory of close double degenerates did yield one system with total mass larger than the Chandrasekhar limit and close enough to merge within a Hubble time, showing that potential SNIa progenitors can be found among DDs as well. These evidences suggest that in Nature both channels may well provide SNIa events with comparable efficiency.

7.3.2
The Distribution of Delay Times

One way to derive a theoretical DTD consists in running Monte Carlo simulations in which a large set of binary systems, each identified with the initial masses of the components and separation (M_1, M_2, A_0), are followed during their evolution to determine the final outcome. This is realized with binary population synthesis (BPS) codes, which, given the distribution of the initial binary parameters, determine the partition of the population in the various channels as well as the delay time of each SNIa event, hence the whole DTD. BPS codes need specification of many items in order to follow the evolution of each system. For example, prescriptions are needed to decide whether at RL overflow a CE occurs, and, in case it does, which fraction of the orbital energy is lost, and the ensuing orbital shrinkage. If a CE phase is not realized, prescriptions should fix how much mass is accreted by the companion; when the accretor is a WD, how the nuclear processing proceeds in the external layers, its radius evolution, and its mass loss. When explosion conditions are met, one needs to specify the ultimate fate of the object, for example whether a nova or a supernova, or if ignition leads to an accretion induced collapse, rather than to an explosion, a likely possibility when carbon ignition occurs off-center. In addition, the realization of these various possibilities in the BPS flow will depend on how some critical quantities are defined, that is, the size of the RL, orbit syncronization, magnetic braking, and ultimately the evolution of the individual components, influenced by convection, mass loss and overshooting. Many of the recipes needed to describe these processes are uncertain, undermining the merit of the BPS codes results. As an alternative, the shape of the DTD can be characterized on general arguments, examining the clock of the explosion and the range of masses and separations which can provide successful events, as is presented in the next sections. In this approach we move from describing the complexities of close binary evolution in favor of a more transparent derivation of the DTD, which relies on a few, robust concepts. The treatment may be oversimplified, but given the very different results from BPS codes in the literature, there's no firm consensus on the numerical DTDs. In addition, the analytic approach provides parametric functions which can easily be used to compute rates and chemical evolution models, thereby gaining constraints from the corresponding data.

7.3.3
The SD Channel

The various paths in Figure 7.7 are labeled with the evolutionary lifetimes of the progenitors at specific points. It can be noticed that for SD explosions the clock is very close to the time it takes for the secondary star to evolve off the MS ($\tau_n(M_2)$). In fact, for the H-rich SD explosion the extra time spent to reach RL dimensions, and the accretion timescale during which the CO WD grows to ignition conditions, are both much smaller than $\tau_n(M_2)$. For the helium star channel, some correction (of about 10%) should be applied to account for the He burning lifetime of the former secondary; in practice, however, $\tau_n(M_2)$ represents fairly well the total delay time also in this case. It should be noticed that if the first RL overflow is conservative, the secondary star may grow in mass, speeding up its evolution. According to literature BPS renditions, the fraction of systems which go through this path and provide a successful supernova is very uncertain, ranging from *negligibly small* to *most of the systems*, depending on the author. Clearly, the efficiency of this path depends on details of the computation. Neglecting this complication we assume that for the SD channel there is a one to one correspondence between the mass of the secondary and the delay time, and the number of events with delay time between t_D and $t_D + dt_D$ is proportional to the number of systems with secondary with mass between M_2 and $M_2 + dM_2$, such that $\tau_n(M_2) = t_D$:

$$n_{Ia}(t_D)|dt_D| \propto n(M_2)|dM_2|, \qquad (7.16)$$

where $n(M_2)$ is the distribution of the secondary masses in systems, which provide successful explosions through the SD channel. From Eq. (7.16) the distribution of the delay times for the SD channel is derived as:

$$f_{Ia}(t_D) \propto n(M_2) \cdot |\dot{M}_2|. \qquad (7.17)$$

The factor $|\dot{M}_2|$, that is, the rate of change of the mass of stars evolving off the main sequence, is a robust prediction of the stellar evolution theory (e. g., from Eq. (2.3)); the factor $n(M_2)$ can be derived analytically by convolving the distribution of primary masses with the distribution of the mass ratios $f(q)$, with $q = M_2/M_1$, in binary systems:

$$n(M_2) \propto \int_{M_{1,i}}^{8} \frac{f(q)}{M_1} dM_1, \qquad (7.18)$$

where the integral extends over the range of mass for the primaries in systems which provide explosions with a delay time $t_D = \tau_n(M_2)$. The lower limit of this mass range ($M_{1,i}$) results from three constraints: (i) the primary must be more massive then the secondary, (ii) it must also be more massive than 2 $M_\odot$, so as to provide a CO, rather than a helium WD, and (iii) its WD remnant must be massive enough to secure that the Chandrasekhar limit is reached in combination with the mass accreted from the secondary. Indicating with $M_{2,env}$ the envelope mass of the donor at RL overflow, and with ϵ the fraction of it which is accreted and burned on top of the CO WD, the third condition becomes:

$$M_{WD}(M_1) + \epsilon M_{2,env} \geq 1.4. \qquad (7.19)$$

In general, the lower the mass of the secondary, the lower $M_{2,env}$; thus, as the delay time increases, M_2 decreases and so does $M_{2,env}$; correspondingly the minimum M_{WD} from Eq. (7.19) increases, and so does the minimum value for M_1 for Chandra explosions, by virtue of the initial-final mass relation (e. g., Eq. (1.11)). Indicating this minimum value with $M_{1,n}$ the lower limit to the integration in Eq. (7.18) is:

$$M_{1,i} = \max[M_2; 2; M_{1,n}]. \qquad (7.20)$$

In this equation, at short delay times M_2 is large and prevails; when M_2 drops below 2 $M_\odot$ the second limit prevails, so that binaries with $M_2 \leq M_1 \leq 2M_\odot$ do not contribute to the integral in Eq. (7.18); as the delay time grows, $M_{2,\text{env}}$ decreases and at some point the third limit becomes larger than 2 $M_\odot$. From this delay time on, the lower limit $M_{1,i}$ coincides with $M_{1,n}$; it increases with the delay time, and the integration range progressively narrows to eventually vanish when $M_{1,i} = 8\, M_\odot$. The setting in of the third regime depends on the efficiency ϵ.

Figure 7.8a shows the distribution of the secondary masses, as from Eq. (7.18) for a Salpeter distribution of the primaries ($n(M_1) \propto M_1^{-2.35}$) and a flat distribution of the mass ratios ($f(q) = \text{constant}$). The distribution of the secondaries differs from a Salpeter IMF because of the restrictions on M_1: the upper limit of 8 $M_\odot$ causes the deviation at large values of M_2; in the flat part, starting at $M_2 = 2\, M_\odot$, a minimum primary mass of 2 $M_\odot$ is imposed; the final drop is due to the lower limit $M_{1,i} = M_{1,n}$ increasing as M_2 decreases, and corresponds to the progressively lighter secondaries requiring a progressively heavier CO WD to meet the Chandrasekhar limit. The efficiency parameter ϵ is unimportant when M_2 is high, and a large amount of mass is available for the CO WD growth, but becomes crucial when the secondary is a low-mass star, that is, at long delay times. For example, at $t_D = 13$ Gyr the mass of the evolving secondary is $\sim 0.9\, M_\odot$; with a core mass of $\sim 0.3\, M_\odot$, only $\sim 0.6\, M_\odot$ can be donated, and, if only 50% of this is actually accreted and burned on top of the WD, the Chandrasekhar limit can be met only if $M_{\text{WD}} \geq 1.1\, M_\odot$. This implies a minimum primary mass of $M_{1,n} \sim 7\, M_\odot$, that is, a very limited mass range for the primary progenitors. The solid line in Figure 7.8a is computed assuming an efficiency of 100%; the dashed line assumes $\epsilon = 0.5$; in this case the final drop sets in earlier because the need to total the Chandrasekhar mass implies a more severe restriction on $M_{1,i}$.

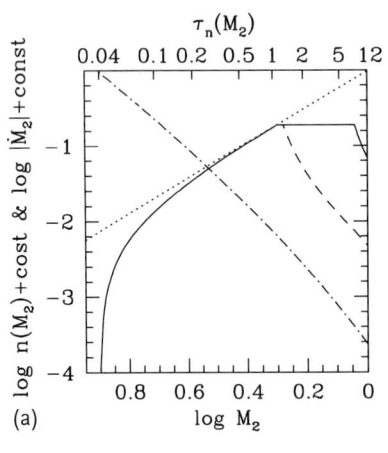

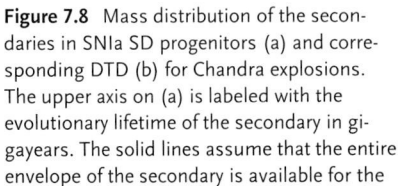

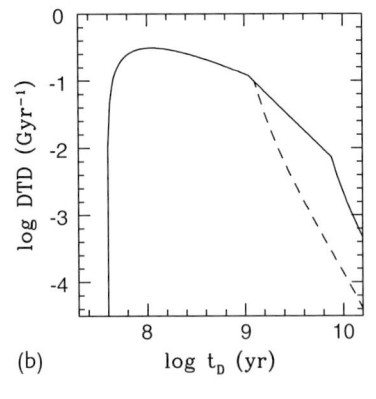

Figure 7.8 Mass distribution of the secondaries in SNIa SD progenitors (a) and corresponding DTD (b) for Chandra explosions. The upper axis on (a) is labeled with the evolutionary lifetime of the secondary in gigayears. The solid lines assume that the entire envelope of the secondary is available for the growth of the CO WD, while the dashed lines assume that only half of the envelope is, the other half being lost in a wind. In panel (a) the dotted line shows a Salpeter distribution; the dot-dashed line shows the derivative of the evolutionary mass, on a logarithmic scale.

Figure 7.8b shows the distributions of the delay times derived from Eq. (7.17) with the functions $n(M_2)$ plotted in Figure 7.8a. The interplay between $n(M_2)$ and the derivative of the evolutionary mass (dash-dot line in Figure 7.8a) shapes the final DTD, which shows a rapid initial rise, a wide maximum up to $t_D \sim 1$ Gyr, followed by a decline which becomes very fast at late epochs, when both factors drop. If only half of the envelope of the secondary is actually accreted and burned on top of the WD, the DTD drops much faster. For the SD Chandra model, the SNIa rate at late times is extremely sensitive to the efficiency parameter ϵ, and this channel has difficulties to account for a sustained SNIa rate in old stellar populations.

As mentioned in Section 3.1 the delay time for the MS + CO WD channel may be shorter than $\tau_n(M_2)$ if RL contact occurs appreciably before core hydrogen exhaustion in the secondary star. The derivation of an analytic DTD for this case is hardly possible, since the delay time depends on the evolution of the binary separation, which in turn is determined by several processes, for example CE evolution, magnetic braking, tidal interactions. Indeed, the results of BPS codes for this evolutionary channel are widely different, both in the mass ranges of the progenitors and in the shape of the DTD. As long as the RL contact occurs in the late stages of the MS evolution the DTD for this channel will be similar to what is shown in Figure 7.8b, obviously limited to the delay times appropriate to the secondaries which feed the MS+CO WD channel.

7.3.4
The DD Channel

In the DD model the first part of the evolution is the same as in the SD model, but, following expansion of the secondary, no accretion on the WD takes place, a CE phase sets in, which leads to the complete ejection of the envelope of the secondary, and the system emerges as a close binary of two WDs. The two components are bound to eventually merge due to the emission of gravitational wave radiation (GWR), and in this case the delay time is:

$$t_D = \tau_n(M_2) + \tau_{GW} , \quad (7.21)$$

where τ_{GW} is the time it takes to the two WDs to merge due to the emission of GWR, given by:

$$\tau_{GW} = \frac{0.15 A^4}{(M_{WD1} + M_{WD2}) M_{WD1} M_{WD2}} \text{ Gyr}, \quad (7.22)$$

where A, M_{WD1} and M_{WD2} are the separation and masses of the components of the DD system at birth in solar units. In the DD model the derivation of the DTD is more complicated than in the SD model, because there's no one to one correspondence between the initial binary parameters and the delay time; rather, a range of initial configurations can end up exploding with the same t_D, or systems with the same $(M_1; M_2)$ but different initial separations can explode at vastly different delay times. Still, some general trends can be outlined. In the following we consider only DD systems composed of two CO WDs. Both primordial components, then, should have masses between 2 and 8 $M_\odot$, and the (nuclear) delay time due to the evolution of the secondary off the MS will not exceed 1 Gyr. All events taking place at a delay time longer than 1 Gyr do so because of the GWR delay.

With simple algebra the denominator in Eq. (7.22) becomes:

$$F = M_{DD}^2 M_{WD2} - M_{DD} M_{WD2}^2 , \qquad (7.23)$$

where $M_{DD} = M_{WD1} + M_{WD2}$. This function is plotted in Figure 7.9a for various values of M_{WD2}, which is assumed to range between 0.6 and 1.2 $M_\odot$ following Eq. (1.11). The dashed portion of the lines highlights the allowed region in M_{DD}, which, for each value of M_{WD2} has to lie within a range corresponding to the minimum and maximum values of M_{WD1}. It can be noticed that the allowed values of F are well described by a unique relation $F = 0.25\, M_{DD}^3$, which is plotted as a solid line. To a very good approximation the GWR delay can then be written as:

$$\tau_{GW} = \frac{0.6 A^4}{M_{DD}^3} \text{ Gyr} , \qquad (7.24)$$

with A and M_{DD} in solar units. This function is plotted on Figure 7.9b for various values of M_{DD}. The GWR delay ranges from virtually zero, for the closest systems, up to many Hubble times for the widest ones. Noticeably, the DD systems ought to be born with separations on the order of a few $R_\odot$ in order to merge within a

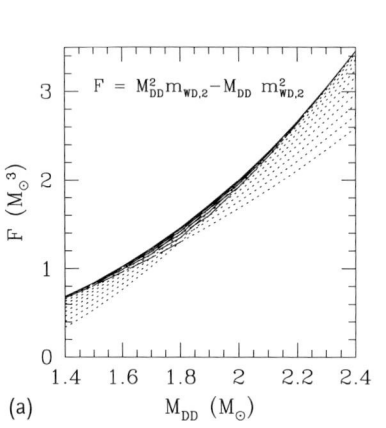

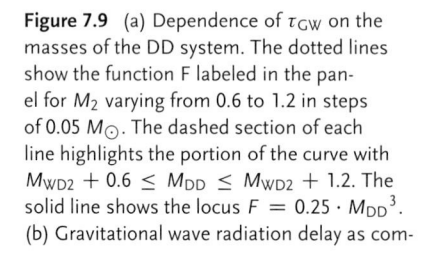

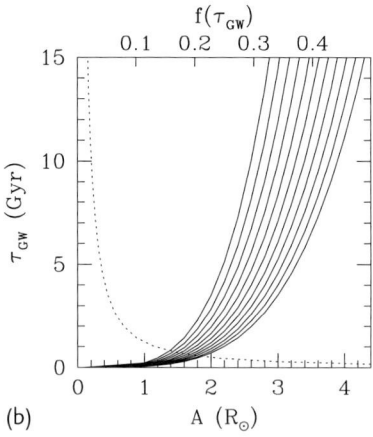

Figure 7.9 (a) Dependence of τ_{GW} on the masses of the DD system. The dotted lines show the function F labeled in the panel for M_2 varying from 0.6 to 1.2 in steps of 0.05 $M_\odot$. The dashed section of each line highlights the portion of the curve with $M_{WD2} + 0.6 \leq M_{DD} \leq M_{WD2} + 1.2$. The solid line shows the locus $F = 0.25 \cdot M_{DD}^3$. (b) Gravitational wave radiation delay as computed from Eq. (7.24). The solid lines show the approximate expression for M_{DD} ranging from 1.4 (lowest curve) to 2.4 (uppermost curve) in steps of 0.1 $M_\odot$. The dotted line shows the distribution of τ_{GW} (labeled on the upper axis) function of τ_{GW} (labelled on the left axis), assuming that the distribution of the separations A is flat.

Hubble time. Since the primordial systems should be born with a much wider separation (from several 10 s to several 100 s $R_\odot$) in order to avoid premature merging, the CE phases must produce a considerable orbital shrinkage for this path to be relevant. It is also worth pointing out that most combinations of the parameters A and M_{DD} provide short τ_{GW}, as can be appreciated in Figure 7.9b. This implies that, unless most DD SNIa progenitors are born with low mass and with large separations, the distribution of the gravitational wave radiation delays will be skewed at short delays. In fact, at given M_{DD}, the number of systems with delay between τ_{GW} and $\tau_{GW} + d\tau_{GW}$ is proportional to the number of systems with separations between A and $A + dA$ such that $A = (1.67 \tau_{GW} M_{DD}^3)^{0.25}$:

$$n(\tau_{GW}) \propto n(A) \frac{\partial A}{\partial \tau_{GW}} \propto n(A) M_{DD}^{0.75} \tau_{GW}^{-0.75} . \tag{7.25}$$

For a flat distribution of A, the distribution of τ_{GW} falls off as $\tau_{GW}^{-0.75}$ for any value of M_{DD}. The shape of this distribution is shown as a dotted line in Figure 7.9b.

Given the sensitivity of the GWR clock to the separation of the DD system, it is very important to examine the effect of the CE evolution, but hydrodynamical models of binary stars spiraling within a common envelope are very difficult to compute, and the results depend on various parameters and assumptions. For this reason the effect of the CE evolution is usually described through a parametric equation in which a fraction (α_{CE}) of the orbital energy of the binary system goes into unbinding the common envelope:

$$\frac{M_{i,d}(M_{i,d} - M_{f,d})}{R} = \alpha_{CE} \left[\frac{M_{f,d} M}{2 A_f} - \frac{M_{i,d} M}{2 A_i} \right] , \tag{7.26}$$

where $M_{i,d}$ and $M_{f,d}$ are the masses of the donor before and after the CE, respectively; M is the mass of the companion that does not change during the CE phase; A_i and A_f are the separations before and after the CE; and R is the radius of the donor at contact, well represented by the average Roche lobe radius. The parameter α_{CE} describes the efficiency of the CE phase to the end of unbinding the CE itself: for α_{CE} much smaller than unity, a big variation of the orbital energy is needed to expel the envelope, and the process is inefficient. In this case, the binary system will emerge from the CE as a much closer system than when the CE phase started, and, following Eq. (7.24) the system will merge in a short time. It should be noticed that the mass distribution in the envelope is not taken into account in the left hand side term of Eq. (7.26). In addition, this equation neglects the thermal energy of the envelope, which helps in lifting it as the RL overflow proceeds. To some extent these complicacies can be included in the parameter α_{CE}, allowing for values larger than unity if other energy sources, besides the orbital one, are used to expel the CE. Typical values of α_{CE} considered in the literature range between 0.5 and 2. If Eq. (7.26) applies to two successive CE phases with the same α_{CE} parameter, a correlation between the binary parameters A and M_{DD} can be produced: the more massive primordial systems, with heavier remnants (M_{DD}), also end up with a smaller A/A_0 ratio, because their envelope is more massive and more energy is needed to unbind it from the system. This has a consequence on the distribution of the gravitational wave radiation delays, because the more massive systems populate only the short τ_{GW} portion of the distribution, and long gravitational delays are provided only from the less massive systems, which manage to remain sufficiently wide after the CE phases. Such a correlation between A and M_{DD} reflects into a DTD, which is particularly skewed at the short τ_{GW} end. On the other hand, there is no guarantee that Eq. (7.26) describes adequately each CE phase, nor that the parameter α_{CE} is necessarily the same during the two CE events. As an alternative, one can consider that binary systems could emerge from the second CE with separations up to several $R_\odot$, irrespective of M_{DD}. In this case also massive systems populate the whole range of GWR delays, up to a Hubble time and M_{DD} and A are unrelated.

We name the first scenario, which produces a larger fraction of close systems, CLOSE DD, to distinguish it from the second scenario (WIDE DD), in which also massive systems end up at wide separations. The main difference between the two scenarios concerns the distribution of the GWR delays: for the CLOSE DDs the more massive the binary the smaller the upper limit to their spanned range of τ_{GW}; for the WIDE DD option any system can end up with a GWR delay up to the Hubble time, and the distribution of τ_{GW} follows from the distribution of M_{DD} and A, regarded as independent variables. In this latter case, it is found that the distribution of τ_{GW} is very sensitive to the distribution of the separations, rather than to that of the DD masses.

Figure 7.10a,b shows examples of the DTD for these two scenarios, having adopted a vanishingly small lower limit to τ_{GW} for any M_{DD} in both cases, and a power law either for the distribution of the GWR delays (CLOSE DDs), or for the distribution of A (WIDE DDs). Figure 7.10a illustrates how the DTDs for the DD model can be viewed as a modification of the SD case, where systems with given τ_n populate a wider range of delay times due to the extra delay τ_{GW}. The WIDE DD case is the least populated at the short delay times, the CLOSE DD the most, since in this case the great majority of systems have a short τ_{GW}. The cusp at $t_D = 1$ Gyr in Figure 7.10a is due to having adopted a minimum M_2 of 2 $M_\odot$ for successful SNIa progenitors, so that delay times in excess of 1 Gyr are obtained only with systems whose τ_{GW} is longer than $(1 - \tau_n)$. In other words, at this delay time one hits a fixed limit on the allowed parameter (τ_n) space. Although this feature in the analytic DTDs is artificial, we do expect an abrupt change in the distribution at a delay time equal to the MS lifetime of the least massive secondary in successful SN Ia progenitors. Some BPS realizations indicate that the minimum M_2 is higher than 2 $M_\odot$, perhaps because if the initial binary mass is too low the final M_{DD} is not likely to reach the Chandrasekhar limit. The effect of varying such minimum M_2 is shown in Figure 7.10b for the WIDE DD scenario. Another important parameter is the distribution of the separations A of the DD systems at birth: the steeper this distribution, the more populated the distribution at the short delay times (see Figure 7.10b). The DTD for the CLOSE DD scenario shows similar trends: the shorter the maximum τ_n and/or the steeper the distribution of τ_{GW} the larger the fraction of systems with short delay times.

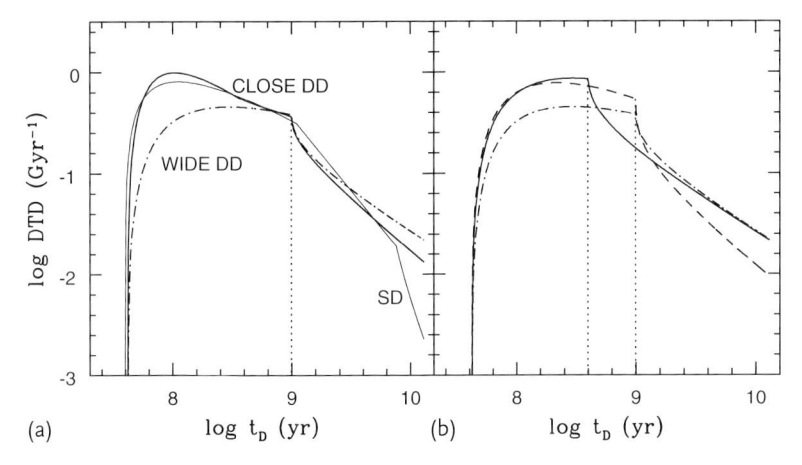

Figure 7.10 Distribution of the delay times for DD models (thick lines). (a) DD CLOSE (solid) and DD WIDE (dot-dashed) models, both with a minimum $M_2 = 2\,M_\odot$, and distributions $n(\tau_{GW}) \propto \tau_{GW}^{-0.75}$ and $n(A)$ = constant, respectively. For comparison, the thin solid line shows the SD Chandrasekhar model with $\epsilon = 1$. (b) The effect on the DD WIDE DTD of the minimum M_2 and of the distribution of the separations. The dot-dashed line is the same as in (a); the solid line assumes a flat distribution of A, but a minimum M_2 of 3 $M_\odot$ in SNIa progenitor systems; the dashed line assumes a minimum M_2 of 2 $M_\odot$, but $n(A) \propto A^{-0.9}$. Notice that, since $n(A_0) \propto A_0^{-1}$, the latter option corresponds to assuming that the CE has little impact on the shape of the distribution of the separations. All the plotted DTDs, normalized to unity in the range $0 \leq t_D \leq 13$ Gyr, assume a maximum M_2 of 8 $M_\odot$, a Salpeter distribution of the primaries, a flat distribution of the mass ratios in the primordial systems.

Figure 7.11 shows the cumulative fraction of events as a function of the delay time for selected model DTDs. For an instantaneous burst of SF, this diagram illustrates the time scale over which SNIa products build up in the interstellar medium. Since SNIa contribute a substantial fraction of iron, but a negligible fraction of α elements, to the chemical cycling of stellar populations, the [α/Fe] ratio can be regarded as a clock to estimate the star formation timescale, modulo the DTD for the SNIa. The tuning of the clock requires fixing the timescale over which SNIa products are released to the interstellar medium. Figure 7.11 shows that such tuning depends on the DTD model. In the literature, it is generally agreed that systems showing a supersolar [α/Fe] abundance ratio have a formation timescale shorter than ~ 1 Gyr: this stems from the wide use of a SD Chandra model, according to which 50% of the SNIa explosions occur within this time span (long dot-dashed line). A similar timescale is obtained with a moderately flat DD CLOSE model (dotted line), but substantially shorter (~ 0.3 Gyr) or longer (~ 1.6 Gyr) timescales are predicted by a particularly steep DD CLOSE (solid line) or a particularly flat DD WIDE (dot-dashed line) models. Actually, the formation timescale of a stellar system showing supersolar [α/Fe] ratios also depends of the SF law; however, the calibration of this chemical clock requires a choice for the DTD, which is difficult to make from a theoretical point a view. More robust constraints can be obtained

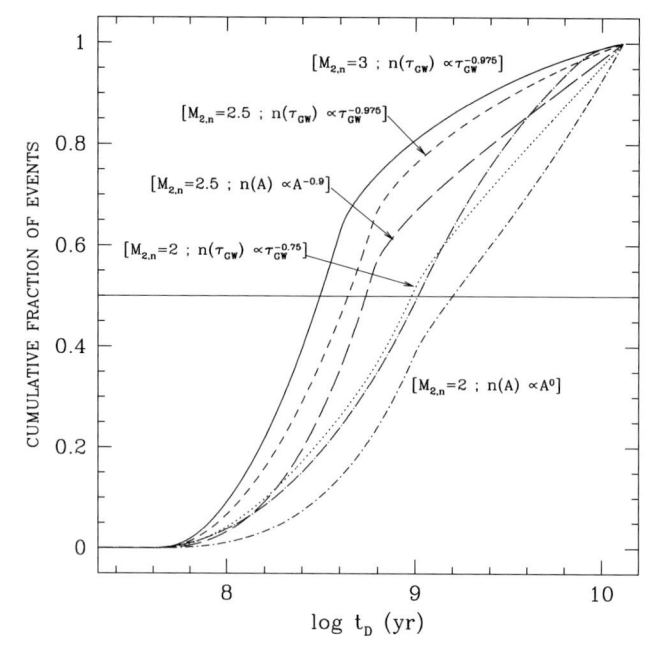

Figure 7.11 Cumulative distribution of the delay times for SD Chandra (long dot-dashed) and a selection of DD WIDE (dot-dashed and long dashed) and DD CLOSE (dotted and short dashed) models. The solid line refers to a particularly steep DD CLOSE model. The choice of the parameters for the DD models is labeled, with $M_{2,n}$ indicating the minimum M_2 in the SNIa progenitor systems in $M_\odot$.

by sorting the DTD that best accounts for the SNIa rate in different astrophysical environments, as we discuss next.

To summarize, all in all SD and DD models provide broadly similar distributions: the DTD features an early maximum and over $\sim 50\%$ of the events occur within ~ 1 Gyr from formation. At later epochs the DTD exhibits a steady decline with a more pronounced downturn for the SD Chandra model. All distributions accommodate delay times as short as ~ 40 Myr (the evolutionary lifetime of a $\simeq 8\,M_\odot$ star), as well as delays up to a Hubble time and more, with continuity over the whole range of delay times; this follows from the assumption of continuous distributions over the relevant parameter space of masses and separations of SN Ia progenitors. Among the many parameters that influence the shape of the DTD, we emphasize the following characteristics of SNIa progenitor systems as crucial:

- The maximum M_2, determining the shortest delay time for both SD and DD channels.
- The minimum M_2, determining the longest delay time for SD models, or the position of the cusp on the DTD of DD models.
- The accretion efficiency ϵ (i.e., the fraction of the envelope of the secondary, which is effectively accreted and burned on top of the CO WD) for the SD model.

- The shape of the distribution of the separations (A) of the DD systems at birth, and whether the CE induces a correlation between M_{DD} and A (CLOSE DD) or not (WIDE DD).

7.3.5
Constraining the DTD and the SNIa Productivity

If the SNIa productivity is constant in time, Eq. (7.1) can be written as

$$\dot{n}_{Ia}(t) = k_{Ia}\mathcal{M}(t)\langle f_{Ia}\rangle_{\psi(t)}, \tag{7.27}$$

where $\langle f_{Ia}\rangle_{\psi(t)}$ is the average of the DTD weighted with the age distribution of the stellar population, or the star formation history, and $\mathcal{M}$ is the total mass turned into stars up to the epoch t. Since the DTD is more populated at short delay times, younger stellar populations (either because of the spanned age range, or because of a SFR increasing with time, or both) will have higher SNIa rate per unit mass (or specific SNIa rate) with respect to older systems. The flatter the DTD, the shallower this predicted trend, and if $f_{Ia}(t_D) = $ constant, with all delay times equally probable, the SNIa rate per unit mass is independent of the age distribution of the stellar population. The observed correlations of the SNIa rate with tracers of the age of the parent galaxy support the notion of a DTD decreasing with increasing delay time, and they can be used to constrain the shape of the DTD.

From Eq. (7.27) the rate appears to be proportional to the galaxy mass, through a factor, which depends on the DTD and on the SFH. Therefore, the best constraint on the DTD can be obtained from the SNIa rate per unit mass binning the galaxies with similar SFH and determining the rate in SNuM in each bin. This procedure requires to evaluate the stellar mass in each galaxy, whereas the galaxy luminosity can be measured directly. The *purely* observed specific rate is then derived in SNu or in SNuK; since the K-band luminosity traces mass better than the B-band luminosity, the latter option is preferable.

As an example, Figure 7.12 shows the comparison of models with data, having chosen the rate in SNuK and the $B - K$ color as tracers of the specific SNIa rate and the parent galaxy age. A galaxy sample of about 10 000 galaxies has been binned by morphological type; for each bin the SNIa rate per unit L_K has been determined as from Eq. (7.7), and the corresponding $B - K$ calculated as the color of the whole bin; this is equivalent to considering a typical galaxy in each bin, which consists of the sum of all individual objects. The theoretical counterpart is computed with model SFHs described by the Eq. (5.6) in the range $0\,\text{Gyr} \leq t \leq 13\,\text{Gyr}$ for all galaxies. The relation in Eq. (5.6) peaks at progressively older ages as the parameter τ_{SF} decreases, thereby providing progressively redder colors. A short τ_{SF} also implies a narrow age distribution, akin to early-type galaxies, while as τ_{SF} grows longer, the age distribution becomes wider and wider, until, for $\tau_{SF} \geq 13\,\text{Gyr}$ $\psi(t)$ monotonically increases up to the present. Three models with an exponentially increasing SFR have been added in order to match the bluest $B - K$ colors. The lines in Figure 7.12 show theoretical relations obtained with selected DTD models (see

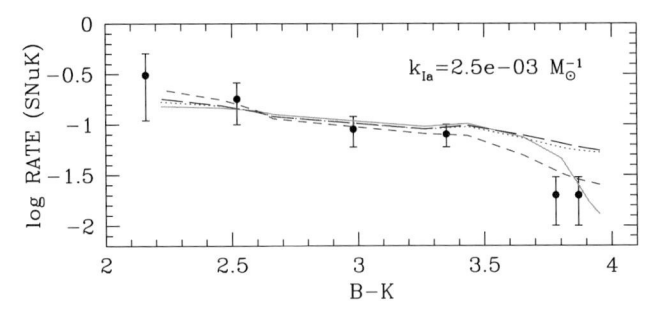

Figure 7.12 SNIa rate per unit K-band luminosity as function of the $B - K$ color of the parent galaxy. Black dots show the rates determined separately for E, S0, Sa, Sb, Sc and Irregular galaxies; lines connect models calculated with the SFR given by Eq. (5.6) and τ_{SF} ranging from 0.1 to 20 Gyr, plus three models with $\psi(t) \propto e^{t/\tau}$ and $\tau = 5, 3$ and 1 Gyr (right to left). The linetypes encode the DTD: SD Chandra (solid), DD CLOSE (dashed), and DD WIDE (dotted) models. The DD CLOSE models adopt a minimum secondary mass of 2.5 $M_\odot$ and a distribution of the GWR delays proportional to $\tau_{GW}^{-0.975}$ (short dashed) and $\tau_{GW}^{-0.75}$ (long dashed). The DD WIDE model adopts a minimum secondary mass of 2 $M_\odot$ and a distribution of the separations proportional to $A^{-0.9}$. The value of the productivity assumed is indicated (updated from Greggio, L. and Cappellaro E. (2009, in *Probing Stellar Populations out to the Distant Universe*, AIPC, 1111, 477) with M05 stellar population models). For a color version of this figure please see in the front matter.

caption), meant to illustrate the dependence on the main parameters. The theoretical rates are computed as the product of the rate per unit mass and the mass-to-light ratio of the galaxy models:

$$\frac{\dot{n}_{Ia}(t)}{L_K} = k_{Ia} \frac{\int_0^t \psi(t-\tau) f_{Ia}(\tau) d\tau}{\int_0^t \psi(t) dt} \times \frac{\int_0^t \psi(t) dt}{\int_0^t \psi(t-\tau) \mathcal{L}_K(\tau) d\tau}, \quad (7.28)$$

where $\mathcal{L}_K$ is the light-to-mass ratio of simple stellar populations in the K-band (function of age). When going from early- to late-type galaxies, the increase of the first factor (rate per unit mass) competes with the decrease of the second factor (mass-to-light ratio); this explains the relatively flat trend of the rate with $B - K$ color, especially for the DD WIDE model. The comparison in Figure 7.12 shows that models, which predict a substantial drop of the rate at the late delay times account better for the observed trend (e.g., the SD Chandra model or the DD CLOSE model with a distribution of the GWR skewed toward short delays).

The comparison between models and observations requires an assumption for the productivity k_{Ia}. For a stellar population formed with a constant SFR up to the current epoch t, Eq. (7.1) can be written as:

$$k_{Ia} = \frac{\dot{n}_{SN}(t)}{\mathcal{M}} F_{Ia}(t_D \leq t), \quad (7.29)$$

if the SNIa productivity is constant in time. The last factor is the fraction of Ia events with delay times shorter than the current age of the stellar population, and as t grows the fraction F approaches unity. Therefore k_{Ia} can be directly evaluated as the current SNIa rate per unit mass in systems formed out of a constant SFR for about a Hubble time, and the result is independent of the shape of the DTD. For this reason, the model rates in Figure 7.12 have been scaled to match the observed value at $B - K \simeq 3$, corresponding to Sb galaxies. For a Salpeter-diet IMF, the normalization yields $k_{Ia} = 2.5 \times 10^{-3}\ M_\odot^{-1}$, that is, 2.5 SNIa every 1000 $M_\odot$ of stars formed in one stellar generation, virtually independent of the DTD model. Another choice for the normalizing bin would imply a substantial dependence of the resulting k_{Ia} on the DTD, particularly strong if we chose the earliest galaxy types.

How does this value for k_{Ia} compare with the abundance of potential progenitors in one stellar population? As previously noted, the upper limit for the mass of the primary is 8 $M_\odot$; the lower limit may in principle be $\sim 2\ M_\odot$, that is, the minimum mass providing a CO WD in a close binary; however, a 2 $M_\odot$ star leaves a light WD (only 0.6 $M_\odot$) and it is more reasonable to consider a lower limit of $\sim 3\ M_\odot$. In this mass range a Salpeter-diet IMF has $\simeq 0.03$ stars $M_\odot^{-1}$. Then, the derived productivity requires that a little less than 10% of all stars in this mass range are the primaries of close binary systems, which end up in a SNIa explosion. If we restrict the primary progenitor masses to the range between 5 and 8 $M_\odot$, then $\sim 30\%$ of them should be evolving into SNIa. This is a very strong constraint on the progenitors. Since only a fraction of stars are born in close binaries, and only a fraction of these will fulfill the conditions that lead to the successful explosion, we conclude that the level of the SNIa rate in spirals requires that the mass range of the primaries in SNIa progenitor systems is relatively wide.

Similar indications on the shape of the DTD and on the productivity of SNIa are derived when considering other correlations between the specific SNIa rate and the average age of the parent stellar populations, as for example the rate in SNu as a function of the parent galaxy color, or of the parent galaxy morphological types, or the rate per unit mass as a function of the sSFR (shown in Figure 7.2b). This validates the approach of Eq. (7.27) and the modeling of the light from the stellar populations. However, at the time of writing, the best fitting DTDs slightly depend on the particular correlation considered, and robust conclusions can be drawn only with bigger datasets and more accurate characterization of the SFH in the various galaxies. In this context it is instructive to consider the evolution of the cosmic SNIa rate as a function of redshift.

Figure 7.13 shows the cosmic SNIa rate, that is, the rate per unit volume, as a function of redshift. The data, derived from many sources in the literature, adopt the same cosmology ($H_0 = 70$ km/(s Mpc), $\Omega_M = 0.3$, $\Omega_\Lambda = 0.7$). The lines are models obtained by convolving the DTDs used for Figure 7.12 with the cosmic SFR, having adopted for it (from Hopkins, A.M. and Beacom, J.F. (2006, Astrophys. J., 651, 142)):

$$\psi(t) = 0.7 \frac{0.017 + 0.13z}{1 + \left(\frac{z}{3.3}\right)^{5.3}} \quad (M_\odot\ \text{yr}^{-1}\ \text{Mpc}^{-3}). \tag{7.30}$$

The different models run very close to each other: the cosmic rate does not seem to provide a good test for the DTD, as one may expect, since at the various red-

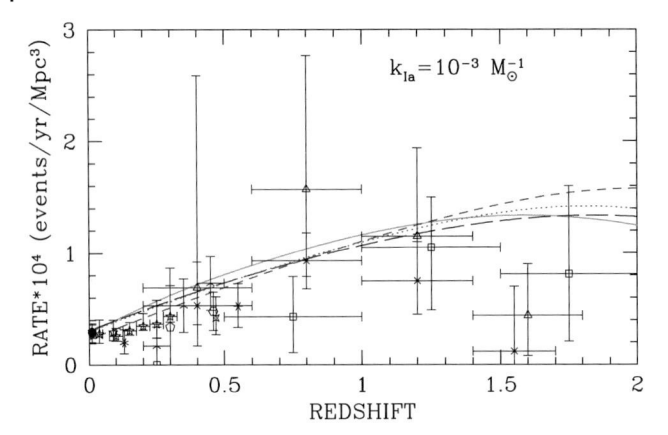

Figure 7.13 SNIa rate per unit volume as a function of redshift. Points are literature data. Lines are theoretical rates computed with Eq. (7.30) and for the same DTDs as in Figure 7.12 with the same line encoding. A productivity of 1 event every 10^3 $M_\odot$ of parent stellar population matches the level of the observed rates (updated from Blanc, G. and Greggio, L. (2008, *New Astron.*, 13, 606)). For a color version of this figure please see in the front matter.

shifts the total stellar population has a wide age spread. At high redshift the models overestimate the SNIa rate: the strong component at short delay times provides a high rate in the young universe, which is not detected. The empirical rates were derived assuming a typical extinction, independent of redshift. However, surveys conducted on a fixed band (e. g., the z band) probe spectral regions which, as the redshift increases, progressively move into the UV, where extinction is stronger. The cosmic SFR, as well as the SFR per unit mass in individual galaxies, increase by over a factor of ~ 20 by a redshift of 1.5, in parallel with a sizable increase of the typical reddening affecting individual galaxies. Hence, the data points at redshift beyond $\simeq 1$ in Figure 7.13 are likely to progressively underestimate the actual SN rate. The observed SNIa rates in Figure 7.13 require $k_{Ia} \simeq 10^{-3}$ $M_\odot^{-1}$, that is, a SNIa productivity, which is a factor of ~ 2.5 lower than that derived by fitting the rates per unit luminosity in Sb galaxies. Notice that this low value of k_{Ia} perfectly accounts for the data point at redshift zero, which is obtained by multiplying the local luminosity density by the SNIa rate in SNu for all the galaxies in the same sample used in Figure 7.12. Therefore, the discrepancy between the two values of k_{Ia} signals a discrepancy between the integrated cosmic SFR and the local *B*-band luminosity density, in the sense that the latter should be higher by a factor of ~ 2.5 if the SFR was given by Eq. (7.30). The problem may be related to the assumptions on the IMF (shape and temporal behavior) and/or on the mass-to-light ratios of the stellar populations, or to erroneous SFRs. At the time of writing, a similar problem affects the local stellar mass density, which appears to be ~ 2 times lower than what obtained by integrating the cosmic SFR. Although the solution to the problem is still under debate, it is reassuring that the indications from the SNIa rates are consistent with what derived from a totally independent method, and that we can constrain the productivity within a factor of ~ 2.

7.3.6
SNIa Yields

As mentioned in Section 7.3, the light curve and spectra of typical SNIa events are most successfully reproduced by an exploding Chandrasekhar mass CO WD. As for CC supernovae, though, the complicated physical processes occurring at explosion prevent us from constructing self-consistent models from first principles; rather, various models are constructed under different assumptions, and the comparison with observational data, that is, light curves and spectra at various epochs, yields information about the underlying physics. Since carbon burning reactions are extremely sensitive to temperature, ignition is confined to a small zone; the burning front may propagate either through a shock wave (detonation), or through a subsonic flame (deflagration). In the first case the fuel is completely burned to nuclear statistical equilibrium producing virtually only iron peak elements. The presence of a sizable quantity of intermediate mass elements in the spectra of SNIa rules out a complete detonation as explosion mechanism for the bulk of SNIa. Thus, there is a general consensus in the literature that the explosion starts as a deflagration; however, there are still different possibilities for the front propagation, that is, the subsonic flame may remain such throughout the whole explosion, or at some point it may turn into a detonation (delayed detonation models). The nucleosynthetic output is different in the two cases, and it depends on the density at which the delayed detonation sets in. Other parameters affect the explosive yields, among which we mention the thermal structure of the exploding WD and the density at ignition (resulting from the previous evolution, including the accretion history); the chemical composition (C/O ratio) throughout the WD; and the velocity of the deflagration flame. Figure 7.14 illustrates the different ejected masses of oxygen, magnesium, silicon and iron obtained by varying some parameters of the explosion models. Notice that these masses include all the isotopic species. The different explosion models provide quite a different nucleosynthetic output. Figure 7.15 shows the distribution of ^{56}Ni masses measured in a sample of about 150 SNIa events from their light curves. The ^{56}Ni SNIa yields appear almost evenly distributed between 0.4 and 0.7 $M_\odot$ with an average value of 0.58 $M_\odot$, which is 10 times the average yields from CC SNe (see Figure 7.6). While this range can clearly be encompassed by explosion models with different parameters, it is unclear how we can relate the different ^{56}Ni masses to the SNIa progenitors and/or model parameters. Therefore, similar to the case of CC SNe, in what follows we adopt an empirical iron yield from each SNIa event of 0.67 $M_\odot$, that is, the average ^{56}Ni from the sample in Figure 7.15, increased by 15% to account for the ^{54}Fe production, as featured by (most) explosion models.

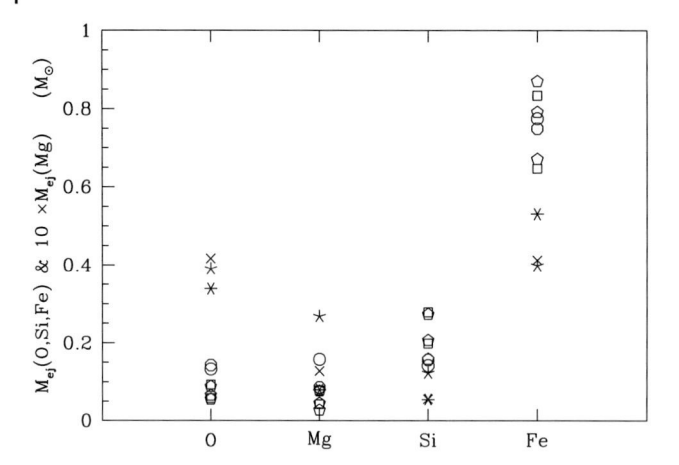

Figure 7.14 Ejected masses of oxygen, magnesium, silicon and iron of SNIa explosion 1D (open symbols) and 3D (starred symbols) models. The open circles refer to deflagration models with slightly different C/O ratios; the other open symbols refer to delayed detonations adopting different transition densities and high (pentagons) or low (squares) central ignition density. The different types of starred symbols show models with different flame geometry (data from Iwamoto, K. *et al.* (1999, *Astrophys. J.* 125, 439) for 1D models; from Travaglio, C. *et al.* (2004, *Astron. Astrophys.*, 425, 1029) for 3D models).

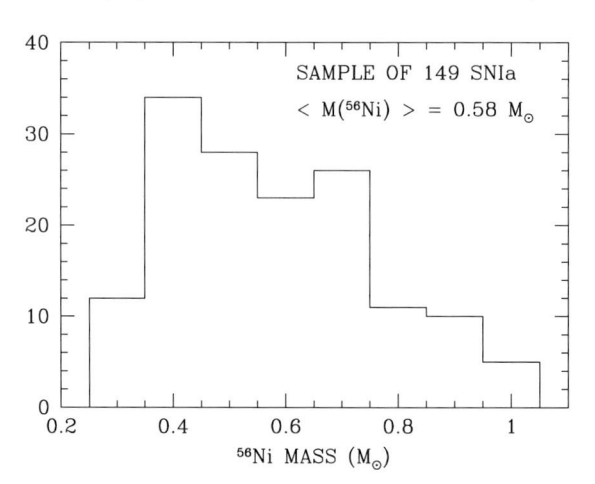

Figure 7.15 Distribution of ejected ^{56}Ni masses in a sample of Type Ia SN events. The masses have been measured from the peak bolometric luminosity and the rise time (*Arnett's Rule*: Arnett, W.D. (1982) *Astrophys. J.*, 253, 785) (adapted from Howell, D.A. *et al.* (2009, *Astrophys. J.* 691, 661)).

7.4
The Relative Role of Core Collapse and Thermonuclear Supernovae

We end this chapter with best guess answers to the initial quest: illustrated in Figure 7.16 we show the temporal evolution of the SN rates following a burst of star

7.4 The Relative Role of Core Collapse and Thermonuclear Supernovae | 203

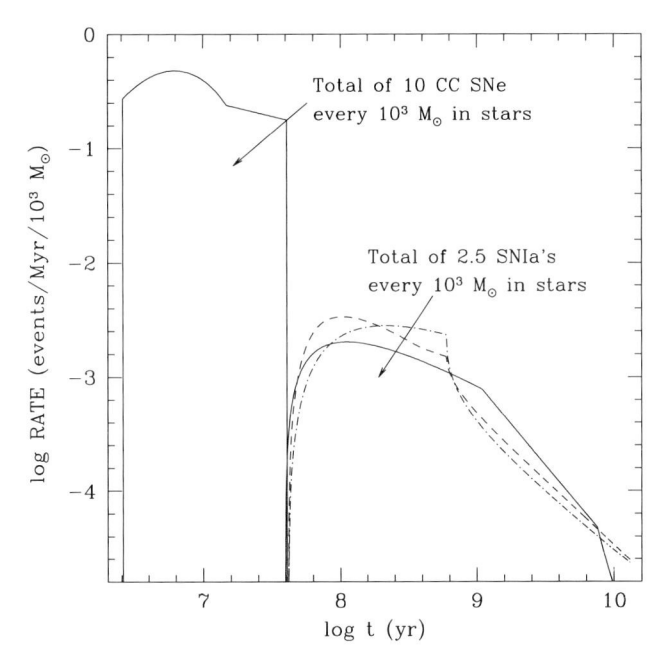

Figure 7.16 Time evolution of the rate of SN explosions following an instantaneous burst of SF of 1000 $M_\odot$. The rate of CC events assumes a *diet* Salpeter IMF and progenitor masses from 8 to 120 $M_\odot$. Three models for the rate of SNIa events are plotted: the SD Chandra (solid) with efficiency $\epsilon = 1$, one DD CLOSE (dashed) with a distribution of $\tau_{GW} \propto \tau_{GW}^{-0.75}$, and one DD WIDE (dot-dashed) with a distribution of $A \propto A^{-1}$. Both DD models assume a minimum $M_2 = 2.5\,M_\odot$. A productivity of $k_{Ia} = 0.0025\,M_\odot^{-1}$ has been adopted.

formation, and in Figure 7.17a,b we plot the time evolution of the ejected masses of three representative α elements and iron. All following figures assume a Salpeter diet IMF.

Following a starburst forming $1.5 \times 10^9\,M_\odot$ ($\sim 2.3 \times 10^9$ stars), CC SNe soon occur at a rate of ~ 1 SN per year; the *fireworks* continue at approximately the same rate for about 40 Myr, to be abruptly replaced by a much lower level of activity from the SNIa channel, at ~ 1 event every 250 yr. After about 400 Myr the explosions start to become less frequent, the time interval between two events grows longer and longer, and at 10 Gyr from the SF burst there is one explosion every $\sim 20\,000$ yr, or more. These figures depend on a few parameters, for example the IMF and the mass range of CC SN progenitors, and the SNIa progenitor model, issues that have been discussed in this chapter; nevertheless they are quite robust, since observations do support this quoted CC rate, and the SNIa rate is calibrated directly on the data.

This same burst of star formation releases metals to the interstellar medium as illustrated in Figure 7.17a. After ~ 40 Myr, about 10 million $M_\odot$ of oxygen, 850 000 $M_\odot$ of magnesium and silicon, and 650 000 $M_\odot$ of iron have accumulated in the interstellar medium. The following SNIa events do not alter the oxygen and mag-

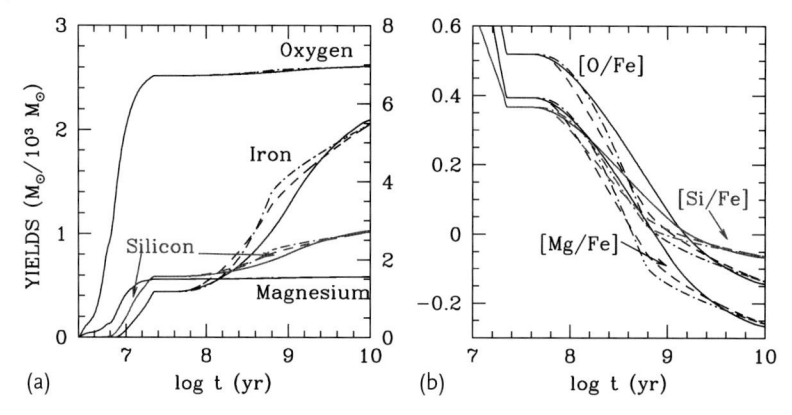

Figure 7.17 (a) Cumulative ejected masses of the various chemical elements from an instantaneous burst of SF of 1000 $M_\odot$ as a function of time which follow from the rates plotted in Figure 7.16. Notice that the oxygen masses are to be read off the right scale. CC yields from the LC set of models have been adopted, while for the type Ia average ejected masses of oxygen, magnesium, silicon and iron of 0.1, 0.01, 0.18 and 0.67 $M_\odot$ have been assumed, consistent with Figure 7.14. (b) Time evolution of the corresponding α elements overabundances, having adopted a solar composition of $(Z_O, Z_{Mg}, Z_{Si}, Z_{Fe}) = (5.41, 0.60, 0.67, 1.17) \times 10^{-3}$.

nesium content, but provide more silicon and iron. At 10 Gyr the total released masses of silicon and iron are $\sim$ 1.5 and 3 million $M_\odot$, respectively. These figures depend on the parameters controlling the SN rates, but, more importantly, on the adopted stellar yields, which have been illustrated in this chapter. We notice in particular that the ejected masses from SNe with progenitor mass between 8 and 11 $M_\odot$ have been neglected in the drawing of Figure 7.17a,b. This may imply an underestimate of the iron yields from CC SNe; alternatively, such yield can be evaluated by multiplying an average ^{56}Ni mass of 0.057 per event times the number of CC SNe, resulting in $\sim$ 880 000 $M_\odot$ of ^{56}Fe. Given the production of $\sim$ 2.4 million $M_\odot$ of iron from SNIa, the ratio of CC to SNIa iron yield from an SSP ranges from 0.27 to 0.37 (for these adopted parameters). The different timescales for the α elements and iron release to the interstellar medium produce the well known trend of the abundance ratios, which can be used as a *nucleosynthetic clock*, as illustrated in Figure 7.17b. The plateau at $\sim$ 30 Myr is due to pollution from CC SNe being completed, while that from SNIa has not started yet. As time progresses, more and more products from SNIa are released into the interstellar medium, its Fe content increases, and correspondingly its α/Fe ratio decreases. The [Si/Fe] ratio decrease is less dramatic than for O and Mg because of the silicon production from the SNIa. Notice that the range of α overabundances covered by the models in Figure 7.17b nicely encompasses the observational data of the Milky Way and external galaxies. The chemical pattern in real stellar systems, though, depends on its SFH. For example, for a constant SFR, the [α/Fe] ratios will remain higher than in the burst case, due to the continuous α-elements replenishment from CC SNe;

the solar ratios (i.e., [α/Fe] = 0) will then be attained at a later epoch in galaxies with a continuous SF.

Acknowledgments

The authors thank Maria Teresa Botticella and Luca Zampieri for helpful discussions on the Core Collapse Supernovae progenitor masses.

Thermonuclear and CC supernovae differ in very many respects, from the mechanism of the explosion, to the mass and nature of their progenitors, to their distribution of the delay times, and so on. Yet, four remarkable coincidences are worth mentioning:

1. The kinetic energy release is in both cases $\sim 10^{51}$ erg.
2. The photon energy release (the time integral of the SN light curve) is $\sim 10^{49}$ erg for CC as well as for SNIa, in spite of the energy source being radically different: the ^{56}Ni decay in one case, the thermal energy stored in the shocked envelope in the other.
3. The total number of events of the two kinds following an episode of star formation is quite similar, differing only by a factor ~ 4.
4. The rates of the CC and of thermonuclear supernovae are quite similar in the local Universe, although one depends on the *present* global SFR density, whereas the other depends on the whole *previous* star formation history. These coincidences are indeed quite intriguing, because the physical processes at work are so radically different in the two cases. For example, the SN productivities (k_{CC} and k_{Ia}) turn out to be similar because the factor ~ 100 in peak rate seen in Figure 7.16 between CC and thermonuclear supernovae is compensated for by the factor ~ 100 in the time interval during which supernova events take place. Because of this, the two kinds of SNe have a similarly important impact on the evolution of our Universe.

Further Reading

Analytic Derivations of the DTD of SNIa

Greggio, L. and Renzini, A. (1983) *Astron. Astrophys.*, **118**, 217.

Greggio, L. (2005) *Astron. Astrophys.*, **441**, 1055.

Greggio, L. (2010) *Mon. Not. R. Astron. Soc.*, **406**, 22.

Physics of Explosions

Hillebrandt, W. and Niemeyer, J.C. (2000) Annu. Rev. Astron. Astrophys., **38**, 191.
Sim, S.A. et al. (2010) Astrophys. J., **714**, L52.
Woosley, S.E. and Weaver, T.A. (1986) Annu. Rev. Astron. Astrophys., **24**, 205.
Woosley, S.E. and Kasen, D. (2011) Astrophys. J., **734**, 38.

Observed SN Rates

Barbary, K. et al. (2010) Astrophys. J. arXiv:1010.5786.
Cappellaro, E., et al. (1999) Astron. Astrophys., **351**, 459.
Li, W. et al. (2011) Mon. Not. R. Astron. Soc., 412, 1473, (Lick Observatory Supernova Survey, LOSS).
Maoz, D. and Gal-Yam, A. (2004) Mon. Not. R. Astron. Soc., **347**, 951.
Mannucci, F., et al. (2005) Astron. Astrophys., **433**, 807.
Sullivan, M. et al. (2006) Astrophys. J., **648**, 868, (Supernova Legacy Survey, SNLS).

Binary Population Synthesis Codes

Belczynski, K. et al. (2008) Astrophys. J. Suppl., **174**, 223.
Han, Z. et al. (1995) Mon. Not. R. Astron. Soc., **272**, 800.
Nelemans, G. et al. (2000) Astron. Astrophys., **360**, 1011.
Mennekens, N. et al. (2010) Astron. Astrophys., **515**, 89.

SNIa Progenitor Models

Hachisu, I. et al. (1999) Astrophys. J., **522**, 487.
Iben Jr., I. and Tutukov A.V. (1984) Astrophys. J. Suppl., **54**, 335.
Iben Jr., I. and Tutukov A.V. (1987) Astrophys. J., **313**, 727.
Livio, M. (2000) in Type Ia SNe. Theory and Cosmology (eds J.C. Niemeyer and J.W. Truran), Cambridge Univ. Press, p. 33.
Webbink, R.F. (1984) Astrophys. J., **277**, 355.
Whelan, J., and Iben Jr., I. (1973) Astrophys. J., **186**, 1007.

Supernova Yields

Iwamoto, K. et al. (1999) Astrophys. J., **125**, 439.
Kobayashi, C. et al. (2006) Astrophys. J., **653**, 1145.
Limongi, M., and Chieffi, A. (2003) Astrophys. J., **592**, 404.
Woosley, S.E., and Weaver T.A. (1995) Astrophys. J. Suppl., **101**, 181.

Progenitors and Taxonomy of CC Supernovae

Smartt, S.J. (2009) Annu. Rev. Astron. Astrophys., **46**, 63.
Turatto, M. (2003) in Supernovae and Gamma-Ray Bursters. (ed. K. Weiler), Lecture Notes in Physics, Springer, **598**, 21.

8
The IMF from Low to High Redshift

At all redshifts much of galaxy properties depend on the IMF, including mass-to-light ratios, derived galaxy masses and star formation rates, the rate of the luminosity evolution of the constituent stellar populations, the metal enrichment, and so on. With so many important issues at stake, we still debate as to whether the IMF is universal, that is, the same in all places and at all cosmic times, or whether it depends on local conditions such as the intensity of star formation (starburst versus steady star formation), or on cosmic time for example, via the temperature of the microwave background. As is well known, we do not have anything close to a widely accepted theory of the IMF, and this situation is likely to last much longer than desirable. Again, star formation is an extremely complex (magneto)hydrodynamical process, indeed much more complex than stellar convection or red giant mass loss, for which we already noted the absence of significant theoretical progress over the last 40–50 years. Thus, the IMF is parameterized for example, as one or more power laws or as a lognormal distribution, and the parameters are fixed from pertinent observational constraints. Wherever the IMF has been measured from statistically significant stellar counts, a *Salpeter* IMF has been found, that is, $\phi(M) \propto M^{-s}$, with $s = 1 + x \simeq 2.35$, however with a flattening to $s \simeq 1.3$ below $\sim 0.5\,M_\odot$. Specifically, where possible, this has been proved for stellar samples including the solar vicinity, open and globular clusters in the Galaxy and in the Magellanic Clouds, actively starbursting regions, as well as the old galactic bulge. Nevertheless, this does not prove the universality of the IMF, as – with one exception – more extreme environments have not been tested yet in the same direct fashion. The exception is represented by the very center the of the Milky Way, in the vicinity of the supermassive q, a very extreme environment indeed, where very massive stars seem to dominate the mass distribution. In this chapter we discuss a few aspects of the IMF, using some of the stellar population tools that have been illustrated in previous chapters, and exploring how specific integral properties of stellar populations depend on the IMF slope in specific mass intervals. In particular, the dependence on the IMF of the mass-to-light ratio of stellar populations is illustrated, along with its evolution as stellar populations passively age. Then the M/L ratios of synthetic stellar populations, and their time evolution, are compared to the dynamical M/L ratios of local elliptical galaxies, as well as to that of ellipticals up to redshift ~ 1 and beyond. These comparisons allow us to set some constraint on the low-mass portion of the IMF, from ~ 0.1 to $\sim 1.4 \, M_\odot$. A strong constraint of the IMF slope from $\sim 1 \, M_\odot$ up to $\sim 40 \, M_\odot$ and above is then derived from considering the metal content of clusters of galaxies together with their integrated optical luminosity.

8.1
How the IMF Affects Stellar Demography

For a fixed amount of gas turned into stars, different IMFs obviously imply different proportions of low-mass and high-mass stars. This is illustrated in Figure 8.1 showing three different IMFs, all with the same slope below 0.5 $M_\odot$, that is, $s = 1 + x = 1.3$, and three different slopes above:

$$\phi(M) = AM^{-s} \qquad \text{for} \quad M \geq 0.5 \, M_\odot$$
$$= 0.5^{1.3-s} AM^{-1.3} \quad \text{for} \quad M < 0.5 \, M_\odot , \tag{8.1}$$

where the factor $0.5^{1.3-s}$ ensures the continuity of the IMF at $M = 0.5 \, M_\odot$. The normalization of the three IMFs corresponds to a fixed amount $\mathcal{M}$ of gas turned into stars, that is, for fixed

$$\mathcal{M} = \int_{0.1}^{120} M \phi(M) dM . \tag{8.2}$$

Here the case $s = 2.35$ corresponds to the Salpeter-diet IMF already encountered in previous chapters. Thick lines in Figure 8.1b show the cumulative distributions, de-

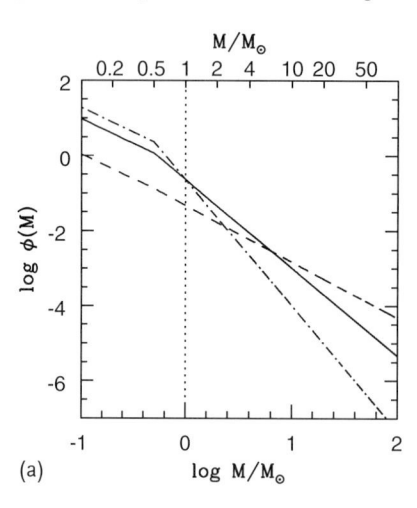

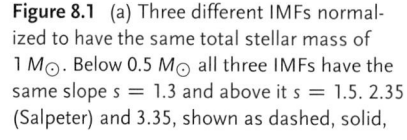

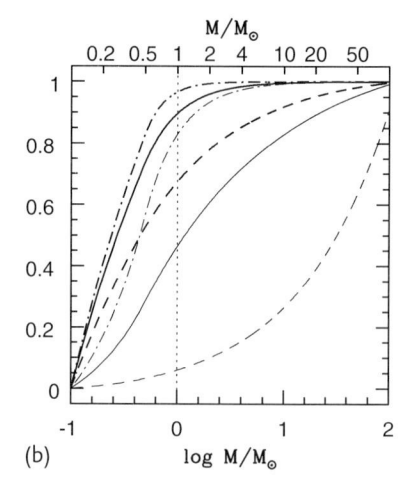

Figure 8.1 (a) Three different IMFs normalized to have the same total stellar mass of 1 $M_\odot$. Below 0.5 $M_\odot$ all three IMFs have the same slope $s = 1.3$ and above it $s = 1.5$. 2.35 (Salpeter) and 3.35, shown as dashed, solid, and dot-dashed lines, respectively. (b) Cumulative distributions of the number (thick lines) and one of the stellar mass (thin lines) for the three IMFs with the same line encoding as in (a).

fined as the number of stars with mass less than M, $N(M' < M) = \int_{0.1}^{M} \phi(M')dM'$, while thin lines show the fraction of mass in stars less massive than M. In a Salpeter-diet IMF ∼ 0.6% of all stars are more massive than $8\,M_\odot$, while for $s = 3.35$ and 1.5 these fractions are 0.03% and 9% respectively. The mass in stars heavier than $8\,M_\odot$ is 20% for a Salpeter diet IMF; it drops to 1% for $s = -3.35$, and is boosted to 77% for $s = 1.5$, a *top-heavy* IMF. Figure 8.1 wants to convey the message that IMF variations have a drastic effect on stellar demography, and therefore on several key properties of stellar populations. Suffice it to say that most of nucleosynthesis comes from $M > 10\,M_\odot$ stars, whereas the light of an old population (say, $t > 10$ Gyr) comes from stars with $M \simeq 1\,M_\odot$, and therefore is proportional to $\phi(M = M_\odot)$.

Figure 8.2 shows the variation of scale factor A (cf. Chapter 2) as a function of the IMF slope, again for a fixed amount $\mathcal{M}$ of gas turned into stars. The scale factor A has a maximum for $s \simeq 2.75$, pretty close to the Salpeter's slope. Since by construction $A = \phi(M = 1\,M_\odot)$ and the luminosity of a $\gtrsim 10$ Gyr old population is proportional to $\phi(M = 1\,M_\odot)$, an IMF with the Salpeter's slope has the remarkable property of almost maximizing the light output of an old population, for fixed mass turned into stars. A flat IMF ($s = 1.35$) is much less efficient in this respect, indeed by a factor of ∼ 8 compared to the Salpeter's slope, as shown by Figure 8.2. This figure also shows the mass-to-number conversion factor K_ϕ, giving the number of stars N_T formed out of a unit amount of gas turned into stars, that is, $N_T = K_\phi \mathcal{M}/M_\odot$. Thus, for a Salpeter-diet IMF $K_\phi \simeq 1.5$, saying that ∼ 150 stars are formed out of $100\,M_\odot$ of gas turned into stars.

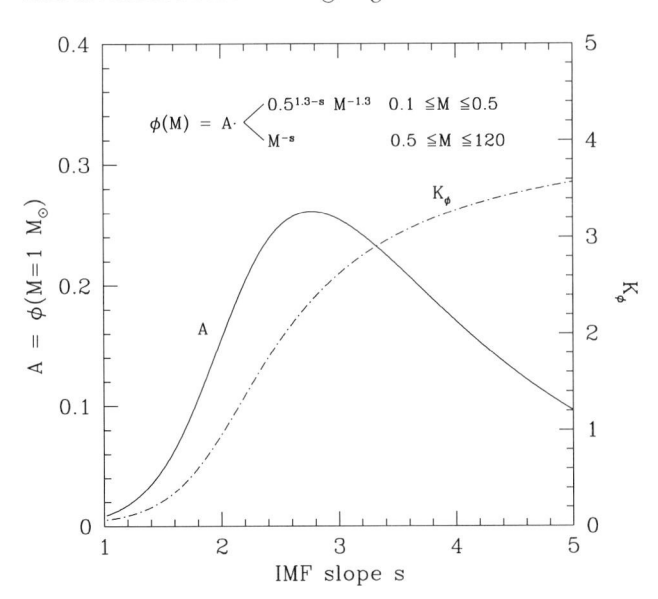

Figure 8.2 The scale factor A_2 as a function of the IMF slope. All IMFs are normalized to a unitary total mass. Also shown is the corresponding mass-to-number conversion factor K_ϕ, such that the total number of stars is given by K_ϕ times the mass turned into stars (in solar units).

An empirically motivated, broken-line IMF such as that shown in Figure 8.1 is widely adopted in current astrophysical applications, yet Nature is unlikely to make an IMF with this number of cusps. Perhaps a more elegant rendition of basically the same empirical data is represented by a Salpeter+lognormal distribution in which a lognormal IMF at low masses joins smoothly to a Salpeter IMF at higher masses, that is:

$$M\phi(M) = A_1 \exp\left[\frac{-(\mathrm{Log} M - \mathrm{Log} M_c)^2}{2\sigma^2}\right], \quad \text{for } M \leq 1\, M_\odot$$

$$= A_2 M^{-x}, \quad \text{for } M > 1\, M_\odot, \quad (8.3)$$

where $A_1 = 0.159$, $M_c = 0.079$, $\sigma = 0.69$, $A_2 = 0.0443$ and $x = 1.3$. Thus, this *Chabrier* IMF is almost identical to the Salpeter IMF above 1 $M_\odot$, and smoothly flattens below, being almost indistinguishable from the Salpeter-diet IMF.

Explorations of variable IMFs can be made by either changing its slope, or by moving to higher/lower masses the break of the IMF slope with respect to Eq. (8.1), or allowing the characteristic mass M_c in Eq. (8.3) to vary. Figure 8.3 shows examples of such evolving IMFs. The two slope IMF with $M_{\mathrm{break}} = 0.5\, M_\odot$ and the Chabrier IMF with $M_c = 0.079\, M_\odot$ (lines a and c in Figure 8.3) fit each other extremely well and both provide a good fit to the local empirical IMF. By moving the break/characteristic mass to higher values one can explore the effects of such evolving IMF, for example mimicking a systematic trend with redshift. The cases with

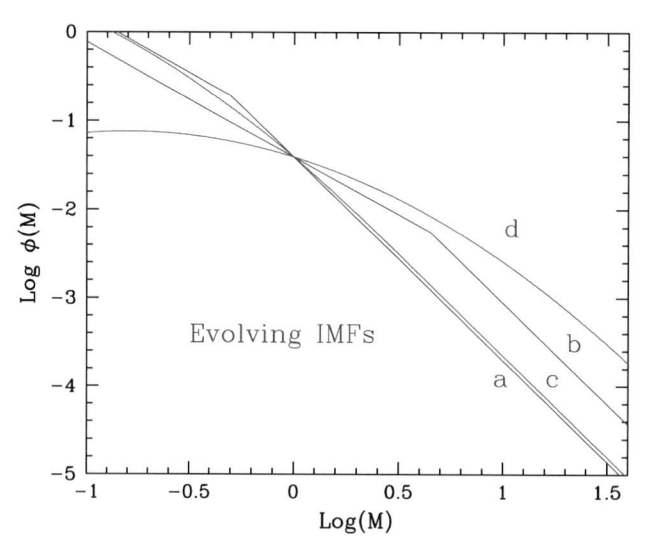

Figure 8.3 Examples of evolving IMFs, for a two-slope IMF and a Chabrier-like IMF. Lines (a) and (c) represent the local IMF. The other lines show modified IMFs to explore a hypothetical evolution with redshift, with the break mass and the characteristic mass M_c having increased to $\sim 4\, M_\odot$, lines (b) and (d) respectively for the two-slope and the Chabrier-like IMF. All IMFs have been normalized to have the same value for $M = M_\odot$. For a color version of this figure please see in the front matter.

$M_c \simeq M_{break} \simeq 4\, M_\odot$ are shown in Figure 8.3 (lines b and d). Having normalized all IMFs to the same value of $\phi(M = 1\, M_\odot)$, Figure 8.3 allows one to immediately gauge the relative importance of massive stars compared to solar mass stars, with the latter ones providing the bulk of the light from old ($\gtrsim 10\,\text{Gyr}$) populations.

8.2
The M/L Ratio of Elliptical Galaxies and the IMF Slope below 1 $M_\odot$

Figure 8.4 shows as a function of age the M_*/L_B ratio (where M_* is the stellar mass) for SSPs with solar composition, and different IMFs each with a single slope s over the whole mass range $0.1\, M_\odot < M < 100\, M_\odot$. Very large mass-to-light ratios are produced by either very flat ($s = 1.35$) or very steep ($s = 3.35$) IMFs, whereas the Salpeter's slope gives the lowest values of the M_*/L_B ratio. This is a result of the different stellar demography already illustrated in Figures 8.1 and 8.2, such that a steep IMF is *dwarf dominated*, that is, most of the mass is in low-mass stars, whereas a flat IMF is *remnant dominated* and most of the mass is in dead remnants.

Measurements of the structure (e. g., half light radius) and stellar velocity dispersion of elliptical galaxies provide estimates of their *dynamical* mass, hence their dynamical mass-to-light ratio can be compared to the stellar M/L ratio. This is shown in Figure 8.4 for a sample of local elliptical galaxies with detailed dynamical model-

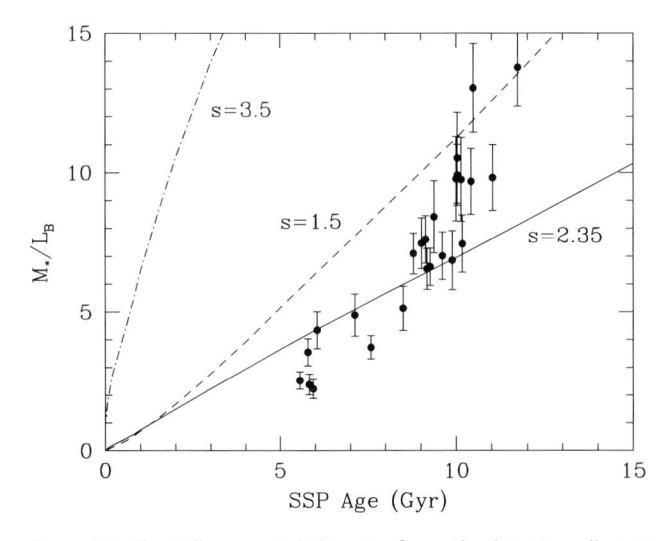

Figure 8.4 The stellar mass-to-light ratio of solar metallicity SSPs as a function of age, for three different single slope IMFs, from 0.1 to 100 $M_\odot$. Also plotted are values of the dynamical M/L ratio for a sample of local elliptical galaxies with detailed dynamical modeling (source: M/L ratios for models: Maraston, C. (1998, Mon. Not. R. Astron. Soc., 300, 872); for the data: Cappellari, M. et al. (2006, Mon. Not. R. Astron. Soc., 366, 1126), van Der Marel, and van Dokkum, P. (2007, Astrophys. J., 668, 756), van Dokkum, P. and van der Marel (2007, Astrophys. J., 655, 30); ages: from Eq. (1) in Thomas, D. et al. (2010, Mon. Not. R. Astron. Soc., 404, 1775)). For a color version of this figure please see in the front matter.

ing, having adopted a relation between the luminosity-weighted age of their stellar populations and velocity dispersion, namely $\text{Log}(\text{Age/Gyr}) = -0.11 + 0.47\text{Log}(\sigma_v)$, consistent with Eq. (6.16). Clearly very steep ($s = 3.5$) and very flat ($s = 1.5$) slopes of the IMF appear to be excluded by the data, whereas the intermediate (Salpeter) slope is quite consistent with the data, apart from the older galaxies which have a higher M/L ratio than the SSP models. However, besides an increase of age also the average metallicity is likely to increase with σ_v, with the galaxies in Figure 8.4 spanning a range from $\sim 1/2$ solar to ~ 2 times solar. Thus, the same galaxies are displayed again in Figure 8.5, together with model M/L ratios for a straight Salpeter IMF and three different metallicities. The trend in M/L ratio exhibited by the data appears to be consistent with a the trend resulting from the metallicity trend with σ_v, and with a straight Salpeter IMF. However, things may not be as simple as they appear. Dark matter may contribute to the dynamical M/L ratios, and the IMF may not be straight Salpeter. A Salpeter-diet IMF such as that shown in Figure 8.1 would give M_*/L_B ratios systematically lower by $\sim 40\%$ than shown in these figures, thus opening some room for a dark matter contribution to the dynamical mass of these galaxies. Alternatively, an IMF slightly flatter than Salpeter at high masses, with its larger contribution by stellar remnants, would reproduce the high dynamical M/L ratios of the oldest galaxies, without dark matter contribution. It is quite difficult to circumvent this dark-matter/IMF degeneracy on the dynamical M/L ratios of elliptical galaxies.

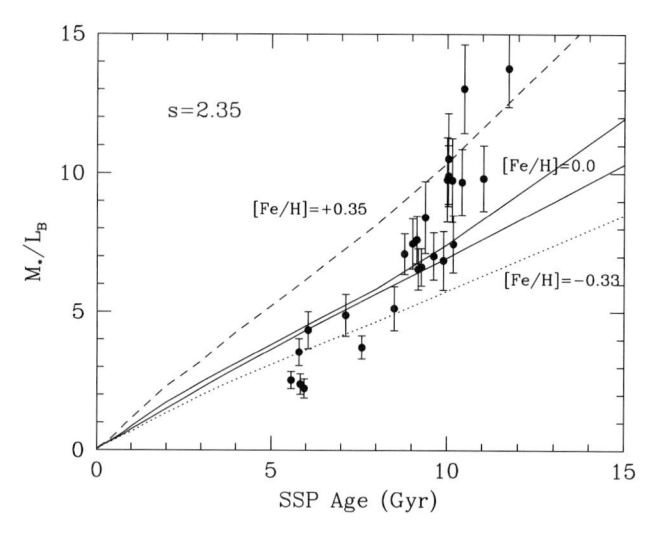

Figure 8.5 The M/L ratio of SSPs with a straight Salpeter IMF, for subsolar (dotted), supersolar (dashed) and solar metallicity (solid), as indicated. The two solid lines refer to two release of the same set of SSP models. (source: model M/L ratios are from: Maraston, C. (1998, *Mon. Not. R. Astron. Soc.*, 300, 872; 2005, *Mon. Not. R. Astron. Soc.*, 362, 799); Data points are the same as in Figure 8.4). For a color version of this figure please see in the front matter.

8.3
The Redshift Evolution of the M/L Ratio of Cluster Ellipticals and the IMF Slope between ~ 1 and $\sim 1.4 M_\odot$

The slope of the IMF controls the rate of luminosity evolution of a SSP, as shown by Figure 2.6 for the bolometric light. The flatter the IMF the more rapid the luminosity declines past an event of star formation. On the contrary, the steeper the IMF the slower such decline, as the light from many low-mass stars compensates for the progressive death of the rarer, more massive and brighter stars. Having identified and studied passively evolving elliptical galaxies all the way to $z \sim 2$ and even beyond, one expects that their M/L ratio must systematically decrease with increasing redshift, and do so by an amount that depends on the slope of the IMF. This test is particularly effective if undertaken for cluster ellipticals, as clusters provide fairly numerous samples of ellipticals at well defined redshifts. Besides the IMF, the rate of luminosity (M/L) evolution also depends on the age of a SSP, being much faster at young ages than at late epochs. Thus, the rate of M/L evolution of elliptical galaxies with redshift must depend on both the IMF slope and the luminosity-weighted age of their stellar populations, or, equivalently on their formation redshift.

We know that the bulk of stars in local massive ellipticals are very old, and for an age of ~ 12 Gyr the light of such galaxies comes from a narrow range around the turnoff mass $M_{\rm TO}$ at $\sim 1\, M_\odot$. Their progenitors at $z \sim 1.5$ must be younger by the corresponding lookback time, that is, ~ 9 Gyr younger, and from the $M_{\rm TO}$-age relation (Eq. (2.2)) we see that the bulk of light of such progenitors has to come from stars of mass around $M_{\rm TO} \simeq 1.4\, M_\odot$. Thus, the evolution of the M/L ratio of old stellar populations from redshift zero all the way to redshift ~ 1.5 is controlled by the IMF slope in the narrow interval between ~ 1 and $\sim 1.4\, M_\odot$. The IMF slope below $\sim 1\, M_\odot$ has no influence on the luminosity evolution, and that above $\sim 1.4\, M_\odot$ was in control of the luminosity evolution at redshifts beyond ~ 1.5. Therefore, the evolution of the M/L ratio of elliptical galaxies from $z = 0$ to ~ 1 allows us to measure the slope of the IMF just near $M \sim 1\, M_\odot$.

Figure 8.6 shows the evolution with redshift of the $M_*/L_{\rm B}$ ratio of solar composition SSPs, for various IMF slopes and different formation redshifts. Also plotted is the average $M_*/L_{\rm B}$ ratio of cluster ellipticals from the literature, from local clusters at $z \sim 0$ all the way to clusters at $z \sim 1.3$. A Salpeter slope ($s = 2.35$) fits the data for a formation redshift $z_{\rm F}$ between ~ 2 and ~ 3, which is in pretty good agreement with both the formation redshift derived from age dating local ellipticals in various ways, and with the observed rapid disappearance of quenched galaxies beyond $z \sim 2$. Assuming that the IMF at the formation redshift of ellipticals was like line b in Figure 8.3, then with $s = 1.3$ at $M = 1\, M_\odot$ this IMF would require a formation redshift well beyond 3 in order to fit the data. Line d instead, with $s = 0.8$ at $M = 1\, M_\odot$ would fail to match the data even assuming $z_{\rm F} = \infty$, as shown in Figure 8.6. One can conclude that the evolution of the M/L ratio of cluster elliptical galaxies up to redshift ~ 1.3 does not favor any significant departure from the Salpeter value $s = 2.35$ of the slope of the IMF in the vicinity of $M \sim 1\, M_\odot$, all the way to a formation redshift beyond ~ 2.

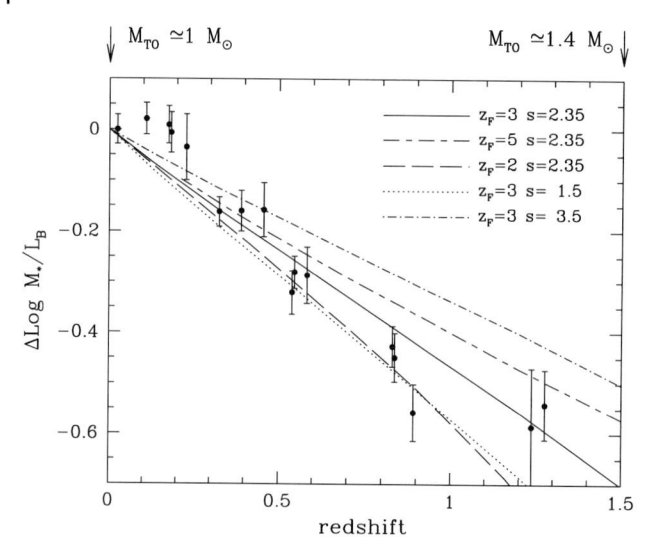

Figure 8.6 The differential redshift evolution (with respect to the value at $z = 0$) of the M_*/L_B mass-to-light ratio of solar composition SSPs, for various choices of the IMF slope between ~ 1 and $\sim 1.4\,M_\odot$, and for various assumed formation redshifts z_F, as indicated. The data points refer to the M_*/L_B ratio of elliptical galaxies in clusters at various redshifts, from $z \sim 0$ up to $z \simeq 1.3$. (Updated from Renzini, A. (2005) *The initial mass function 50 years later*, (ed. E. Corbelli et al.), Ap. Sp. Sci. Library, 327, 221) For a color version of this figure please see in the front matter.

8.4
The Metal Content of Galaxy Clusters and the IMF Slope between ~ 1 and $\sim 40\,M_\odot$, and Above

In its youth, a stellar population generates lots of UV photons, core collapse supernovae and metals that go to enrich the ISM. In its old age, say ~ 12 Gyr later, the same stellar population radiates optical-near-IR light from its $\sim 1\,M_\odot$ stars, while all more massive stars are dead remnants. The amount of metals (M_X) that are produced by such populations is proportional to the number of massive stars $M \gtrsim 8\,M_\odot$ that have undergone a core collapse supernova explosion, whereas the luminosity (e.g., L_B) at $t \simeq 12$ Gyr is proportional to the number of stars with $M \sim 1\,M_\odot$. It follows that the metal mass-to-light ratio M_X/L_B is a measure of the number ratio of massive to $\sim M_\odot$ stars, that is, of the IMF slope between ~ 1 and $\sim 40\,M_\odot$. Clusters of galaxies offer an excellent opportunity to measure both the light of their dominant stellar populations, and the amount of metals that such populations have produced in their early days. Indeed, most of the light of clusters of galaxies comes from ~ 12 Gyr old, massive ellipticals, and the abundance of metals can be measured both in their stellar populations and in the intracluster medium (ICM). Iron is the element whose abundance can be most reliably measured both in cluster galaxies and in the ICM, but its production is likely to be dominated by Type Ia supernovae whose progenitors are binary stars. As extensively discussed in Chapter 7, a large fraction of the total iron production comes from Type Ia supernovae, and the contribution from CC supernovae is uncertain; therefore the M_{Fe}/L_B ratio of clusters is less useful to set constraints on the IMF slope between ~ 1 and $M \gtrsim 10\,M_\odot$. For this reason, we focus on oxygen and silicon, whose production is indeed dominated by core collapse supernovae.

Following the notations in Chapter 2, the IMF can be written as:

$$\phi(M) = a(t,Z) L_B M^{-s}, \tag{8.4}$$

where $a(t,Z)$ is the relatively slow function of SSP age and metallicity shown in Figure 2.10, multiplied by the bolometric correction shown in Figure 3.1. Thus, the metal mass-to-light ratio for the generic element "X" can be readily calculated from:

$$\frac{M_X}{L_B} = \frac{1}{L_B} \int_8^{120} m_X(M) \phi(M) dM = a(t,Z) \int_8^{120} m_X(M) M^{-s} dM, \tag{8.5}$$

where $m_X(M)$ is the mass of the element X, which is produced and ejected by a star of mass M. From stellar population models one has $a(12\,\text{Gyr}, Z) = 2.22$ and 3.12, respectively, for $Z = Z_\odot$ and $2Z_\odot$ and we adopt $a(12\,\text{Gyr}) = 2.5$ in Eq. (8.5). Using the oxygen and silicon yields $m_O(M)$ and $m_{Si}(M)$ from theoretical nucleosynthesis (cf. Figure 7.5), Eq. (8.5) then gives the M_O/L_B and M_{Si}/L_B metal mass-to-light ratios that are reported in Figure 8.7 as a function of the IMF slope between ~ 1 and $\sim 40\,M_\odot$. As expected, the M_O/L_B and M_{Si}/L_B are extremely sensitive to the IMF slope. The values observed in local clusters of galaxies, from X-ray observations of the ICM and assuming stars are near-solar on average, are ~ 0.1 and $\sim 0.008\,M_\odot/L_\odot$, respectively, for oxygen and silicon, as documented in Chapter 10. These empirical values are also displayed in Figure 8.7. A comparison with the calculated values shows that with the Salpeter IMF slope ($s = 2.35$) the standard explosive nucleosynthesis from core collapse supernovae produces just the right amount of oxygen and silicon to match the observed M_O/L_B and M_{Si}/L_B ratios in clusters of galaxies, having assumed that most of the B-band light of clusters comes from ~ 12 Gyr old populations. Actually, silicon may be even somewhat overproduced if one allows a $\sim 40\%$ contribution from Type Ia supernovae (cf. Figure 7.17).

Figure 8.7 also shows that with $s = 1.35$ such a top heavy IMF would overproduce oxygen and silicon by more than a factor of ~ 20. Such a huge variation with $\Delta s = 1$ is actually expected, given that for a near Salpeter slope the typical mass of metal producing stars is $\sim 25\,M_\odot$. By the same token, the IMF labeled (b) in Figure 8.3 would overproduce metals by a factor of ~ 4 with respect to a Salpeter-slope IMF (lines (a) and (c)), whereas the IMF labeled (d) in Figure 8.3 would do so by a factor of ~ 20. Thus, under the assumptions that the bulk of the light of local galaxy clusters comes from ~ 12 Gyr old stars, that current stellar theoretical nucleosynthesis is basically correct, and that no large systematic errors affect the reported empirical values of the M_O/L_B and M_{Si}/L_B ratios, then one can exclude

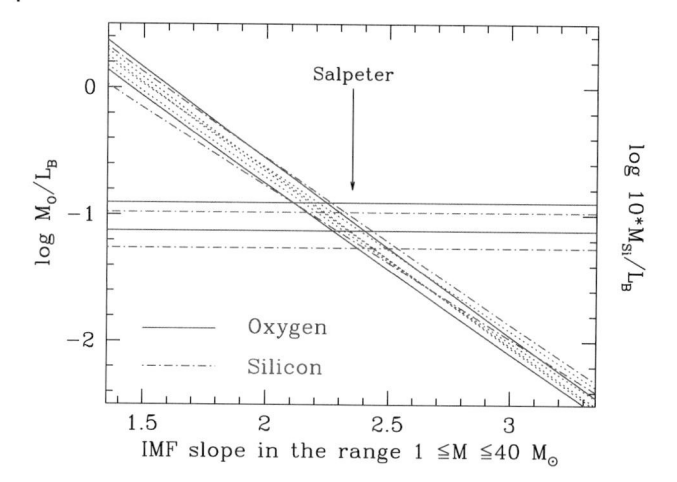

Figure 8.7 The oxygen and silicon mass-to-light ratios as a function of the IMF slope for a ~ 12 Gyr old, near-solar metallicity SSP with oxygen and silicon yields from standard nucleosynthesis calculations. Different lines refer to the different theoretical yields shown in Figure 7.5a,b. The horizontal lines show the uncertainty range of the observed values of these ratios in clusters of galaxies, with central values as reported in Chapter 10, and allowing for a $\sim \pm 25\%$ uncertainty. For a color version of this figure please see in the front matter.

a significant evolution of the IMF with cosmic time, such as for example, one in which the IMF at $z \sim 3$ would be represented by line (b) or (d) in Figure 8.3, and by line (a) or (c) at $z = 0$.

A variable IMF is often invoked as an *ad hoc* fix to specific discrepancies that may emerge here or there, which however may have other origins. For example, an evolving IMF with redshift has been sometimes invoked to ease a perceived discrepancy between the cosmic evolution of the stellar mass density, and the integral over cosmic time of the star formation rate. In other contexts it has been proposed that the IMF may be different in starbursts as opposed to more steady star formation, or in disks versus spheroids. Sometimes one appeals to a top-heavy IMF in one context, and then to a bottom-heavy one in another, as if it was possible to have as many IMFs as problems to solve. Honestly, we do not know whether there is one and only one IMF. However, if one subscribes to a different IMF to solve a single problem, then at the same time one should make sure the new IMF does not destroy agreements elsewhere, or conflicts with other astrophysical constraints.

While it is perfectly legitimate to contemplate IMF variations from one situation to another, it should be mandatory to explore all consequences of postulated variations, well beyond the specific case one is attempting to fix. This kind of sanitary check is most frequently neglected in the literature appealing to IMF variations. A few examples of such checks have been presented in this chapter.

Further Reading

This chapter expands and updates the paper: Renzini, A. (2005) in *The Initial Mass Function 50 Years Later* (ed. E. Corbelli *et al.*), Ap. Sp. Sci. Library, Vol. 327, p. 221.

Most Popular Initial Mass Functions

Chabrier, G. (2003) *Publ. Astron. Soc. Pac.*, **115**, 763.

Kroupa, P. (2002) *Science*, **295**, 82.

Salpeter, E.E. (1955) *Astrophys. J.*, **121**, 161.

Scalo, J.M. (1986) *Fund. Cosm. Phys.*, **11**, 1.

Recent Review

Bastian, N. *et al.* (2010) *Annu. Rev. Astron. Astrophys.*, **48**, 339.

The IMF at High Redshift

Davé, R. (2008) *Mon. Not. R. Astron. Soc.*, **385**, 147.

Tacconi, L.J. (2008) *Astrophys. J.*, **680**, 246.

van Dokkum, P.G. (2008) *Astrophys. J.*, **674**, 29.

9
Evolutionary Links Across Cosmic Time: an Empirical History of Galaxies

In the 1980s and 1990s a bold, fully theoretical approach to galaxy formation and evolution dominated the scene. The cold dark matter (CDM) theory proposed an attractive scenario, in which galaxies start to form from primordial perturbation seeds, and grow via hierarchical merging of dark matter halos. The simplicity of the underlying physics of collisionless particles has allowed rapid progress, and the theory had a great cultural impact due to its direct link to cosmology. N-body simulations reached great sophistication and impressive success in predicting (from first physical principles and observationally established initial conditions) the development of CDM structures on grand scales, that is, the underlying backbone without which the visible galaxies would have never assembled. Thus, for almost two decades theory has enjoyed an almost unconstrained expansion, being confronted by insufficient data on galaxy evolution, which barely extended more than a lookback time of a few gigayears. Then, new generations of observing facilities of unprecedented power came into play, and discrepancies between the simulated galaxy populations and observations started to emerge.

In the most common approach to galaxy formation and evolution the results of semianalytic models or hydrodynamical simulations are compared to observations, checking to which extent the models fit or do not fit the data. Usually they do not fit so well, and therefore parameters in the theoretical models are adjusted in the attempt to improve the fit, thus venturing into the next iteration. In this endeavor one (perhaps predictable) problem has progressively emerged, that baryonic physics is far more complex than dark matter dynamics, and theory cannot have the same predictive power: there are too many physical processes that certainly play an important role in shaping galaxies and their evolution. As more of them are incorporated in the simulations, in the attempt to reach a more comprehensive description of the events, more postulates, and assumptions and parameters are also introduced, and the original elegance, beauty and simplicity of the CDM scenario gets lost. In Chapter 1 we emphasized that the only major difficulties encountered in modeling the evolution of stars come from *macrophysics*, that is, from treating matter bulk motions. When modeling the formation and evolution of galaxies the same kind of physics difficulties are encountered to an exacerbated degree and on much larger scales.

Stellar Populations, First Edition. Laura Greggio and Alvio Renzini.
© 2011 WILEY-VCH Verlag GmbH & Co. KGaA. Published 2011 by WILEY-VCH Verlag GmbH & Co. KGaA.

Multiwavelength surveys of galaxies are now delivering an unprecedented wealth of data, with fairly representative slices of the Universe at all redshifts. In a sense, the situation has radically changed compared to 20 years ago: now telescopes are mapping the Universe at such a pace that simulations soon become obsolete, forcing adjustments in assumptions and model parameters over and over again. Meanwhile, another, fully empirical and phenomenological approach becomes possible for the first time: one in which evolutionary links and trends are derived directly from the data, without dealing at all with the complexities of the many physical processes that indeed must govern galaxy evolution. This is quite a different approach, compared to the traditional one, and requires a sort of *change of reference frame* with respect to our previous attitude, restraining temporarily from searching for the underlying physics, and just exploring what the data demand.

At first sight the population of galaxies appears to host an infinite variety of types and subtypes, of *peculiar* objects making it a taxonomist's nightmare. However, when large samples of galaxies are culled and studied, it appears that the vast majority of galaxies, at all redshifts, follow simple scaling relations, while outliers represent a tiny minority. In particular, three *remarkable simplicities* have emerged, namely:

- The tight relation between the star formation rate (SFR) of *star-forming* galaxies, their mass, and the cosmic epoch as described by Eq. (6.17) or Eq. (6.18) up to $z \simeq 2$–2.5, whereas beyond this redshift the specific SFR flattens as shown in Figure 6.20.
- The M^* and α parameters in the Schechter mass function of *star-forming* galaxies are remarkably constant, while the normalization ϕ^* increases slightly with time, that is, the mass function of star-forming galaxies is almost independent of redshift (see Figure 6.23).
- The fraction of passively evolving (red) galaxies is a strong function of both stellar mass and environment (local overdensity), but the differential effect of mass and environment on the quenching of star formation in galaxies are fully separable, that is, the relative effect of mass in the quenching of star formation is independent of environment, and the relative effect of environment is independent of mass (see Figure 6.21). Mass matters for quenching, as does the environment, but they do so independently of one another.

In this chapter a phenomenological model based on these three pillars is illustrated. It is meant to describe just the broad features of galaxy populations and evolution, such as the growth in mass and the quenching of star formation (i.e., the star formation histories), and do so as a function of redshift/cosmic time and environment. The data to be used are primarily the stellar mass and SFR of galaxies, with stellar population ages and metallicities providing further checks for the emerging scenario. Thus, the phenomenological model is almost exclusively based on the stellar population diagnostics that has been illustrated in the previous chapters, and represents its most ambitious incarnation.

9.1
The Growth and Overgrowth of Galaxies

As documented in Chapter 6, at all redshifts star-forming galaxies cluster along the main sequence in the SFR–$M_\star$ diagram, and this main sequence evolves nearly parallel to itself, steadily declining as a function of cosmic time (see Figure 6.17). The tightness of the main sequence implies that galaxies are bound to remain close to it, insofar as they keep forming stars, whereas they can leave it in two, possibly related ways:

1. Either experiencing a strong starburst to become temporarily a submillimeter galaxy (SMG), being driven to it for example by a merging event or some other global instability.
2. Or, by quenching star formation altogether and becoming a passively evolving galaxy.

It is also conceivable that the SMG phase immediately precedes and promotes the quenching. If not, as the SMG burst subsides the galaxy must return to the main sequence, resuming its evolution along it. Thus, star-forming galaxies must follow Eq. (6.17) (or Eq. (6.18)) most of the time, and therefore the time evolution of its stellar mass can be obtained by integrating the equation (from Renzini, A. (2009), *Mon. Not. R. Astron. Soc.*, 398, L58)):

$$\frac{dM_\star}{dt} = \langle \text{SFR} \rangle \simeq 270 \times \eta\, M_{11} \left(\frac{t}{3.4\,\text{Gyr}}\right)^{-2.5} \quad (M_\odot\,\text{yr}^{-1})\,, \qquad (9.1)$$

or the equivalent relation using Eq. (6.18). Here M_{11} is the stellar mass in units of $10^{11}\,M_\odot$, and t is the cosmic time in years. The parameter η is meant to describe two independent aspects of the SFR–$M_\star$ relation:

1. Exploring a possible systematic offset of the empirically derived SFRs with respect to reality, which certainly cannot be currently excluded.
2. Explore the effect of systematic departures of the SFRs of individual galaxies from the average, that is, for sustaining systematically higher/lower than the average by a factor η, through all its evolution.

The consequences are now explored, assuming that this relation adequately describes the evolution of the SFR and mass growth of galaxies on the main sequence, from $z \sim 3$ ($t \sim 2\,\text{Gyr}$) all the way to $z \sim 0$ ($t \sim 13.7\,\text{Gyr}$). The integration of Eq. (9.1) with fixed η leads to a galaxy growth with time from $t = 2\,\text{Gyr}$ onward, which is described by the equation:

$$M_\star(t) \simeq M_\star(2\,\text{Gyr}) \times \exp(13.53\eta)\,\exp(-38.26\eta\, t^{-1.5})\,. \qquad (9.2)$$

Similarly, combining Eqs. (9.1) and (9.2) one derives the corresponding evolution with time of the SFR of individual galaxies:

$$\text{SFR}(t) = \text{SFR}(2\,\text{Gyr}) \times 5.66 \times \exp(13.53\eta) \exp(-38.26\eta t^{-1.5}) \times t^{-2.5}, \quad (9.3)$$

where, as usual, the SFR is in $M_\odot\,\text{yr}^{-1}$. Notice that these equations describe only part of the possible mass growth, hence provide only lower limits to it, because strong starbursts (such as the SMG phases) or growth by merging are not taken into account.

Focusing first on the $\eta = 1$ case, Figure 9.1 shows the extremely rapid growth of the stellar mass, amounting to more than a factor $\sim 10^5$, if Eq. (9.1) were to hold true from $t = 2\,\text{Gyr}$ to the present (from $z \sim 3$ to 0). Clearly, observations do not support such a dramatic overgrowth: certainly $\sim 10^{10}\,M_\odot$ galaxies at $z \sim 3$ do not grow to become $M_\star = 10^{15}\,M_\odot$ galaxies by $z = 0$! Perhaps, even more dramatic is the discrepancy between how much galaxies would grow according to Eq. (9.2) and the established fact that the mass function of star-forming galaxies is almost constant in time, as shown in Figure 6.23. How can galaxies grow so rapidly in mass, and yet the mass function remains almost unchanged?

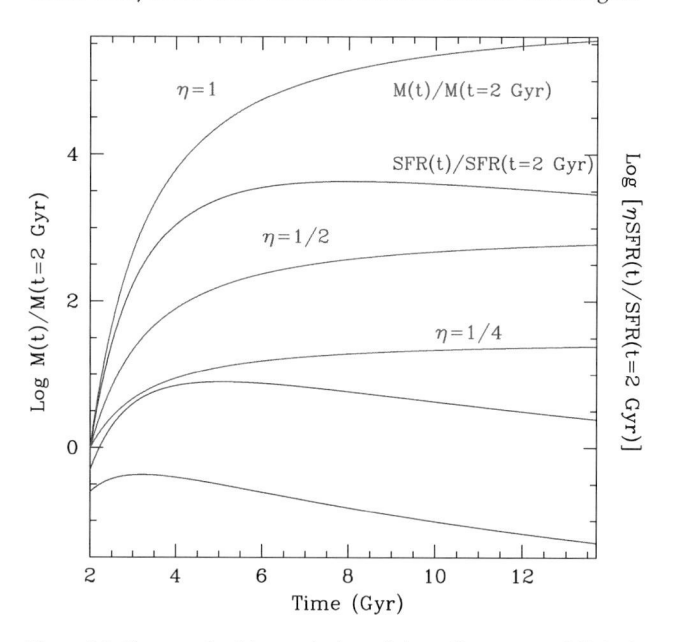

Figure 9.1 The growth with cosmic time of the stellar mass normalized to its initial value at $t = 2\,\text{Gyr}$ ($z \sim 3$), following Eq. (9.2), and for three values of η as indicated. Also shown is the corresponding evolution with time of the SFR, following Eq. (9.3), for the same values of η. The three curves are initially offset by a factor η to show the initial difference in SFR (i.e., at $t = 2\,\text{Gyr}$). One can appreciate that SFRs for a given mass and time that differ by only a factor of 4 lead to vastly different evolutionary paths (source: Renzini, A. (2009, Mon. Not. R. Astron. Soc., 398, L58)) (Reproduced by permission of the Royal Astronomical Society.). For a color version of this figure please see in the front matter.

Clearly, such extremely fast mass growth cannot be sustained for long: at some epoch the SFR, as given by Eq. (9.1), no longer applies, and galaxies must leave the main sequence of star-forming galaxies, which is described by this equation. The only way this can happen is by quenching star formation, thereby joining the passive branch of *red and dead* galaxies. Indeed, galaxies have no other place to go in SFR–$M_\star$ plots such as those shown in Figure 6.14 or 6.17. The apparent contradiction between the high SFRs and the invariance of the mass function, *demands* quenching. If we were blind to passive galaxies, those two empirical facts would be sufficient to demonstrate to us the existence of quenched galaxies(!).

Before exploring further the quenching issue, to which the rest of this chapter is dedicated, there is another aspect of the mass growth implied by Eq. (9.1) worth emphasizing: the extreme dependence of the growth on the normalization of the SFR($M_\star$, t) relation, here expressed by the fudge factor η. One intriguing aspect of the SFR as given by Eq. (9.1) is that upon integration its normalization η appears at the exponent in Eqs. (9.2) and (9.3). Hence, the effects of relatively small differences in η dramatically amplify as time goes by, as illustrated in Figure 9.1 where the cases with $\eta = 1, 1/2$ and $1/4$ are compared. Thus, the true value of the time averaged specific SFR critically determines the subsequent growth rate of the stellar mass of galaxies, a factor that varies slightly making enormous differences in the subsequent evolution. This means that the average specific SFR should be measured with great accuracy in order to precisely predict the subsequent mass growth. Current SFR estimates are indeed affected by systematic uncertainties that may amount to a factor of ~ 2, hence η in Eq. (9.1) will have to be used as an adjustable parameter within uncertainties affecting the SFR measurements.

A second possible interpretation of the η parameter is also quite intriguing: galaxies whose specific SFR systematically differs by even a relatively small factor ($\Delta\eta/\eta \lesssim 1$) experience radically different mass evolutions. Some galaxies can enjoy a rather modest mass growth, with secularly declining SFRs, while others suffer a runaway, quasiexponential mass growth, which certainly cannot be sustained for more than ~ 1 Gyr. Indeed, Eq. (9.1) with $\eta = 1$ refers to the *average* SFR, hence one may expect some galaxies to have SFRs systematically lower than the average, and others higher than the average. However small this dispersion can be, it naturally tends to dramatically amplify in the course of time, as demanded by Eq. (9.2) and illustrated in Figure 9.1, but still galaxies would remain on the main sequence.

The possible origin of such a dispersion is environment. As mentioned above, the tightness of the main sequence of star-forming galaxies indicates that they experience a (quasi-)steady star formation. If star formation is fueled by quasisteady gas accretion from the environment, it is conceivable that galaxies in different environments may experience different rates of gas accretion, hence different specific SFRs. Actually, Eq. (9.2), in spite of its simplicity, may capture both *nature* and *nurture* aspects of galaxy evolution, which to some extent undoubtedly must coexist. Indeed, the stellar mass, certainly a main driver in galaxy evolution, can stand for *nature*, and a dispersion of η may result from a dispersion in the physical properties of the local environment of individual galaxies (*nurture*). Moreover, the $t^{-2.5}$

factor in Eq. (9.1) describes the global, cosmological evolution of the environment, a combination of cosmic expansion and the progressive consumption of the cold-gas reservoir, as more baryons are shock heated to virial temperatures, or even above it by feedback effects (galactic winds).

But, in closing this speculative parenthesis, one has to emphasize that observations do not support the expectation of a specific SFR dependence on environment (overdensity). Indeed, this dependence does not appear to exist either in the local Universe, as shown in Figure 6.14, nor up to $z \sim 1$. Beyond this redshift the galaxy density field remains to be mapped, and therefore at present one cannot exclude that a dependence of the specific SFR may eventually emerge. In the following, it is assumed that the specific SFR of star-forming galaxies is independent of environment even beyond redshift 1.

9.2
A Phenomenological Model of Galaxy Evolution

The major features and results of a simple phenomenological model for the evolution of star-forming and quenched populations of galaxies are now presented. The model is based on the following assumptions:

1. The specific SFR of star-forming galaxies is described up to $z = 2$ by Eq. (6.18), having set $\beta = 0$ for simplicity (i.e., the SFR is set proportional to the stellar mass):

$$\text{SFR} \simeq 25 \left(\frac{M_\star}{10^{10} \, M_\odot} \right) \left(\frac{t}{3.5 \, \text{Gyr}} \right)^{-2.2} \quad (M_\odot \, \text{yr}^{-1}) \, , \tag{9.4}$$

   assumed to hold independent of environment. Beyond this redshift the specific SFR remains constant at the value $2.5 \, \text{Gyr}^{-1}$, as shown in Figure 6.20.
2. The Schechter characteristic mass $M_\star^*$ of star-forming galaxies is constant in time, at the value $\sim 4 \times 10^{10} \, M_\odot$, as illustrated by Figure 6.23.
3. The relative effect on star-formation quenching of mass and environment are completely separable at all redshifts.

This is deliberately a simplified model, as for example, β may not be strictly zero and might even evolve with redshift, and other assumptions have not been verified beyond a certain redshift. Still, the model has the merit of providing a global vision of the growth of galaxies through cosmic times and the progressive emergence of the population of passively evolving galaxies and meets a remarkable series of other observational constraints. Besides those listed above, a few additional assumptions will be introduced when appropriate.

9.2.1
How Mass Quenching Operates

Once one assumes that low-mass seed galaxies appear at very high redshift, their mass growth due to star formation is initially perfectly exponential, because $dM_\star/dt \propto M_\star$. Then, below $z = 2$, when cosmic time exceeds 3.5 Gyr, the specific SFR starts to drop, whereas the mass may still grow quasiexponentially for a while. This is illustrated in Figure 9.2, showing the mass growth demanded by assumption (1) above. In order to maintain an almost stationary mass function of star-forming galaxies, the quenching probability must increase with stellar mass, hence with SFR following Eq. (9.4). In particular, the requirement to maintain fixed $M_\star^*$ for star-forming galaxies dictates a remarkably simple mass quenching law, in which the mass quenching rate is proportional to SFR alone, independent of cosmic epoch and of environment. In fact, for galaxies at the top of the mass

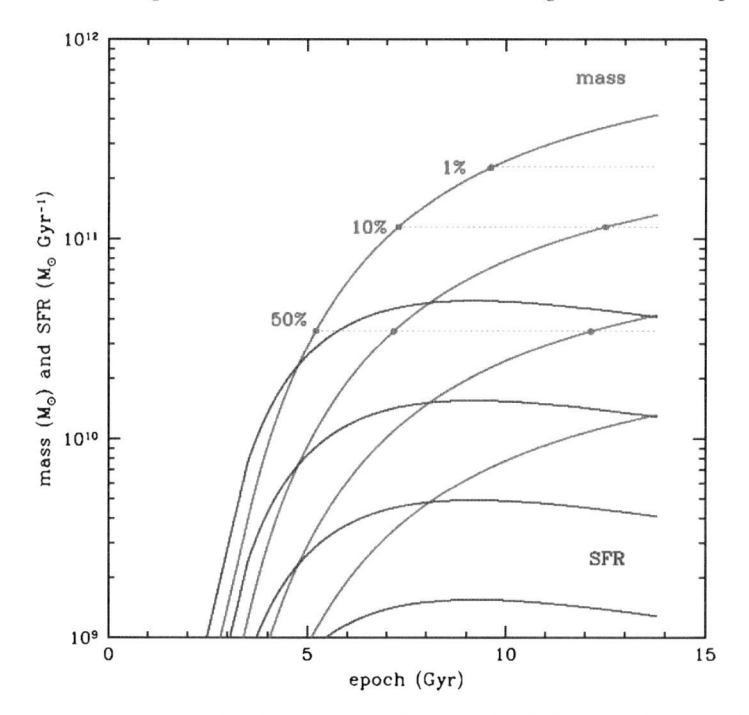

Figure 9.2 The growth with cosmic time of the stellar mass and SFR of galaxies following Eq. (9.4) for cosmic time $t \geq 3.5$ Gyr, and constant specific SFR for cosmic time $t \leq 3.5$ Gyr. The specific SFR rapidly declines between $z = 2$ and 0 but the actual SFR predicted by Eq. (9.4) remains nearly the same because the decrease in specific SFR is compensated by the secular increase in mass. The dots along the mass lines indicate at which mass the survival probability (avoiding quenching of star formation) has to drop below 50, 10 and 1% in order for the mass function of star-forming galaxies to remain nearly stationary (see text). As the mass of a galaxy grows via star formation, the chance to be "mass quenched" rapidly increases (source: Peng, Y. et al. (2010, Astrophys. J., 721, 193)) (Reproduced by permission of the AAS.). For a color version of this figure please see in the front matter.

function ($M_\star \gtrsim M_\star^*$) the rate of increase in $\log(M_\star)$ is proportional to the specific SFR, and therefore the mass function of star-forming galaxies would shift a solid body by an amount $\Delta \log(M_\star)$ proportional to the specific SFR. Thus, if this shift must be contrasted and vanished by quenching, the quenching rate must also be proportional to the specific SFR. Moreover, in order to maintain $M_\star^*$ fixed, the quenching rate must also be proportional to $M_\star$. This is because quenching must be faster where the mass function is steeper, and where an untamed mass growth would otherwise lead to a dramatic increase in the number of objects (i.e., of the mass function of star-forming galaxies), which is not observed. Thus, the mass quenching rate in the high-mass regime ($M_\star \gtrsim M_\star^*$) must be proportional to the product of $M_\star$ times the specific SFR, that is, to just the SFR.

Figure 9.3 illustrates the case: horizontal arrows show the effect on a Schechter mass function of a constant $\Delta \log M_\star$ increase (= 0.3 dex in the specific example) resulting from applying to all galaxies a mass increase with constant specific SFR. The vertical arrows show the amount of quenching as a function of mass that is required to restore the original mass function, and in particular to keep $M_\star^*$ unchanged. The dashed line shows in full this $\Delta \log \phi$ (difference between the two mass functions) that is required to bring back the dotted mass function to the original one. This $\Delta \phi / \phi$ is proportional to $M_\star / M_\star^*$, and therefore the quenching rate must be proportional to the specific SFR times $M_\star / M_\star^*$. In summary, the mass

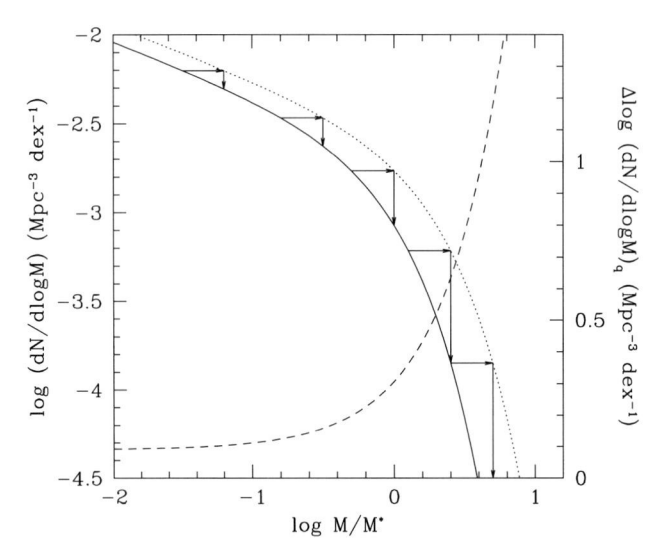

Figure 9.3 The combined effect of star formation and quenching on a Schechter mass function. Horizontal arrows show the effect of a constant $\Delta \log M_\star$ (= 0.3 dex) increase in the mass of all star-forming galaxies, as demanded by a constant specific SFR acting for a fixed time interval. The vertical arrows show the amount of quenching that is required to restore the initial mass function and in particular to keep $M_\star^*$ constant. This is also shown as a dashed line (to be read on the right axis scale), which is also proportional to $M_\star / M_\star^*$.

quenching rate is given by (from Peng, Y., et al. (2010, Astrophys. J., 721, 193)):

$$\left(\frac{dN_q}{dt}\right)_m \simeq \frac{\text{SFR}}{M_\star^*} N_{\text{sf}} = 2.5 \left(\frac{t}{3.5\,\text{Gyr}}\right)^{-2.2} \frac{M_\star}{M_\star^*} \phi_{\text{sf}}(M_\star) \quad (\text{dex}^{-1}\,\text{Gyr}^{-1}), \tag{9.5}$$

where the suffix "m" stands for mass quenching, N_q and N_{sf} are the numbers per unit logarithmic mass of the mass quenched and of the star-forming galaxies, respectively, and where Eq. (9.4) has been used for the SFR. Note that the mass quenching rate is obviously proportional to the number of galaxies that could be quenched, via the quenching *probability* (SFR/$M_\star^*$). In this simple model, it is then postulated that the Eq. (9.5) holds true for the mass quenching rate at all masses, redshifts and environments, an assumption that at this stage can be justified only by the soundness of the results.

Having replaced N_{sf} with the corresponding mass function of star-forming galaxies, this equation allows us to reveal two important aspects of the mass quenching rate. The first is that if ϕ_{sf} is a Schechter function, then the mass quenching rate also has the shape of a Schechter function, with the same $M_\star^*$, but with an effective α that differs by $+1$ with respect to the α of the mass function of star-forming galaxies. Thus, this holds for the resulting mass function of mass-quenched galaxies as well. But notice that this is so only insofar as $\beta = 0$, as assumed in this simple model. Being the product of a quenching probability times the mass function of star-forming galaxies, the quenching rate is small at low masses because of the small quenching probability (the $M_\star/M_\star^*$ factor) and increases with $M_\star$. At high mass the quenching probability is high, but there are few star-forming galaxies to quench, because of the exponential cutoff. Thus, the resulting mass function of mass-quenched galaxies must have a maximum for $M_\star$ near $M_\star^*$.

The second notable property of this mass quenching rate is its evolution with cosmic time. This is governed by the secular evolution of the specific SFR, and therefore the mass quenching rate must be highest at high redshifts, and decrease as cosmic time goes by. However, for the mass quenching rate to reach its peak one has to wait until many star-forming galaxies have grown to a mass near and above $M_\star^*$. But before illustrating further these trends it is worth bringing into play the other quenching actor, the environment.

9.2.2
How Environmental Quenching Operates

To illustrate the workings of environment in quenching star formation in galaxies one has to explicitly introduce the *relative environmental quenching efficiency*, defined as a function of local density ρ (from Peng, Y. et al. (2010, Astrophys. J., 721, 193)):

$$\varepsilon_\rho(\rho, \rho_\circ, M_\star) = \frac{f_{\text{red}}(\rho, M_\star) - f_{\text{red}}(\rho_\circ, M_\star)}{f_{\text{blue}}(\rho_\circ, M_\star)}, \tag{9.6}$$

where f_{red} and f_{blue} are the fractions of quenched and star-forming galaxies, respectively, and $\rho_\circ$ is the density of a reference low-density environment (such as that of voids), where environment effects on galaxies must be minimal. Thus, ε_ρ varies from ~ 0 in the lowest density environments ($\rho \sim \rho_0$) where most galaxies are star-forming ($f_{blue} \simeq 1$), and reaches near 1 in the highest density environments, where virtually all galaxies are quenched ($f_{red} \simeq 1$). The behavior of ε_ρ can be grasped by looking at Figure 6.21, and is further illustrated in Figure 9.4a,b. Empirically, ε_ρ is found to be independent of $M_\star$ up to $z \sim 1$, and as stated above such separability of environmental quenching and mass quenching is assumed to hold at all redshifts. Figure 9.4a,b shows another remarkable simplicity, that is, that the environmental quenching efficiency as a function of overdensity is also independent of redshift, at least up to $z \sim 1$, and again in this simple model it is assumed that this remains true at all redshifts.

Figure 9.4b shows the empirical cosmic evolution of the median overdensity up to $z \simeq 1$, for the highest and lowest density quartiles, as well as for all galaxies. In the lowest density quartile overdensity barely evolves at all, and in the model it is assumed to be strictly constant at all redshifts. Dashed lines in Figure 9.4b show the fitting functions adopted in the model to describe the evolution of the overdensity, which are then extrapolated beyond $z = 1$. Such fitting functions flatten towards high redshifts, describing the fact that high overdensity structures are relatively late products of the evolution of the Universe, and so too must the environmental quenching be. Thus, environmental quenching develops with cosmic time, as the galaxy populations come to inhabit higher and higher density regions. Therefore, contrary to mass quenching that is stronger at early times, environmental quenching must be a relatively late process in the evolution of galaxy populations.

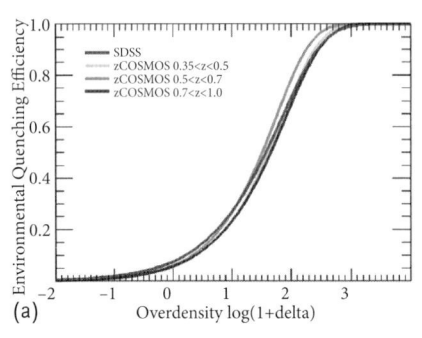

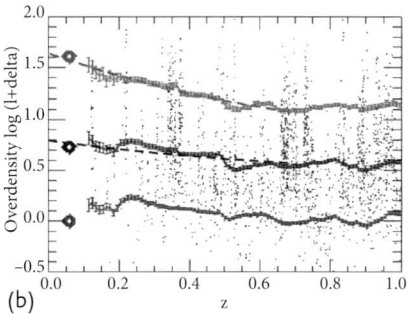

Figure 9.4 (a) The relative environmental quenching efficiency ε_ρ as a function of overdensity $\log(1 + \delta)$, and for several redshift bins. At low density virtually all galaxies are star forming, whereas in the highest density regions, with overdensity exceeding ~ 1000, virtually all galaxies are quenched. (b) The growth of structure through cosmic times from $z = 1$ to 0.1, as described by the evolution of the median overdensity of all galaxy inhabited regions (middle line), as well as for the highest and lowest density quartiles (upper and lower line, respectively). Dashed lines show the corresponding best fit relations. Large circles refer to the local ($z \sim 0.1$) Universe. Small points represent the galaxies used to map the density field (source: Peng, Y. et al. (2010, Astrophys. J., 721, 193)) (Reproduced by permission of the AAS.). For a color version of this figure please see in the front matter.

The ingredients to describe environmental quenching are now in place. With some algebra (not reported here) it can be shown that the environmental quenching rate is given by:

$$\left(\frac{dN_q}{dt}\right)_{env} \simeq \frac{1}{(1-\varepsilon_\rho)} \frac{\partial \varepsilon_\rho}{\partial \log \rho} \frac{\partial \log \rho}{\partial t} \times N_{sf} = \lambda_\rho \phi_{sf}(M_\star), \qquad (9.7)$$

which is the environmental analog to Eq. (9.5) for the mass quenching rate, and where λ_ρ lumps together the three density-depending terms and is independent of mass. Notice that the environmental quenching must be large at intermediate overdensities (cf. Figure 9.4a), where the density derivative of ε_ρ is maximum, and becomes more prominent at late epochs, compared to mass quenching that declines following the declining specific SFR. At low redshifts, in the highest density structures (where $\partial \varepsilon_\rho / \partial \log \rho \simeq 0$) the quenching rate is low just because virtually all galaxies are already quenched (see Figure 9.4a). The total quenching rate is therefore derived by combining Eqs. (9.5) and (9.7), thus obtaining:

$$\left(\frac{dN_q}{dt}\right)_{tot} \simeq \left[25 \left(\frac{t}{3.5\,\text{Gyr}}\right)^{-2.2} \frac{M_\star}{M_\star^*} + \lambda_\rho + \kappa\right] \phi_{sf}(M_\star), \qquad (9.8)$$

which also includes a term κ meant to explore the quenching effect of merging events. This is the only ingredient in the model not based on stellar population diagnostics, and for this reason here it is not discussed.

9.2.3
The Evolving Demography of Galaxies

Equations (9.4) and (9.8) are then integrated together, one describing the mass growth of galaxies resulting from star formation, the other the quenching of star formation via mass, environment and merging. The value of $M_\star^*$ in Eqs. (9.5) and (9.8) is set to the observed value $4 \times 10^{10}\,M_\odot$. As initial conditions, a distribution of seed star-forming galaxies is assumed at $z = 10$, following a power-law mass function ϕ_{sf}° with a slope $\alpha = -1.4$, the same as the low-mass end slope of the Schechter mass function for star-forming galaxies at $z \lesssim 2$ (shown in Figure 6.23). This primordial mass function needs an arbitrary high-mass cutoff to prevent already very massive galaxies from overgrowing, but its precise value is immaterial provided it is lower than $\sim M_\star^*$. Indeed, the way the mass quenching law operates ensures that the correct value of $M_\star^*$ is set up automatically, no matter where the initial cutoff is placed.

The results of putting this simulation into motion are finally shown in Figure 9.5, where Figures 9.5a,c and 9.5b,d refer to the low- and high-density quartiles, respectively. Figure 9.5a,b show the mass functions of star-forming and quenched galaxies at successive redshifts, distinguishing for the latter ones those which have been mass quenched from those which have been environmentally quenched. Figure 9.5c,d show instead the red fraction as a function of stellar mass, for several redshifts. The actual data for the local Universe are shown as solid squares, and

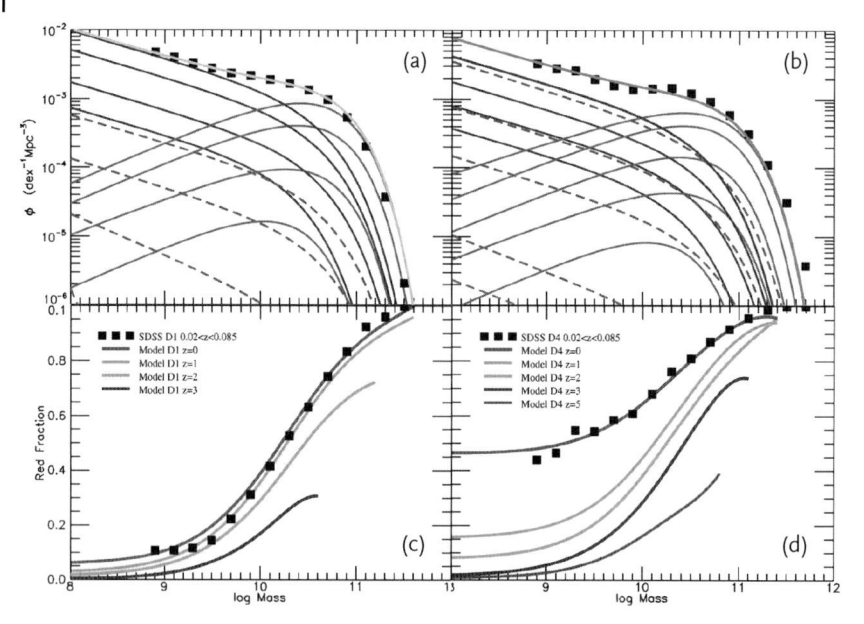

Figure 9.5 The phenomenological model evolution of the mass functions (a,b) and the quenched (red) fraction (c,d), respectively, in the low-density quartile (a,c) and high-density quartile (b,d). Red solid lines show the evolution of the mass function of quenched galaxies that are mass-quenched, while red dashed lines refer to galaxies that are environment-quenched. The blue lines show the evolution of the mass function of star-forming galaxies. Mass functions are shown bottom-up for $z = 3, 2, 1, 0$, and also for $z = 5$ in (b,d). The top gray line is the final, cumulative mass function (star-forming plus quenched galaxies) of the model, with overplotted the empirical mass function from SDSS data (black squares), also shown in (c,d) (source: Peng, Y. et al. (2010, Astrophys. J., 721, 193)) (Reproduced by permission of the AAS.). For a color version of this figure please see in the front matter.

the uppermost lines in panels a and b are the resulting global mass function of the simulation (star-forming and quenched galaxies together) at redshift ~ 0. It is *not* a fit to the data(!). Indeed, the mass function ϕ_{sf} progressively evolves from the input primordial mass function ϕ_{sf}°, into the Schechter mass function of star-forming galaxies, until it perfectly meets the observed mass function at $z \sim 0$, as shown in Figure 6.22. Thus, the adopted quenching algorithm works in such a way to quickly establish and then maintain a Schechter mass function for star-forming galaxies. In the meantime, the mass function of quenched galaxies ϕ_q starts being populated, initially by galaxies that have been mass quenched, and later also by others which have been environmentally quenched. Quite naturally, a double Schechter mass function for quenched galaxies progressively emerges, with the two functions sharing among themselves the same value of $M_\star^*$, as well as with the mass function of star-forming galaxies ϕ_{sf}. The two α differ by one unity, while the two normalizations ϕ_{mq}^* and ϕ_{eq}^* (respectively for the mass- and environment-quenched galaxies) secularly increase at a different pace. All this can be clearly seen by just looking at Eq. (9.8). Whereas ϕ^*_{mq} increases rapidly at early times and then slows down, much of the ϕ^*_{eq} increase takes place at late times. Eventually, the sum of the three mass functions generated by the model perfectly matches the observed global mass function at $z \sim 0$, and does so both in the high-density and low-density regions of the local Universe. All these behaviors are clearly seen in Figure 9.5a–d, which is largely self-explanatory, and are even easier to visualize in the corresponding animation that can be retrieved from the Internet. It is also worth noting from Figure 6.13 that in the local Universe most of the stellar mass in quenched galaxies is contained in galaxies more massive than $\sim 10^{10} M_\odot$, and from Figure 9.5 that above this mass it is mass quenching that dominates. Therefore, in terms of bulk mass, one can say that most quenched mass has been mass quenched.

These main features of the model are not qualitatively altered by including merging. Actually, the full model whose results are shown in Figure 9.5a–d does include a treatment of merging, not only via the term κ in Eq. (9.8), but also to take into account the effect of galaxy merging on the evolution of the mass function of star-forming as well as of the quenched galaxies. Globally, the effect of merging is relatively small, for example, producing an average increase in the mass of quenched galaxies by $\sim 15\%$ in the high density quartile. However, the importance of merging between already quenched galaxies (the so-called *dry merging*), increases with stellar mass, because of the steepness of the mass function at its top end. In the model, the most massive galaxies ($M_\star \simeq 10^{12} M_\odot$) all result from dry merging of two near equal mass quenched galaxies. However, the effect of dry merging even in the highest density quartile (where merging is most effective) is quite modest, increasing $M^*_\star$ of quenched galaxies by just 0.1 dex, that is, from $\sim 4 \times 10^{10}$ to $\sim 5 \times 10^{10} M_\odot$. In this respect one can note that the local mass function of quenched galaxies shown in Figure 6.22 is fit with exactly the same value of $M^*_\star$ both in the high-density and in the low-density quartiles. This suggests that indeed dry merging does not have a dramatic effect. The fact that at high masses $\phi_q \gg \phi_{\text{sf}}$ has been sometime interpreted as evidence that massive ellipticals must be formed by dry merging. Actually, this is only partially true. Indeed, high-mass star-forming galaxies are quite rare, but they do exist, and they are so few just because the probability of having them mass quenched is very high, hence they do not live much before being quenched. Thus, direct transitions from the *blue cloud* to the *red sequence* not only exist also at high masses, but represent the very process that establishes the $M^*_\star$ cutoff of the Schechter mass function.

In summary, based entirely on observationally established facts, the phenomenological model provides a fairly successful scenario for the evolution of the galaxy populations through cosmic times, and its main merits include the following:

- The model establishes the Schechter function of star-forming galaxies, and the constancy of its M^* and α, as demanded by the observations from at least $z \sim 2$ to the present.
- It also establishes the double Schechter mass function, which is characteristic of passive (quenched) galaxies, resulting from having two independent quenching processes, one controlled only by mass and one controlled only by environment,

with the slope of the low-mass end of the mass function that differs by $\Delta\alpha = 1$ between mass-quenched and environment-quenched galaxies, as demanded by observations.
- It produces precisely the local set of values of the Schechter parameters (M^*, ϕ^*, α) for the mass functions of both star-forming and quenched galaxies, and the combination of the single Schechter function for star-forming galaxies with the double Schechter function for quenched galaxies results precisely in the observed double Schechter function for the whole population.
- It shows that the value of M_*^*, common to star-forming and quenched galaxies alike, and irrespective of how they were quenched, is set uniquely by the mass-quenching process.
- It reproduces the observed fraction of quenched galaxies in different environments, as well as the relative weight ϕ_{mq}/ϕ_{eq} of the mass-quenched and environment-quenched galaxies as a function of overdensity.
- It reproduces the observed trend with mass of stellar population ages in quenched galaxies, with the more massive galaxies being statistically quenched at earlier times than less massive ones, hence hosting older stellar populations.
- For most of the galaxies that are mass quenched, the SFR accelerates with time up to quenching epoch, therefore quenching takes place near their maximum SFR. This is the natural consequence of the mass dependence of the SFR, and of a mass-quenching probability that increases with SFR. Thus, most star-forming galaxies at $z \gtrsim 1$ are likely to be caught at their maximum SFR, opposite to what is implicit in the assumption of an exponentially declining SFR, for which they would all be caught at their minimum SFR.
- Star formation being confined to a shorter time interval in massive (predominantly mass-quenched) galaxies, compared to less massive (predominantly environment-quenched) galaxies, naturally reproduces the observed trend of the α-element enhancement with mass (cf. Eq. (6.16)), with the latter being established during the quasiexponential mass growth of galaxies.

One may argue that this remarkable series of successful *predictions* is built-in by having adopted the observed values of α and M_*^* for star-forming galaxies. This is certainly partly true, but it is not the whole story: the star formation rates and quenching algorithms have indeed the effect of keeping these Schechter parameters constant in time, while driving the emergence of the correct mass function of the quenched galaxies. Moreover, as the mapping of galaxy demography and density field proceeds to higher redshifts, it becomes possible to submit to closer and closer observational scrutiny the predicted evolution of the mass functions of star-forming and quenched galaxies, and their relation to environment.

9.2.4
Caveats

As emphasized many times, the model deliberately makes several simplifying assumptions, such as setting $\beta \equiv 0$ in Eq. (6.18) and extrapolating various relations to all redshifts, whereas so far they have been established only below $z \sim 1$. The question is therefore how robust are the results, given that several ingredients are affected by uncertainties, and some of the assumptions may not be fully valid.

One first issue concerns the specific SFR, whose empirical values may be affected by a systematic uncertainty perhaps as large as a factor of ~ 2. This effect is explored in Figure 9.1, showing that indeed a factor of two would matter quite dramatically as far as the mass growth is concerned. By increasing (decreasing) systematically the SFR, that is, by introducing the parameter η after the factor 25 in Eq. (9.8), one can modify at leisure the growth rate of all galaxies. Note that increasing this coefficient in Eq. (9.5) is equivalent to reducing $M_\star^*$, and therefore the quenching law should be modified to counter this effect and ensure that the model produces the observed $M_\star^*$. With a higher SFR ($\eta > 1$) all galaxies would grow faster, thus reaching the mass-quenching condition at an earlier time: quenched, passively evolving galaxies would then appear at a higher redshift. Conversely, reducing the SFR ($\eta < 1$) galaxy growth would be slower, quenching would be delayed, and quenched galaxies would appear at a lower redshift. Thus, the specific SFR and its evolution with redshift works as a *cosmic clock* in this model, accelerating or delaying the unfolding of the events if it is changed, but not altering the outcome, with just one exception. Since in the baseline model with the nominal SFRs ($\eta = 1$) the massive mass-quenched galaxies appear at approximately the right time ($z \sim 2$) one can conclude that the adopted specific SFRs cannot be far from reality. The exception is the relative proportion of mass-quenched and environment-quenched galaxies: increasing the specific SFR at all times would promote more mass quenching while leaving the rate of environmental quenching the same, hence $\phi_{\rm mq}$ would increase with respect to $\phi_{\rm eq}$. Once more, the excellent fit to the local mass function shown in Figure 9.5a–d argues for the specific SFR adopted in the model being close to reality.

The same reasoning applies to the possibility that the specific SFR depends on environment (overdensity), that is, having η as a function of ρ: $\eta(\rho)$. For example, an η that increases with ρ would have the effect of accelerating the mass-quenching rate in high-density regions and of slowing it down in low-density regions, again degrading the match with the observed mass functions. Indeed, the high-mass hump in the double Schechter mass function would become more prominent in high-density regions, and would diminish in low-density regions.

It is certainly possible that the specific SFR of star-forming galaxies is not completely independent of mass, that is, unlike that adopted in the baseline model, β may differ from zero. It can be seen from Eq. (9.5), that in such a case the difference in the low-mass slope of the mass function of mass-quenched and star-forming galaxies becomes $\Delta \alpha = 1 + \beta$. It can also be shown that with $\beta < 0$ the low-mass slope α of the mass function of star-forming galaxies would gradually steepen with time, whereas observations indicate that it does not evolve at all between $z \sim 2$ and 0 (see Figure 6.23). This argues for β being indeed close to zero.

In spite of the model meeting so many observational constraints, departures from the baseline assumptions ($\eta \equiv 1$, $\beta \equiv 0$, etc.) are certainly possible, or to some extent even likely. However, the model is sufficiently flexible that changes in one ingredient could be balanced by relatively small adjustments in others, and one can conclude that the main drivers of galaxy evolution are indeed incorporated in the model, and put at work in ways that produce quite sound results. In any event, since the flow of observational data on high-redshift galaxies is ever increasing, all assumptions of the model can be progressively checked and refined, and so the model along with them.

9.2.5
The Physics of Quenching

The physical processes feeding star formation in galaxies and quenching it are left deliberately out of this fully empirical, phenomenological approach. So, the model provides an intriguing description of what happens, but it does not say why. Yet, certainly one would also like to know what mass quenching and environmental quenching mean in physical terms, that is, what are the actual physical processes at work. A discussion of such processes is beyond the scope of this book, which is fully and exclusively dedicated to stellar populations, and what can be inferred on galaxy evolution by using only the stellar population diagnostics. Nevertheless, it may be worth opening here a side discussion just to allude to the physics, which may be responsible for the observed and simulated phenomenology. If the mass-quenching rate is proportional to the SFR, it is the physics of star formation that should be understood first. This is a very complex issue, but here one can address just one aspect of it: the tightness of the main sequence of star-forming galaxies (cf. Chapter 6). Such tightness implies that within individual galaxies the SFR can fluctuate by typically a factor of ~ 2 up and down with respect to the mean. As already mentioned in Chapter 6, this means that the vast majority of star-forming galaxies, at all redshifts, are making stars at a quasisteady rate, even in the case of high-redshift galaxies with SFRs of $\sim 100\ M_\odot\ \mathrm{yr}^{-1}$ or more, which in the local Universe are exhibited only by starburst galaxies. To sustain such high, and yet quasisteady SFRs fresh gas needs to be continuously fed to galaxies from the environment, and the notion is currently entertained according to which feeding takes place in the form of *cold streams* reaching galaxies from large distances. Indeed, without such a continuous supply, the observed amount of gas inside high-z galaxies would be consumed in a very short time at the observed rates of star formation and mass outflows.

In this scenario the rate of gas accretion via cold streams increases in parallel with the mass growth of galaxies, massive disks at $z \sim 2$ become more and more gas-rich until fragmentation into separate clumps sets in, thus triggering disk instabilities. Then massive clumps migrate and coalesce at the center to form a bulge in a last major star formation event, and, in parallel with it, possibly feeding an AGN. The ensuing feedback from supernovae, or from the AGN. or from both combined, would then be responsible for expelling all the gas from the galaxy, hence quenching star formation. A process of this kind would quite naturally maintain some vestiges of the disk even after quenching, whereas merging may be needed to destroy such vestiges entirely. It may be that mass quenching not followed by merging leads to *disky* elliptical galaxies, whereas merging (and especially dry merging among massive galaxies) may result in *boxy* ellipticals.

Turning now to environmental quenching, two possible processes are being considered: ram-pressure stripping and *strangulation*, sometimes also called *suffocation*. As a galaxy enters a high-density region, such as a cluster, it can be stripped of its gas interacting with the intracluster medium (ICM). For this process to be efficient one needs a dense medium and high velocities relative to it, two conditions that are met in clusters but usually not elsewhere. Thus, it is believed that ram pressure stripping can be responsible for only a fraction, and probably a small one, of all the environmental quenching events. The alternative process (strangulation/suffocation, sometimes called satellite quenching in the CDM parlance), instead of stripping gas from galaxies, is cutting off the cold stream supply of gas, and works once a galaxy enters a higher-density region, where all the gas has already been shock-heated and cold streams have evaporated. One intriguing aspect of environmental quenching is that it appears to affect the SFR only to the extreme, that is, quenching it altogether, and yet unable to affect SFRs at all insofar as galaxies continue to be star forming.

These scenarios for the physics of quenching remain largely speculative, and hard to model without introducing many free parameters. Nevertheless it is likely that the invoked processes do effectively operate and quench star formation in galaxies. It is actually quite ironic to realize that mass and environment may quench star formation in two drastically opposite ways: one by feeding excessive fuel to galaxies, the other by starving them to death.

Observations of large samples of galaxies from low to high redshifts compose a picture of remarkable simplicity:

1. The star formation rate of star-forming galaxies scales almost linearly with mass, strongly declines with cosmic time, and exhibits very small scatter around the average relation.
2. Due to the high observed SFRs the mass of galaxies at high redshifts must increase very rapidly, and yet the mass function of star-forming galaxies evolves only very slightly with redshift.
3. At all redshifts the fraction of quenched (passively evolving) galaxies increases with galactic stellar mass and with local overdensity, with the remarkable property that the relative efficiency of "mass quenching" is independent of environment, and that of "environment quenching" is independent of mass.

The simple phenomenological model summarized in this chapter demonstrates that these three empirical facts suffice to account for the observed evolution of the galaxy mass function and naturally generate the "double Schechter" mass function for quenched galaxies.

We believe that this simple, perfectible model reproduces fairly well the population dynamics of galaxies, that is, their evolving demography, and represents a major step forward towards a systematic understanding of galaxy evolution. Yet, this is still a purely phenomenological model, where there is no physics. At the onset of the era of precision (concordance) cosmology someone once said *"Now that the proper stage has been set up, what we need is a good play."* With this phenomenological model we do not have the play yet, just the plot, but a fairly good one. For the full play we need to introduce actors, that is, the real physical processes at work, and force them to strictly follow the plot. The company for the play includes many actors, and several have prima donna ambitions. Among them are star formation, galactic winds, cold streams, AGN feedback, the IMF, clump physics, supernova feedback, chemical evolution, black hole formation, strangulation, mergers, starbursts, disk instabilities, ram pressure, and more. Quite a turbulent assembly of actors for the theoretical impresarios to orchestrate! We can only wish them good luck.

Acknowledgments

The model heavily relies of SDSS data, in particular from Baldry, I.K. *et al.* (2006, *Mon. Not. R. Astron. Soc.*, 373, 469) and Brinchmann, J. *et al.* (2004, *Mon. Not. R. Astron. Soc.*, 351, 1151), and represents the top products of the zCOSOMS-Bright survey described in Lilly, S. *et al.* (2009, *Astrophys. J. Suppl.*, 184, 218), with particular reference to Pozzetti, L. *et al.* (2010, *Astron. Astrophys.*, 523, A13) and Bolzonella, M. *et al.* (2010, *Astron. Astrophys.*, 524, A76).

Further Reading

This whole chapter is based on the paper:

Peng, Y., Lilly, S. *et al.* (2010) Mass and Environments as Drivers of Galaxy Evolution in SDSS and zCOSMOS and the Origin of the Schechter Function, *Astrophys. J.*, **721**, 193.

It provides a description of its main assumptions, procedures and results.
Results of the original paper are most effectively visualized by looking at the animation www.exp-astro.phys.ethz.ch/zCOSMOS/MF_simulation_d1_d4.mov.

Theory and Physics of Galaxy Formation

Mo, H., van den Bosch, F., and White, S. (2010) *Galaxy Formation and Evolution*, Cambridge University Press.

10
The Chemical Evolution of Galaxies, Clusters, and the Whole Universe

The chemical evolution of a stellar system, be it the solar neighborhood, a whole galaxy or a cluster of galaxies, is governed by relatively simple equations. These equations describe the enrichment of the ISM out of which new stellar generations form, given the star formation history, the rates of mass return for each relevant element, and the rates of mass exchange (in and out) with the surroundings. The robust core of chemical evolution models is represented by stellar evolution prescriptions, giving stellar lifetimes, rates of mass return, and chemical yields. But this is hardly sufficient to construct models, which instead depend heavily on the other, far less robust ingredients: the star formation history and the rates of gas and metal exchange with the surroundings, which all tend to be poorly constrained by the observations.

Rather than introduce the chemical evolution equations and describe some of the resulting models, this chapter aims to provide the reader with a few high level concepts and constraints, without venturing into the detailed chemical history as modeled for specific stellar systems.

Besides shining, stellar populations return mass to the ISM along with newly synthesized elements and isotopes, and the two products are tightly related to each other. The previous chapters have dealt with both aspects, and now the two ingredients are put in play together.

10.1
Clusters of Galaxies

Clusters of galaxies are the largest bound systems in the Universe, and possibly offer the best example of a closed system, that is, a system in which all the actors are present from the beginning to the end. This is to say that present-day clusters are now assumed to contain all the baryons that have contributed to star formation, all the stars that have formed out of them, and all the metals produced by the successive stellar generations. It is also assumed that the baryonic fraction of clusters is in the cosmic ratio, $\Omega_{\rm b}/\Omega_{\rm m} \simeq 0.17$ – a hypothesis that can be subjected to observational testing, and appears to be verified at least for the rich clusters.

Stellar Populations, First Edition. Laura Greggio and Alivio Renzini.
© 2011 WILEY-VCH Verlag GmbH & Co. KGaA. Published 2011 by WILEY-VCH Verlag GmbH & Co. KGaA.

10.1.1
Iron in the Intracluster Medium and the Iron Mass-to-Light Ratio

In one of the rare cases in which theory anticipates observations, the existence of large amounts of heavy elements in the intracluster medium (ICM) was predicted shortly before it was actually observed. This came from (now old-fashioned) *monolithic collapse* models for elliptical galaxy formation, in which the observed color-magnitude relation was reproduced in terms of a metallicity trend. In turn, such a trend was established by supernova-driven galactic winds being more effective in low-mass galaxies with shallow potential wells, compared to more massive galaxies harbored in deeper potential wells. Thus, in this class of models the mass-metallicity relation is established because lower mass galaxies eject a higher fraction of the metals they have produced, compared to more massive galaxies; hence, the ejected metals must be found somewhere in the ICM, if clusters have evolved as closed systems.

These models are now inadequate in many respects, but this prediction was soon confirmed by the discovery of a strong iron-K line in the X-ray spectrum of galaxy clusters, and with time and more potent X-ray satellites, a wealth of other metal lines have been discovered, as illustrated by the sample spectrum shown in Figure 10.1. The figure displays the X-ray spectrum of the inner and outer regions of the cluster A496, whose best fit temperatures are ~ 3.4 and $\sim 4.2\,\text{keV}$, respective-

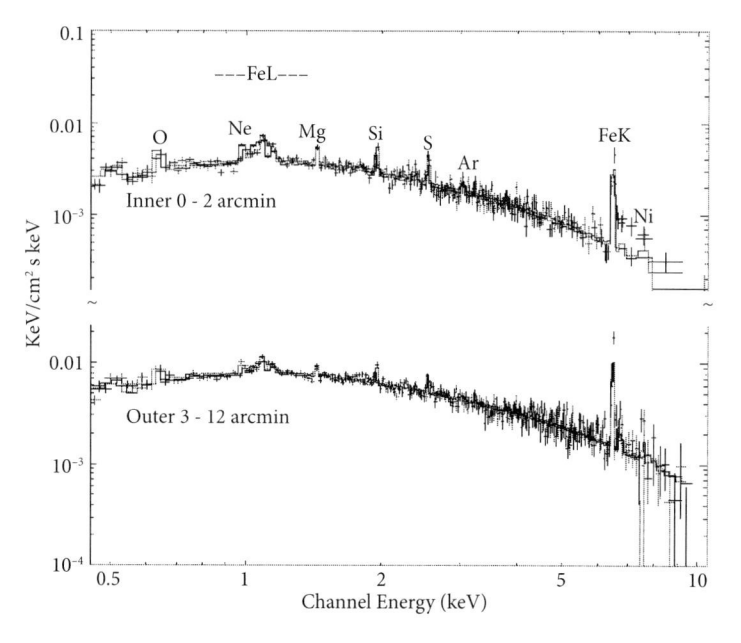

Figure 10.1 The X-ray spectra of the inner and outer parts of the cluster A496 with the main emission lines having been marked. Data are from the ASCA satellite (source: Dupke, R.A. and White, R.E. III (2000, *Astrophys. J.*, 537, 123)) (Reproduced by permission of the AAS.).

ly. Emission lines of several elements are clearly visible, with the most prominent feature being the iron-K line at ~ 6.7 keV. The continuum X-ray emission is due to electron bremsstrahlung, while the lines come from decays from higher energy levels which were populated by collisional excitations, plus collisional ionization and subsequent recombination and cascade transitions down to the ground level. Thus, for example, the Fe-K line comes from the transitions down to the K shell of H-like and He-like iron ions, whereas the Fe-L complex at ~ 1 keV comes from transitions down to the L shell of iron ions with 3 or even many more bound electrons. Getting abundances from such spectra is universally done using packages, which are based on theoretically calculated collisional excitation probabilities. Therefore, the derived abundances of simple ions (e. g., H-like and He-like) should be regarded as more reliable than those of ions still with many bound electrons.

Iron is the best studied element in clusters of galaxies, as ICM iron emission lines are present in all clusters and groups, either warm or hot. Figure 10.2 shows the iron abundance in the ICM of clusters and groups as a function of ICM temperature from a fairly dated compilation, which however can serve to illustrate some of the problems one can encounter. For $kT \gtrsim 3$–4 keV the derived ICM iron abundance is constant at $Z_{Fe} \simeq 0.3 Z_{Fe}^{\odot}$, independent of the ICM temperature. Abundances for clusters in this *horizontal* sequence come from the iron-K complex. At lower temperatures the situation is much less straightforward. Figure 10.2 shows also data for cooler groups, with the iron abundance having been derived with both one-temperature and two-temperature fits. The one-temperature (1T) fits give iron abundances for these cool groups which are more or less in line with those of the hotter clusters. The two-temperature fit (2T) abundances, instead, form an almost vertical sequence, with a great deal of dispersion around a mean ~ 0.75 solar. Ear-

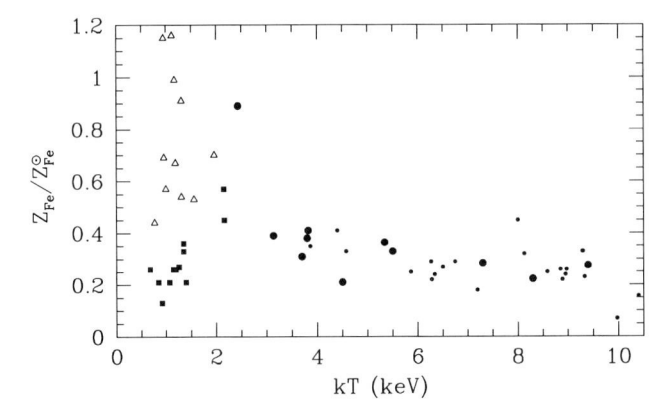

Figure 10.2 A compilation of the iron abundances in the ICM as a function of the ICM temperature for a sample of local clusters and groups. Clusters at moderately high redshift ($\langle z \rangle \simeq 0.35$) are represented by small filled circles. For $kT \lesssim 2$ keV 11 groups are shown, whose temperatures and abundances are determined from 1- and 2-temperature fits (filled squares and open triangles, respectively) (source: Renzini, A. (2004, Clusters of Galaxies: Probes of Cosmological Structures and Galaxy Evolution, eds. J.S. Mulchaey, A. Dressler, and A. Oemler, Cambridge: CUP, p. 261)).

lier estimates gave extremely low values for cooler groups, $kT \lesssim 1\,\text{keV}$, as well as for the iron abundance in the hot ISM of elliptical galaxies, much lower than what is expected from the stellar abundance in the same galaxies plus the ongoing iron enrichment by Type Ia supernovae. Compiling values from the literature a strong dependence of the abundance on ICM temperature is apparent, being very low at low temperatures, steeply increasing to a maximum around $kT \sim 1.5$–$2\,\text{keV}$, then decreasing to reach ~ 0.3 solar by $kT \gtrsim 3$–$4\,\text{keV}$.

Such a strong correlation of the iron abundance with ISM/ICM temperature was highly suspicious, suggesting that some sort of bias was affecting the abundance determinations. The atomic physics of iron ions with more than 2 bound electrons (essential for the iron-L diagnostics) was initially pointed to as the suspect culprit, but now it appears that the problem is more in the 1T fits, whereas the X-ray gas in real galaxies, groups and clusters is unlikely to be isothermal. Indeed, the iron-L emission is extremely sensitive to the plasma temperature for $0.7\,\text{keV} \lesssim kT \lesssim 4\,\text{keV}$, being very low at low temperature, rising steeply to a maximum and then falling steeply to vanish at $kT \gtrsim 4\,\text{keV}$. This is due to iron dominant ionization stages having many bound electrons at low temperatures and emitting in the far UV rather than in the X-ray band, whereas at higher temperatures only H-like and He-like ions remain, and they emit the iron-K lines, rather then in the iron-L complex. Thus, if a multitemperature plasma is fit with an isothermal model, the resulting iron abundance is too low or too high, depending on whether the compromise temperature of the fit is below or above the temperature at which the iron-L emission is maximum. The former effect got the name of *iron bias* and the latter that of *inverse iron bias*, though one may just call it a temperature bias. Thus, this effect accounts for much of the dispersion of the reported iron abundances when they are based on the iron-L complex. Yet, the 2T fits shown in Figure 10.2 exhibit very high iron, much higher than that of hotter clusters. If 2T fits are more realistic, should one conclude that the groups are more iron-rich than clusters? Well, yes and no. The fact is that much of the X-ray emission in poor groups actually comes from their central galaxy, which often gives its name to the whole group. Thus, the resulting iron tends to be a compromise between the iron-rich ISM of the central galaxy, and the lower iron of the genuine intragroup medium, if any. Given all these complications and discrepancies, it is better to concentrate on the hotter clusters, which all exhibit an uncontroversial iron abundance ~ 0.3 solar.

The chemical abundances of the ICM are generally reported in solar units. Unfortunately, the solar units are not carved on rock, but change from time to time, and can be different depending on what one means by the sun. One option is the present *photospheric* abundances as derived from the solar spectrum and model-atmosphere analysis. Another option is the solar-system abundances as derived from the analysis of meteorites (the *meteoritic* abundances). Yet another option refers to *protosolar* abundances, which aims at taking into account that helium and heavy elements have partly diffused out of the photosphere since our sun formed 4.7 Gyr ago. All these abundances change with time as better models and measurements come into being, and the result is that we do not know what for example, *0.3 solar* precisely means until we go to look at which solar abundances have been used as units. This is causing unwanted confusion, as different authors assume a different sun of their choice, whereas all the plasma codes in use calculate an absolute measure of the element-to-hydrogen ratio, *not* a measure relative to the sun. After this plea in favor of absolute, as opposed to relative abundances for the ICM, one can finally reveal that 0.3 solar for the iron abundance in the hot, rich clusters shown in Figure 10.2 actually means an abundance by mass $Z^{Fe}_{ICM} = 0.0006$, or that each kilogram of ICM contains ~ 0.6 grams of iron.

Next question is how many kilograms of iron are contained in the ICM of galaxy clusters, or, better, how many solar masses of iron are contained in the ICM for every solar luminosity of the stellar populations inhabiting the clusters. But, before getting to that one has to face another complication: the abundances shown in Figure 10.2 refer to the cluster central regions, whereas radial gradients in the iron abundance have been reported for several clusters. The results of a systematic study of the radial distribution of iron are shown in Figure 10.3a,b. Clusters break up in two distinct groups: the so-called *cool core* clusters (CC, formerly known as *cooling flow* clusters before the failure of the stationary cooling flow model was generally acknowledged) are characterized by a steep iron gradient in the core, with the abundance reaching ~ 0.6 solar near the center. These clusters are characterized by the presence of a strong ICM temperature and density gradient towards the center, after which they are named. Instead, in non-CC clusters (where no temperature gradient is found) no metallicity gradient appears either. The origin of the dichotomy remains to be fully understood. The fact that metallicity gradients are found in association with large temperature gradients in the central regions may look suspicious, as noted for the strong dependence of Z^{Fe} on ICM temperature, but the existence of such gradients now appears to be well established. Since the temperature structure of non-CC clusters is definitely simpler than that of CC clusters, their iron abundances appear to be less prone to the temperature bias mentioned above. As shown in Figure 10.4, some of the most recent measurements give slightly lower values for non-CC clusters, $Z^{Fe}_{ICM} \simeq 0.0004$, hence in the following this value is

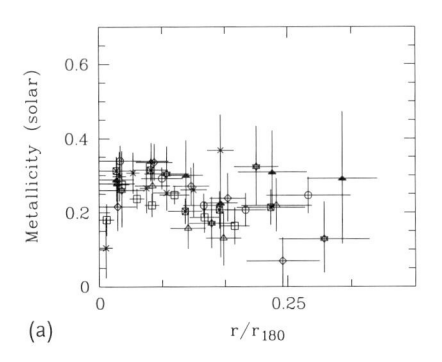

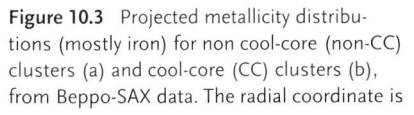

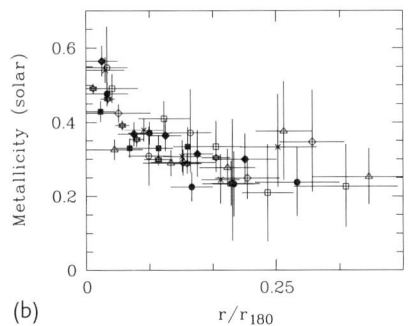

Figure 10.3 Projected metallicity distributions (mostly iron) for non cool-core (non-CC) clusters (a) and cool-core (CC) clusters (b), from Beppo-SAX data. The radial coordinate is normalized to the radius with overdensity factor 180 (source: De Grandi, S. and Molendi, S. (2001, *Astrophys. J.*, 551, 153)) (Reproduced by permission of the AAS.).

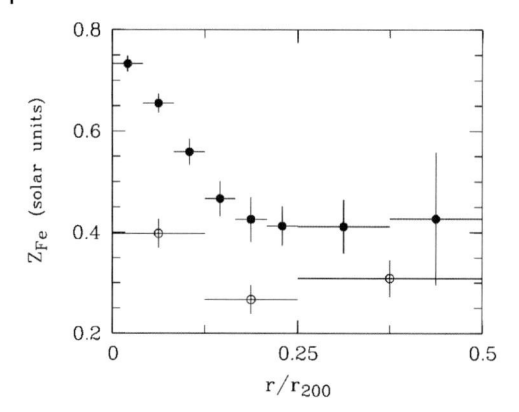

Figure 10.4 Deprojected iron abundance relative to solar as a function of radial distance, averaged over a set of CC clusters (filled circles) and non-CC clusters (open circles). The radial coordinate is normalized to the radius with overdensity factor 200 (source: De Grandi, S. et al. (2004, Astron. Astrophys., 419, 7)) (Reproduced by permission of © ESO.).

taken as the lower limit for the likely abundance of iron in the ICM of rich clusters, and 0.0006 is taken as the upper limit of its range.

The iron-mass-to-light-ratio (FeM/L) of the ICM is defined as the ratio M_{Fe}^{ICM}/L_B of the total iron mass in the ICM over the total B-band luminosity of the galaxies in the cluster. This is a quite useful quantity, because it relates two products of the stellar populations inhabiting clusters of galaxies, namely the whole mass of iron they have produced and dispersed in the ICM through all their previous history, and their present luminosity. In turn, the total iron mass in the ICM is given by the product of the iron abundance times the mass of the ICM, that is, $M_{Fe}^{ICM} = M_{ICM} Z_{ICM}^{Fe}$, and Figure 10.5 shows the resulting FeM/L from an earlier compilation. The drop of the FeM/L in poor clusters and groups (i.e., for $kT \lesssim 2\,\text{keV}$) can be traced back to a drop in both the iron abundance (which however may not be real, see above) *and* in the ICM mass. Such groups appear to be gas-poor compared to clusters, which suggests that they may have been subject to baryon and metal losses due to strong galactic winds driving much of the ICM out of them. Alternatively, early winds may have *preheated* the gas around galaxies, thus preventing it to *fall* inside groups, or having *inflated* the gas distribution relative to the distribution of galaxies. In one way or another, groups are not discussed further because of the greater uncertainty of their abundances, and of the quite concrete possibility that they have not evolved as a closed system. Thus, here we deal with clusters with $kT \gtrsim 3\,\text{keV}$, for which the interpretation of the data appears more secure. Still, cautionary remarks are in order, even concerning these hotter clusters.

A main concern is that two of the three ingredients entering in the calculation of the FeM/L values shown in Figure 10.5 (namely M_{ICM} and L_B) may not be measured consistently in the various sources used in the compilation. Both quantities come from a radial integration up to an ill-defined cluster boundary, for example,

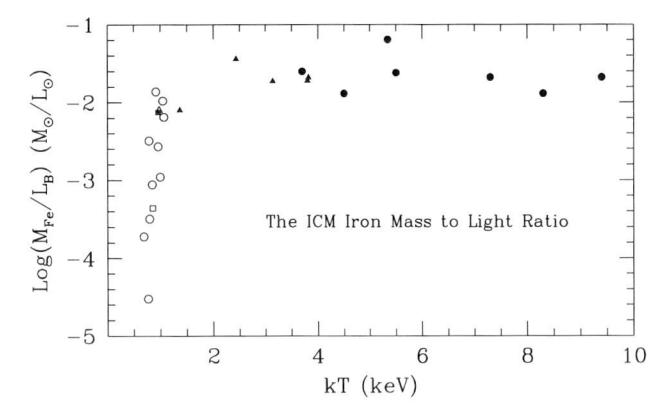

Figure 10.5 The iron mass-to-light ratio of the ICM of clusters and groups as a function of the ICM temperature from an earlier compilation, for which 50 km s^{-1} Mpc^{-1} was assumed for the Hubble constant (source: Renzini, A. (2004, Clusters of Galaxies: Probes of Cosmological Structures and Galaxy Evolution, ed. J.S. Mulchaey, A. Dressler, and A. Oemler, Cambridge: CUP, p. 261)).

the Abel radius, the virial radius, or to a radius at some fixed overdensity. Sometimes it is quite difficult to ascertain what definition has been used by one author or another, with the complication that in general X-ray and optical data have been collected by different groups using different assumptions. There is certainly room for improvement here, and a new compilation, paying attention to analyze all clusters in a homogeneous way, would be highly desirable. Finally, estimated total luminosities L_B refer to the sum over all cluster galaxies, and do not include the population of stars which are diffused through the cluster, and which may account for at least $\sim 10\%$ of the total cluster light, and perhaps more.

Various determinations can be found in the literature for FeM/L, and evaluating the level and uncertainty on this quantity is not so straightforward. Clearly, one should ensure that total luminosity and total ICM mass refer to the same cluster boundary, and when adopting different sources for L_B and M_{ICM} this is not guaranteed. Optical catalogs give coordinates for each individual galaxy, an L_B is given within a certain distance from the center in Mpc, whereas ICM masses from X-ray observations are often given within a certain overdensity, that varies from one case to another (e.g., 150, 200, 500, 1000, ...). These *language* differences between optical and X-ray astronomers often make it quite laborious to derive consistent values for these quantities. A similar problem affects the determination of the iron mass in the ICM, obtained by multiplying its average abundance Z^{Fe}_{ICM} by the ICM mass M_{ICM}, the two factors coming in general from different sources. Moreover, the average abundance may be representative of a region smaller than that sampled to derive the mass of the ICM, as in general the iron abundance is less well measured in the outer, low-density regions with small emission measure.

While keeping these caveats in mind, one sees from Figure 10.5 that FeM/L runs remarkably flat with increasing cluster temperature, for $kT \gtrsim 3$ keV. This constancy of the FeM/L comes from both Z^{Fe}_{ICM} and M_{ICM}/L_B showing very little trend with cluster temperature. Therefore, we adopt here (numerical value from White, S.D.M. *et al.* (1993, Nature, 366, 429)):

$$\frac{M_{\rm ICM}}{L_{\rm B}} \simeq 25 h_{70}^{-1/2} \left(\frac{M_\odot}{L_{\rm B,\odot}}\right), \qquad (10.1)$$

for all rich clusters, where this value was measured on the Coma cluster for which the galaxy census and the mapping of the X-ray emission are fairly complete, and $h_{70} = H_\circ/(70 {\rm km\,s^{-1}\,Mpc^{-1}})$. The resulting $Fe M/L$ in solar units is therefore:

$$\left(\frac{Fe M}{L_{\rm B}}\right)_{\rm ICM} = Z_{\rm ICM}^{\rm Fe} \frac{M_{\rm ICM}}{L_{\rm B}} \simeq (4\text{--}6) \times 10^{-4} \times 25 h_{70}^{-1/2}$$

$$\simeq 0.01\text{--}0.015 h_{70}^{-1/2} \left(\frac{M_\odot}{L_{\rm B,\odot}}\right), \qquad (10.2)$$

that is, in the ICM there are about 0.01 solar masses of iron for each solar luminosity of the cluster galaxies. This value of $Fe M/L$ may be slightly underestimated, as other studies indicate values as high as $0.02\,h_{70}^{-1/2}\,(M_\odot/L_{\rm B,\odot})$, and therefore it seems fair to conclude that the ICM iron mass-to-light ratio is between 0.01 and 0.02 solar masses for each solar B-band luminosity of the stellar populations inhabiting the cluster. This estimate does not change appreciably if one includes in $L_{\rm B}$ also the $\sim 10\%$ contribution by stars free floating within the clusters, for not being bound to any detected galaxy.

The most straightforward interpretation of the constant $Fe M/L$ is that clusters did not lose iron (hence baryons), nor differentially acquired pristine baryonic material, and that the conversion of baryonic gas into stars and galaxies did proceed with the same efficiency and the same stellar IMF in all rich clusters. Otherwise, there should be cluster-to-cluster variations of $Z_{\rm ICM}^{\rm Fe}$ and $Fe M/L$. All this is true insofar as the baryon to dark matter ratio is the same in all $kT \gtrsim 3 {\rm\,keV}$ clusters, and the ICM mass-to-light ratio and the gas fraction are constant. Nevertheless, there may be hints for some of these quantities showing (small) cluster-to-cluster variations, but no firm conclusion has been reached yet.

10.1.2
The Iron Share between ICM and Cluster Galaxies

The metal abundance of the stellar component of cluster galaxies is derived from integrated spectra coupled to synthetic stellar populations. Much of the stellar mass in clusters is confined to passively evolving spheroids (ellipticals and bulges), for which the iron abundance $Z_*^{\rm Fe}$ may range from $\sim 1/3$ solar to a few times solar, see Eq. (6.16). The iron abundance in galaxies coming from integrated spectra of stellar populations, it ultimately reflects the photospheric abundance of the constituent stars which are measured relative to the sun. Therefore, for the solar iron it is appropriate to adopt the latest determination of the solar *photospheric* abundance, that is, $Z_{Fe}^{\odot} = 0.0011$. The Fe M/L of cluster galaxies is then given by:

$$\left(\frac{Fe M}{L_B}\right)_{gal} = Z_*^{Fe}\frac{M_*}{L_B} \simeq 0.006 h_{70} \quad (M_{\odot}/L_{B,\odot}), \tag{10.3}$$

where $Z_*^{Fe} = Z_{Fe}^{\odot}$ and one has adopted $M_*/L_B = 5 h_{70}$. This estimate comes from the M/L_B determinations for the sample of ellipticals shown in Figure 8.4, which have been used to derive an average *dynamical* mass-to-light ratio $\langle M/L_B \rangle = 7 h_{70}$. This value is further reduced to ~ 5 assuming a likely $\sim 30\%$ dark matter contribution to the total mass within the galaxy effective radius. Thus, in rich clusters the fraction of baryons now in stars is assumed to be $\sim 5/(5+25) \simeq 0.16$, in fair agreement with other recent determinations giving $\sim 14\%$. The ratio of the iron mass in the ICM to the iron mass locked into stars and galaxies is given as the ratio of the two Fe M/L:

$$\frac{Z_{ICM}^{Fe} M_{ICM}}{Z_*^{Fe} M_*} \simeq (1.6-3.3) \times h_{70}^{-3/2}. \tag{10.4}$$

So, it appears that there is ~ 2 times more iron mass in the ICM than locked into cluster stars (galaxies), perhaps even more if Z_*^{Fe} is subsolar due to abundance gradients within individual galaxies. In turn, this empirical iron share (ICM versus galaxies) sets a strong constraint to models for the chemical evolution of galaxies, and demonstrates that galaxies do not evolve as chemically closed systems. Actually, it appears that they eject more iron than they are able to retain.

Combining Eqs. (10.2) and (10.3) one then obtains the total cluster Fe M/L:

$$\left(\frac{Fe M}{L_B}\right)_{cl} \simeq (4-6) \times 10^{-4} \times \left(\frac{M_{ICM}}{L_B}\right) h_{70}^{-1/2} + 1.1 \times 10^{-3} \frac{M_*}{L_B} h_{70}, \tag{10.5}$$

or $\sim 0.016-0.021$ $M_{\odot}/L_{B,\odot}$, for the adopted values of the mass-to-light ratio of the ICM and galaxies, and for $h_{70} = 1$.

It is worth cautioning that the values of the total cluster Fe M/L and of the iron share derived in this section strictly depend on the adopted values of M_*/L_B, M_{ICM}/L_B and on the iron abundance in the ICM and galaxies, which may all be subject to change as better estimates become available.

10.1.3
Elemental Ratios

Several X-ray observatories have sufficiently high spectral resolution that, besides those of iron, the emission lines of many other elements can be detected and measured (see Figure 10.1). These include oxygen, neon, magnesium, calcium, silicon, sulfur, argon, and nickel. Most of these are α elements (i.e., made by an integer number of α particles), which are predominantly synthesized in massive stars exploding as core collapse supernovae, whereas silicon can also be contributed by thermonuclear, Type Ia supernovae (see Chapter 7). As is well known, iron-peak elements are also produced by Type Ia supernovae, and 50–75% of iron in the sun may come from them.

Estimates of the α-element to iron ratio $\langle[\alpha/\text{Fe}]\rangle$ in the ICM have fluctuated considerably since the early attempts. Initially substantially supersolar values were reported, which prompted claims that the ICM was enriched predominantly by massive stars, with no detectable contribution from Type Ia supernovae. Then subsequent estimates went down to near solar or even below solar α/Fe ratios. The main difficulty for an accurate measure of α-element to iron ratios comes from the fact that iron is best measured in hot clusters from the iron-K lines, but in these clusters most α elements are fully ionized, and their X-ray lines are very weak. Instead, the lines of α elements are strong in cool clusters, where iron can be measured from the iron-L complex, but it is subject to the temperature bias mentioned above.

Several attempts have been made to constrain the relative role of the two kinds of supernovae in the enrichment of the ICM, or even to choose among different models for the explosion of thermonuclear supernovae. But given the sizable changes in the abundance determinations from one study to the next, it may be worth restraining to less ambitious goals.

Taken from a recent compilation, Figure 10.6a–d shows several abundance ratios relative to iron in the ICM of clusters with a wide range of temperatures. These ratios appear to be essentially constant with temperature, at near-solar or slightly subsolar values, with just the Ni/Fe ratio appearing to be above solar. In particular, the average Si/Fe ratio is ~ 0.7 solar with relatively small scatter, which given the solar abundances adopted to draw Figure 10.6a implies $Z_{\text{ICM}}^{\text{Si}}/Z_{\text{ICM}}^{\text{Fe}} \simeq 0.41$ for the

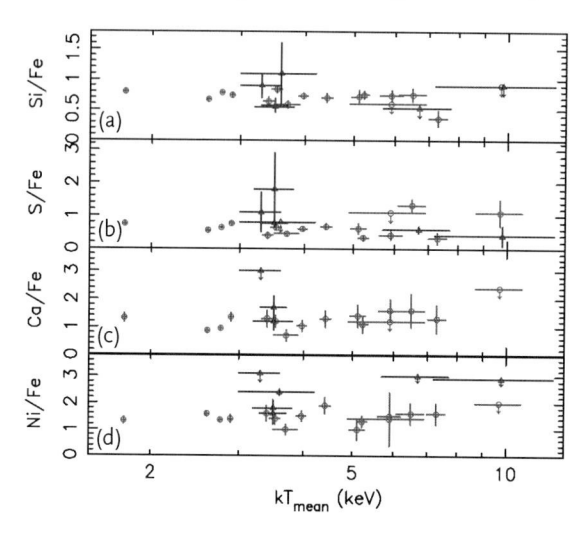

Figure 10.6 The silicon (a), sulfur (b), calcium (c) and nickel-to-iron ratios (d) in the ICM (relative to the solar ratios) as a function of ICM temperature. No appreciable trend with ICM temperature (i. e., cluster richness) is apparent (source: Werner, N. et al. (2008, Space Sci. Rev., 134, 337)). For a color version of this figure please see in the front matter.

silicon-to-iron mass ratio in the ICM. Therefore, using Eq. (10.2) the silicon mass-to-light ratio is here estimated as $M^{Si}_{ICM}/L_B \simeq (0.004$–$0.006)h_{70}^{-1/2}$ $(M_\odot/L_{B,\odot})$.

Oxygen is the most abundant of all α elements, but its X-ray lines can be measured only in relatively cool clusters $kT \lesssim 2$–3 keV, and therefore the O/Fe ratio is affected by somewhat larger uncertainties. The most recent results indicate solar or sightly subsolar values. Adopting a solar O/Fe ratio for the ICM, one gets $Z^O_{ICM}/Z^{Fe}_{ICM} \simeq 4.7$ for the oxygen-to-iron mass ratio in the ICM. Again, from Eq. (10.2) the oxygen mass-to-light ratio in the ICM is here estimated as $M^O_{ICM}/L_B \simeq (0.047$–$0.07)h_{70}^{-1/2}$ $(M_\odot/L_{B,\odot})$.

As far as the abundance of these elements in cluster galaxies is concerned, following Eq. (6.16) we adopt an average [O/Fe] = [Si/Fe] = +0.2, and therefore the oxygen and silicon mass-to-light ratios of cluster galaxies is estimated to be ~ 0.044 and ~ 0.0034 $(M_\odot/L_{B,\odot})$, respectively. Combining with the above values for the ICM, the total cluster oxygen and silicon mass-to-light ratios become ~ 0.09–0.11 and ~ 0.007–0.009, respectively. To generously account for possible systematic errors one can perhaps double or triple this uncertainty range, but the middle values appear to be fairly reasonably well established, and both have been anticipated in Chapter 8, and used to constrain the IMF slope between ~ 1 and $\sim 40\, M_\odot$.

Assuming solar elemental proportions for the ICM, the ICM metal-mass-to-light ratio is therefore $\sim 0.3 \times 0.02 \times 25 h_{70}^{-1/2} = 0.15\, (M_\odot/L_{B,\odot})$, having adopted $Z_\odot = 0.02$ and for $h_{70} = 1$. By the same token, the overall cluster metal-mass-to-light ratio (ICM + galaxies) is $\sim 0.25\, (M_\odot/L_{B,\odot})$, that is, clusters contain $\sim 1/4$ of a solar mass of full metals, for every solar luminosity of its galaxy members.

10.1.4
Metal Production: the Parent Stellar Populations

The constant Fe M/L of clusters obviously means that the total mass of iron in the ICM is proportional to the total optical luminosity of the cluster galaxies. The simplest interpretation of this linearity is that the iron and all the metals now in the ICM have been produced by supernovae belonging to the same stellar population whose surviving low-mass stars now radiate the bulk of the cluster optical light. Since much of the cluster light comes from old spheroids (ellipticals and bulges), one can conclude that *the bulk of cluster metals were produced by the same stellar generations that make up the old spheroids that we see today in clusters.*

It is also worth asking which galaxies have produced the bulk of the iron and the other heavy elements, that is, what has been the relative contributions to the metal production as a function of the present-day luminosity and mass of cluster galaxies. From their luminosity function it is easy to realize that the bright galaxies (those with $L \gtrsim L^*$ and $M_\star \gtrsim M_\star^*$) produce the bulk of the cluster light and account for the bulk of the stellar mass, while the contribution of the dwarfs is negligible in spite of them dominating the galaxy counts by a large margin (cf. Figure 6.13). This is valid for the general galaxy population as well as for the population of cluster galaxies. In practice, most galaxies do not do much, while only the brightest $\sim 3\%$ of all galaxies contribute $\sim 97\%$ of the whole cluster light and mass. Far from being politically correct, giants dominate the scene whereas dwarfs do not count much. Following the simplest interpretation according to which metals were produced by the same stellar population that now shines, one can conclude that the bulk of the cluster metals have also been produced by the giant galaxies that contain most of the stellar mass. The relative contribution of dwarfs to ICM metals may have been somewhat larger than their small relative contribution to the cluster light, since metals can more easily escape from their shallower potential wells. Yet, this is unlikely to alter the conclusion that the giants dominate metal production by a very large margin. Indeed, almost 2/3 of the whole mass in stars in the local Universe is now in spheroids, $\sim 1/3$ in disks, and less than 1% in irregular galaxies, and in clusters the dominance of spheroids is likely to be even higher than in the general field. Therefore, the prevalence of spheroids offers an opportunity to estimate the epoch (redshift) at which (most) metals were produced and disseminated, since one knows quite well when most stars in cluster spheroids were formed. As illustrated in Chapter 6, all evidence converges to indicate that most stars in cluster ellipticals formed at $z \gtrsim 2$, hence at a lookback time of ~ 10–12 Gyr, while only minor residual star formation has occurred later.

With most of star formation in clusters having taken place at such high redshift, most cluster metals should also have been produced and disseminated at $z \gtrsim 2$. Little evolution of the ICM composition is then expected all the way to fairly high redshifts, with the possible exception of iron from SNIa, whose rate of release does not follow as closely the star formation rate as does the core collapse supernovae (CC SNe) rate. Therefore, not much evolution of the ICM composition is expected to fairly high redshifts ($z \sim 1$).

10.1.5
Iron from SNIa

The E/S0 galaxies that dominate cluster light produce only Type Ia SNe at a rate of $\sim (0.16 \pm 0.06) h_{70}^2$ SNu, with 1 SNu corresponding to 10^{-12} SNe yr^{-1} $L_{B\odot}^{-1}$. Assuming such a rate to have been constant through cosmological times (~ 12 Gyr), the number of SNIa exploded in a cluster of present-day luminosity L_B is therefore $\sim 1.6 \times 10^{-13} \times 1.2 \times 10^{10} L_B h_{70}^2 \simeq 2 \times 10^{-3} L_B$. With each SNIa producing $\sim 0.67 M_\odot$ of iron (cf. Chapter 7), the resulting FeM/L of clusters would be:

$$\left(\frac{M_{Fe}}{L_B} \right)_{SNIa} \simeq (1.4 \pm 0.5) \times 10^{-3} h_{70}^2 , \qquad (10.6)$$

which falls short by a factor of ~ 10–20 compared to the observed total cluster FeM/L (0.016–0.021 for $h_{70} = 1$). The straightforward conclusion is that either SNIa did not play any significant role in manufacturing iron in clusters, or their rate in what are now E/S0 galaxies had to be much higher in the past. This argues for a strong evolution of the SNIa rate in E/S0 galaxies and bulges, with the past average being ~ 10 times higher than the present rate.

In Chapter 7 the number of Type Ia supernovae produced by a given amount of gas turned into stars has been semiempirically evaluated as $k_{Ia} \simeq 2.5 \times 10^{-3}$

events/$M_\odot$. Therefore, the total number of events that have occurred through the whole history of clusters per unit B-band light is given by:

$$N_{SNIa} = k_{Ia} \frac{\mathcal{M}}{L_B} L_B, \qquad (10.7)$$

where $\mathcal{M}$ is the mass that went into stars. This can be evaluated as the present mass (at an age of 10–12 Gyr), corrected for the mass return, a correction that amounts to a factor 1.6 for a Salpeter-diet IMF (cf. Chapter 2). Thus, the SNIa contribution to the whole cluster FeM/L is:

$$\left(\frac{M_{Fe}}{L_B}\right)_{SNIa} \simeq 1.6 \times 0.67 \times k_{Ia} \frac{M_\star}{L_B} \simeq 0.013 \left(\frac{M_\odot}{L_{B,\odot}}\right), \qquad (10.8)$$

where $M_\star/L_B = 5$ has been adopted, as in Eqs. (10.3) and (10.5). Thus, Type Ia supernovae alone come close to accounting for all the iron in clusters, whose FeM/L has been estimated to be in the range of 0.016–0.021 ($M_\odot/L_\odot$). Yet, core collapse supernovae may also have a role to play.

10.1.6
Iron and Metals from Core Collapse SNe

In the case of SNIa, we currently believe to have a fairly precise knowledge of the amount of iron produced by each event, while the nature of the progenitors and the run of the rate still remain open issues. The case of CC SNe is quite the opposite: one believes to have unambiguously identified the progenitors (stars more massive than $\sim 8\,M_\odot$), while a great uncertainty affects the amount of iron $M_{Fe}^{CC}(M)$ produced by each CC SN event as a function of the progenitor's mass. This is due to the fundamental difficulty for core-collapse SN models to precisely locate the *mass cut* between the collapsing core going to form the neutron star (or BH) remnant, and the ejected envelope, with such cut often being within the iron-peak layer. As discussed in Chapter 7, the amount of radioactive ^{56}Ni ejected in CC SN events can be measured from the supernova light curve, but then it is very difficult to assign an initial mass to the progenitor star.

The total number of CC SNe – N_{CCSN} – is obtained by integrating the stellar IMF from for example, 8 to 40 $M_\odot$, with the IMF expressed following the formulation introduced in Chapters 2 and 3:

$$\phi(M) = A(t,Z)\,M^{-s} = a(t,Z)\,L_B\,M^{-s}, \qquad (10.9)$$

where $a(t,Z)$ is a (slow) function of the SSP age and metallicity. For a stellar population of age $t = 12$ Gyr, $a(Z) = 2.22$ and 3.12, respectively for $Z = Z_\odot$ and $2Z_\odot$, with L_B and M being in solar units. Clearly, the flatter the IMF slope the larger the number of massive stars per unit present luminosity, the larger the number of CC SNe, and the larger the massive star contribution to the FeM/L. Thus, adopting $\langle M_{Fe}^{CCSN}\rangle = 0.057\,M_\odot$ as documented in Chapter 7, $a = 3$ and integrating over the

IMF one gets:

$$\left(\frac{M_{Fe}}{L_B}\right)_{CC} = \frac{M_{Fe}^{CC} N_{CC}}{L_B} \simeq 0.002\,;\quad \simeq 0.007\,;\quad \simeq 0.03\ \left(\frac{M_\odot}{L_{B,\odot}}\right), \quad (10.10)$$

for $s = 2.7$, 2.35 and 1.9, respectively. Hence, if the Salpeter IMF slope above $\sim 1\,M_\odot$ ($s = 2.35$) applies also to cluster ellipticals, then CC SNe underproduce iron by a factor of ~ 2–3 compared to the total cluster Fe M/L that has been estimated above in the range 0.016–0.021 ($M_\odot/L_{B,\odot}$). Clearly, an IMF somewhat flatter than Salpeter's, but not dramatically so, would suffice to produce all iron in clusters by CC SNe alone.

Combining with the contribution from SNIa, with a Salpeter-slope IMF above $1\,M_\odot$ the total contribution of supernovae to the cluster Fe M/L becomes $\sim 0.013 + 0.007 = 0.020$, well within the range (0.016–0.021) estimated from the observations. Moreover, SNIa appear to contribute $\sim 2/3$ of the iron, and CC SNe $\sim 1/3$, a fairly reasonable share indeed.

Given that iron is produced by both types of supernovae, it is not the best element to constrain the IMF slope in the high mass range. Instead, α elements are produced almost exclusively by CC SNe, and therefore the IMF slope can be better constrained by them. In particular, this is so for oxygen and silicon, although SNIa may also contribute almost as much silicon as CC SNe (cf. Chapter 7). This exercise was anticipated in Chapter 8, using the oxygen- and silicon mass-to-light ratios estimated above. In summary, it appears that with a Salpeter-diet IMF and standard nucleosynthesis prescriptions massive stars can produce the observed amounts of oxygen and silicon present in clusters of galaxies, while perhaps falling short by a factor ~ 2–3 to produce the observed iron. Yet, with an IMF just slightly shallower than Salpeter SNII could also make all the iron, but then there should be clearer evidence for an α-element overabundance in galaxy clusters than currently indicated by the observations.

All in all, nucleosynthesis may well have proceeded in clusters not unlike in the Milky Way, where one currently believes that about $\sim 1/2$–$2/3$ of the iron has been produced by thermonuclear, Type Ia supernovae, and the rest by core-collapse supernovae. With $s \simeq 2.35$ and a past average rate of SNIa in ellipticals ~ 10 times the present rate, the iron content of clusters and the global ICM [α/Fe] ratio are accounted for. Therefore, there appears to be no need to abandon the attractive simplicity of a still viable universal nucleosynthesis process (i.e., IMF and SNIa/CC SN productivity ratio) for embarking towards more complex, multiparametric scenarios.

10.2
Metals from Galaxies to the ICM: Ejection versus Extraction

Having established that most metals in clusters are now out of the parent galaxies, one would like to understand how they were transferred from galaxies to the ICM. There are two main possibilities: (i) extraction by ram pressure stripping as galaxies plow through the ICM, and (ii) ejection by galactic winds powered from inside galaxies themselves. In the latter case the power can be supplied by supernovae (the so-called star formation feedback) and/or by AGN activity (AGN feedback). Notice that here one encounters once more phenomena that may be strictly related to mass quenching and environmental quenching (cf. Chapter 9).

Ram pressure stripping certainly operates in clusters, as clusters are assembled by growing groups and isolated galaxies, which by entering a dense ICM are stripped of their gas and quenched. However, several arguments favor ejection over extraction for doing most of the work:

- There appears to be no trend of either $Z_{\rm ICM}^{\rm Fe}$ or the Fe M/L with cluster temperature or cluster velocity dispersion ($\sigma_{\rm v}$), whereas the efficiency of ram pressure stripping increases steeply with increasing $\sigma_{\rm v}$.
- Field ellipticals appear to be virtually identical to cluster ellipticals. They follow basically the same color-magnitude, line strength-σ, and fundamental plane relations, with just barely noticeable differences. Given that most metals reside in the ICM, if stripping was responsible for extracting metals from galaxies one would expect galaxies in low-density environments to have retained about twice more metals than cluster galaxies, hence showing higher metal indices for given σ, which is not seen.
- Nongravitational energy injection to the ICM seems to be required to account for the break of the self-similar X-ray luminosity-temperature relation for groups and clusters. While galactic winds are an obvious vehicle for such *preheating*, no preheating is associated with metal transfer by ram pressure.
- Strong galactic winds are actually observed in star-forming galaxies at high redshift, and some of them must be like the progenitors of local ellipticals in clusters.
- Most star formation and associated metal production in today's clusters took place at very early times ($z \gtrsim 2$), probably well before the clusters were assembled, hence before the dense ICM was in place, and therefore when there was not much ram pressure that the galaxies could experience.

One can quite safely conclude that most of the metals in the ICM have been *ejected* from galaxies by supernova- or AGN-driven winds, rather than stripped by ram pressure. Moreover, if the empirical model illustrated in Chapter 9 is basically correct, then the bulk of quenched mass was mass quenched, and mass quenching is independent of environment. On the other hand, clusters are the objects with the highest overdensity in the Universe, and therefore where environmental quenching must have been most effective. Therefore, a contribution by ram-pressure stripping must be present, and is likely to be related to the transformation of spirals into S0 galaxies at relatively low redshifts ($\lesssim 1$), as they enter the clusters and start interacting with the ICM.

10.3
Clusters versus Field and the Overall Metallicity of the Universe

To which extent are clusters fair samples of the $z \sim 0$ Universe as a whole? In many respects clusters look much different from the field, for example, in the morphological mix of galaxies, or in the star formation activity which in clusters has almost completely ceased while it is still going on in the field. Yet, when we restrict ourselves to some global properties, rich clusters and field are not so different. For example, the baryon fraction of the universe is $\Omega_b/\Omega_m \simeq 0.17 \pm 0.01$, which compares to $\sim 0.14 \pm 0.01$ as estimated for clusters, adopting $h_{70} = 1$. This tells us that no appreciable baryon versus dark matter segregation has taken place at a cluster scale.

Even more interesting is the case of the star mass over baryon mass in clusters and in the field, which measures the efficiency of baryon to star conversion in these different environments. This star fraction is given by $f_\star^{cl} = M_\star/(M_\star + M_{ICM})$ in clusters, and by $\Omega_\star/\Omega_b$ in the field, where $\Omega_\star = \rho_\star/\rho_c$, $\Omega_b = \rho_{baryon}/\rho_c$ are the contribution of stars and baryons to the cosmic density parameter, with $\rho_c = 1.36 \times 10^{11} h_{70}^2 \, M_\odot \mathrm{Mpc}^{-3}$ being the critical density of the Universe.

Estimates of the stellar mass-to-baryon mass ratio in rich clusters have ranged from ~ 0.10 up to 0.16 (the value that has been adopted above), whereas estimates for the general field have been given in the range from ~ 0.10 down to ~ 0.06 (Fukugita, M. and Peebles, P.J.E. (2004) *Astrophys. J.*, 616, 643). Thus, we are in the somewhat embarrassing situation in which we do not know for sure whether the gas to star conversion in the field was nearly the same as in clusters, or almost 3 times lower. This is a very big difference, with quite diverging ramifications, and efforts to reduce it as much as possible would be most welcome. Much of this difference arises from the different stellar mass-to-light ratios adopted for cluster and field galaxies, where $M_\star/L_B = 5$ has been adopted here for clusters, and $M_\star/L_B = 2.4$ has been used to get the lower, 0.06 value for the star mass-to-baryon mass ratio in the local Universe. A higher $M_\star/L_B$ in clusters with respect to the field is obviously to be expected, because stars in galaxy clusters are on average definitely older than those in the general field. Yet, the difference may not be as large as adopted here.

In any event, we follow the consequences of two extreme scenarios: one in which there is no field versus cluster difference (Case A: with a star fraction 0.1 in both clusters and field), and one in which the star fraction on the general field is locally ~ 3 times lower than in clusters (Case B: with a star fraction 0.06 in the field and 0.16 in clusters).

In *Case A*, with no cluster-to-field difference in the star mass-to-baryon mass ratio, the metallicity of the local Universe has to be virtually identical to that measured in clusters (i.e., $\sim 1/3$ solar), since star formation and the ensuing metal enrichment have proceeded to the same level of *baryon consumption* ($\sim 10\%$). Like in clusters, a majority share of the metals now reside out of galaxies in the intergalactic medium (IGM) containing a major share of all the baryons and metals. Moreover, similar overall star formation activities most likely result not only in similar metal enrichment, but also in similar energy depositions by galactic winds. Hence, the average temperature of the local IGM should be similar to whatever the cluster preheating temperature will turn out to be, the latter corresponding to the nongravitational contribution to the internal energy (and entropy) of the ICM. Such warm/hot IGM, the elusive WHIM, is supposed to hide the still unaccounted fraction of baryons and metals.

In *Case B*, with the stellar mass-to-baryon mass ratio in the local Universe being only $\sim 1/3$ the cluster value, all quantities related to star formation must be scaled down by a factor of ~ 3 with respect to Case A. So, the average metallicity of the local Universe would be just $\sim 1/10$ solar, and the average preheating of the IGM a factor of ~ 3 lower than in Case A.

One may expect that in high-density regions (bound to become clusters in due time) star formation starts at an earlier cosmic epoch compared to low-density regions. However, environmental quenching will also be more efficient, so it is not obvious whether high-density regions will turn more baryons into stars than low-density ones, or vice versa.

Whatever value the global metallicity has in the local Universe, its evolution with redshift/cosmic time can be reconstructed from an empirical star formation history such as that described by Eq. (7.30), here shown in Figure 10.7. Thus, about 10% of all stars appear to have formed by $z = 3$ (i.e., in the first ~ 2 Gyr since the Big Bang), and therefore $\sim 10\%$ of the metals must have been produced and dispersed by that time. This does not apply to elements produced also, or predominantly, by

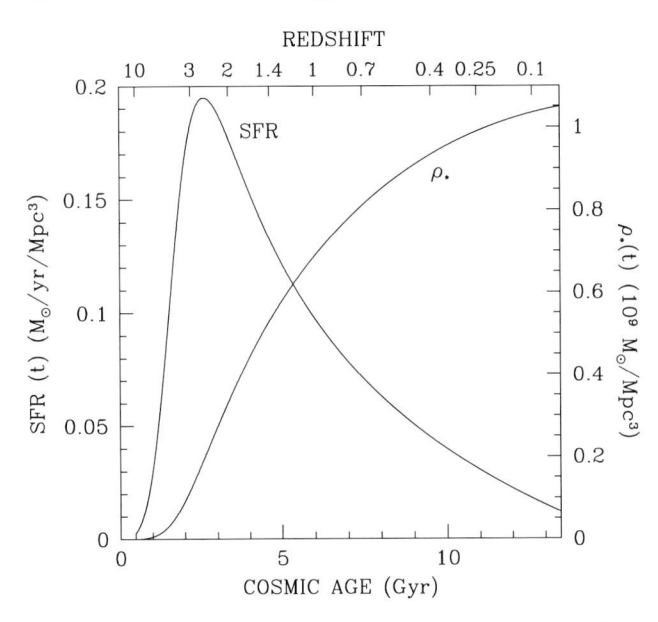

Figure 10.7 The fit to empirical SFR density as a function of redshift as from Eq. (7.30). Also shown is its integral over cosmic time, which gives the mass turned into stars as a function of time.

Type Ia supernovae such as iron and partly silicon. Similarly, one would expect that $\sim 50\%$ of the metals were produced and dispersed during the first ~ 5 Gyr since the Big Bang, that is, at $z \gtrsim 1.5$. This reasoning implies an overall metallicity of the Universe at $z = 3$ of 0.01–0.03 solar, respectively for Case B and Case A as described above, which one can consider as the likely extremes of the allowed range. These values are 10–30 times higher than the average metallicity of the absorbers responsible for the Lyman-α forest at the same redshift ($\sim 10^{-3}$ solar).

The Lyman-α forest has been occasionally seen as representative of the global metallicity of the high-z Universe, as it may contain most of the baryons. Instead, its metallicity is much lower than expected on the basis of the empirical star formation history up to $z = 3$, implying that at this early time most metals were confined to structures that do not qualify as Lyman-α absorbers. Instead, most metals were likely to be locked into stars and dispersed in a shock-heated IGM, and none of these components qualifies as a Lyman-α absorber. Quenched galaxies host little cool gas, and hence do not qualify either. Star forming galaxies have relatively small cross sections, and the probability to intercept the line of sight to a distant quasar is small. Yet, when this happens its optical depth in the UV may be so high that it completely obscures the background quasar: metal-enriched and actively star-forming galaxies may absorb too much! One can conclude that whatever the nature of the Lyman-α forest material, it provides only a very biased and partial census of the cosmic metallicity, and therefore it is of limited interest for the study of the chemical evolution of the Universe.

10.4
Clusters versus the Chemical Evolution of the Milky Way

Having started to investigate chemical evolution on the grand scale of galaxy clusters and the whole Universe, the time has come to look closer at our host galaxy, the Milky Way, and apply to a smaller scale some of the insight gained from galaxy clusters. Specifically, having good reason to assume that rich galaxy clusters have evolved as closed boxes, empirical iron and metal yields of cluster stellar populations have been derived. These empirical yields can then be applied to other, lesser systems to check to what extent they have departed from the closed-box approximation.

For example, a metal-poor 12 Gyr old globular cluster with a present-day mass of one million $M_\odot$ and [Fe/H] $= -2$ contains $0.01 \times 1.1 \times 10^{-3} \times 10^6 M_\odot \simeq 10\,M_\odot$ of iron, or $\sim 200\,M_\odot$ of metals. Such a cluster has a $M_\star/L_B$ ratio about half that of a solar metallicity population with the same age and IMF, therefore for the same L_B it has formed $\sim 1/2$ of the stars. Since stellar and supernova models indicate that the metal yields depends very little on stellar metallicity, using the galaxy cluster values for the FeM/L and for the metal mass-to-light ratio one can infer that the globular cluster has produced $\sim 0.5 \times 0.02 \times 10^6\,M_\odot \simeq 10\,000\,M_\odot$ of iron and $\sim 0.5 \times 0.25 \times 10^6 \simeq 1.2 \times 10^5\,M_\odot$ of metals. Thus, globular clusters have produced far more metals than they presently contain.

One may think that this result depends on having chosen a relatively small object such as a globular cluster, with a fairly shallow potential well. But one can extend the argument to a bigger entity such as the Galactic bulge. The bulge luminosity is $\sim 10^{10}\, L_{K,\odot}$, and $\sim 6\times 10^9\, L_{B,\odot}$, respectively, in the K and in the B band. If we take the cluster empirical yield of metals ($\sim 0.25 \times L_B\, M_\odot$) as universal, it follows that the Galactic bulge has produced $M_Z \simeq 0.25\, L_B^{\text{BULGE}} \simeq 0.25\times 6\times 10^9 \simeq 1.5\times 10^9\, M_\odot$ of metals. Where are all these metals now? One billion solar masses of metals should not be easy to hide. Part of them must be in the stars of the bulge itself, and part was blown out by winds. Indeed, the stellar mass of the bulge follows from its K-band mass-to-light ratio, $M_*^{\text{BULGE}}/L_K \simeq 1$, hence $M_*^{\text{BULGE}} \simeq 10^{10}\, M_\odot$. Its average metallicity is about solar or slightly lower, that is, $Z = 0.02$, and therefore the bulge stars altogether contain $\sim 2 \times 10^8\, M_\odot$ of metals. Thus, only $\sim 1/7$ of the metals produced when the bulge was actively star forming, some 10–12 Gyr ago, are still in the bulge! Hence, over $\sim 80\%$, or over $\sim 10^9\, M_\odot$ of metals were ejected into the surrounding space by an early wind, and were never incorporated into the bulge stars we see today.

One can speculate that at the time of bulge formation, such $\sim 10^9\, M_\odot$ of metals ran into largely pristine ($Z \sim 0$) material, shock-heating it, and experienced Rayleigh–Taylor instabilities leading to chaotic mixing, thus establishing a distribution of metallicities in the IGM surrounding the young Milky Way bulge. For example, this enormous amount of metals would have been able to bring to a metallicity $1/10$ solar about $5 \times 10^{11}\, M_\odot$ of pristine material, several times the mass of the present-day Galactic disk. Therefore, it is likely that for the subsequent ~ 10 Gyr the Galactic disk grew out of such pre-enriched material. Thus, an application to the bulge of the cluster metal mass-to-light ratio leads to the inference that the bulge, and likely the disk as well, did not evolve as closed-box systems. Moreover, if the disk has grown accreting material from the environment, then this material was probably preenriched by bulge ejecta, rather than being pristine and then devoid of metals.

There is still another quite intriguing aspect of chemical evolution that is highlighted by the comparison of the Milky Way to clusters of galaxies. Both for the Galactic disk and for clusters one can estimate the empirical metal yield based on stellar mass, as opposed to stellar luminosity as so far followed in this chapter. This is defined as the ratio of the mass of metals to the mass of stars in the system, which in the case of the Milky Way disk is given by:

$$y^{\text{disk}} \simeq \frac{Z_\odot(M_* + 0.2\, M_*)}{M_*} = 1.2\, Z_\odot, \tag{10.11}$$

assuming that both stars and the ISM are solar on average and the mass of the disk ISM is $\sim 20\%$ the mass of the stars. The cluster yield is instead:

$$y^{\text{clusters}} \simeq \frac{Z_\odot(M_* + 0.3 \times 5 \times M_*)}{M_*} = 2.5\, Z_\odot, \tag{10.12}$$

where stars in clusters are again assumed to be solar, the ICM is assumed 0.3 solar, and 5 times more massive than the stars in galaxies. Thus, the apparent yield in clusters is about twice that of the Galactic disk.

Two opposite solutions of the discrepant yields can be entertained:

Option A. The IMF in galaxy clusters is flatter than in the galactic disk; with a *top-heavy IMF* more massive stars are produced, hence more metals.

Option B. The IMF is the same in the disk as it is in clusters, but the discrepancy arises from not having counted metals produced by the Milky Way disk stars and which have been ejected by disk winds, that is, the disk has lost metals (just like the bulge).

Two arguments can be made in favor of Option A:

1. If B were true then most chemical evolution models of the Milky Way galaxy would have been based on an invalid assumption.
2. There is no evidence for supernova activity in the Milky Way disk and in the disks of nearby spirals causing mass loss. The common wisdom is that materials ejected from the disk in galactic *fountains* sooner or later fall back to the disk.

It is certainly true that Option B would cause some problems to most chemical evolution models, as such models rest on three assumptions that may not be valid:

1. Disks started forming out of pristine ($Z = 0$) material.
2. They grow further by accumulating pristine ($Z = 0$) material.
3. Disks did not lose any mass.

In favor of Option A one can also spend the conjecture that galaxies which formed stars at very high rates in the young Universe, such as the precursors of elliptical galaxies which dominate in clusters, did so with a flatter IMF than disks, which instead formed stars at a much lower rate and over a much more extended period of time.

On the other hand, even if the present SFR in the MW disk is very low ($\sim$ 1–2 $M_\odot$ yr^{-1}), it may have been much higher in the past. Actually, most of the factor of $\sim$ 10 increase in the global SFR between $z = 0$ and $\sim$ 1 seen in Figure 10.7 is due to an increase within individual disks. If so, several gigayears ago the Milky Way disk was forming stars much more violently than it is doing now, and a substantial ejection of metals from the disk by supernova activity is not at all inconceivable. By the same token, if most of the disk build-up was through stronger bursts than today, then the global disk IMF may well be close to that in ellipticals. Whereas the choice between Option A and B is still open, spectroscopic evidence is accumulating that high-redshift disks do support strong winds, hence lose metals.

In summary, the chemistry of galaxy clusters is at the crossroads of many interesting astrophysical and cosmological issues, with important implications for our understanding of galaxy evolution and its driving physical processes. In this respect, the chemistry of clusters offers an opportunity to check and fruitfully use the stellar population tools described in the previous chapters. A number of unexpected inferences are derived starting from a few empirical facts, namely, the iron and metal content of the ICM and cluster galaxies, the fraction of the baryons locked into stars in clusters and in the field, and the age and mass-to-light ratios of stellar populations and galactic spheroids. Such inferences include the following:

- In both clusters and in the general field, there are more metals in the diffused gas out of galaxies (ICM and IGM) than there are locked in the stars inside galaxies. The loss of metals to the surrounding media is therefore a major factor in the chemical evolution of galaxies.
- In clusters and in the general field, $\sim 10\%$ of the baryons are now locked into stars inside galaxies. Yet, given current uncertainties, in clusters this fraction could be as high as $\sim 16\%$ and in the general field as low as $\sim 6\%$. At this global level, the outcome of star formation through cosmic time may be largely independent of environment, possibly just because a major fraction of all stars formed before cluster formation. However, uncertainties are still large, and the baryon-to-star conversion may have been ~ 2 times larger in clusters compared to in the general field.
- Various arguments support the notion that most of the metals now in the ICM/IGM were *ejected* by galactic winds, rather then being *extracted* from galaxies by ram pressure. If so, there must be a close relation between the mass-quenching process of star formation and the spread of metals out of galaxies. Yet, since environmental quenching has been most effective in clusters, also ram pressure stripping should have contributed to the metal enrichment of the ICM.
- Having processed similar fractions of baryons into stars, the global metallicity of the local Universe has to be nearly the same as what one can measure in clusters, that is, $\sim 1/3$ solar. In any event, the metallicity of the Universe scales from this value in proportion to the fraction of baryons turned into stars in the general field over the same fraction in clusters.
- For the same reason, one expects the IGM to have experienced a similar amount of *preheating* from galactic winds as did the ICM.
- On the basis of its empirical star formation history, it appears that the Universe experienced a prompt metal enrichment, with the global metallicity possibly reaching $\sim 0.01\text{--}0.03$ solar already by $z \sim 3$, much higher than the metallicity of the materials responsible for the Lyman-α forest. However, most metals remain unaccounted for, both at low as well as high redshift, and likely reside in a WHIM.
- On the scale of our own Milky Way galaxy, early winds from the forming Galactic bulge can have preenriched to $\sim 1/10$ solar a much greater mass of gas, out of which the Galactic disk grew in the following $\sim 10\,\text{Gyr}$.
- The empirical metal yield of clusters is at least twice that of the Milky Way disk. This signals that either the stellar IMF of the disk is a little steeper above $\sim 1\,M_\odot$ than that of ellipticals, or – perhaps more likely – that the disk has lost at least as many metals as it has retained.
- All this brings us to conclude that galaxies did not chemically evolve in isolation, as closed boxes, but have been fed by the environment, while in the meantime

leaking gas and metals. The necessity to account for this mass and metal exchanges with the environment is a challenge for chemical evolution models.

The iron, oxygen, and silicon mass-to-light ratios measured in clusters of galaxies are fully empirical estimates of the chemical yields of these elements by the stellar populations having generated them and still harbored by the clusters. Building on such empirical yields, simple calculations are sufficient to set fairly precise constraints on the role of the various supernova types in contributing to the establishment of such yields, as well as to constrain the IMF slope between ~ 1 and $\sim 40\,M_\odot$ and over, and to constrain the past SNIa activity in clusters, hence the distribution of SNIa delay times. The attractive aspect of these calculations is that they do not need more than a pocket calculator, and hence can easily be checked by others without having to write and run complex chemical evolution codes. Moreover, they can readily be updated as better measurements become available for the various ingredients that have been put in play. An economy of means can save time and gain clarity.

Acknowledgments

Recent ICM abundance measurements used in this chapter include: Simionescu, A. *et al.* (2009, *Astron. Astrophys.*, 493, 409); Gastaldello, F. *et al.* (2011) *Astron. Astrophys.*, 522, A34). For the star and baryon in rich clusters data from Giodini, S. *et al.* (2009, *Astrophys. J.*, 703, 982) and McGaugh, S.S. *et al.* (2010, *Astrophys. J.*, 708, L14) have been used, together with the cosmic baryon fraction from Komatsu, E. *et al.* (2009, *Astrophys. J. Suppl.*, 180, 330).

Further Reading

This chapter updates and expands over the paper Renzini, A. (2004) in *Clusters of Galaxies: Probes of Cosmological Structures and Galaxy Evolution*, eds. J.S. Murchaey *et al.*, Cambridge, CUP, p. 261.

Recent Reviews on the Chemical Abundances in the ICM

Böhringer, H. and Werner, N. (2010) *Astron. Astrophys. Rev.*, **18**, 127.

Werner, N. *et al.* (2008) *Space Sci. Rev.*, **134**, 337.

For the Role of Type Ia and Core Collapse Supernovae

Blanc, G. and Greggio, L. (2008) *New Astron.*, **13**, 606.

Calura, F. *et al.* (2007) *Mon. Not. R. Astron. Soc.*, **378**, L11.

Chemical Evolution of Galaxies

Edmunds, M.G. (1990) *Mon. Not. R. Astron. Soc.*, **246**, 678.

Matteucci, F. (2001) *The Chemical Evolution of the Galaxy*, Dordrecht: Kluwer Academic Publishers.

Pagel, B.E.J. (1997) *Nucleosynthesis and Chemical Evolution of Galaxies*, Cambridge, CUP.

Portinari, L. *et al.* (2004) *Astrophys. J.*, **604**, 579.

Thomas D. *et al.* (1999) *Mon. Not. R. Astron. Soc.*, **302**, 537.

Index

symbols
4000 Å break 118, 124
η Carinae 27

a
A496 238
accretion 82, 129, 133, 138, 184–191, 223
accretion efficiency 196
accretion induced collapse (AIC) 184, 187
adiabatic gradient 5, 18
AGB manqué 11
age dating 77, 88, 91, 146, 167, 213
age-metallicity anticorrelations 59
age-metallicity degeneracy 58, 91, 146–152
age-metallicity relation 73, 102, 105
AGN 137, 139, 155, 234, 236, 251
α-element 19, 23, 37, 115, 147, 150, 151, 195, 203, 204, 232, 245, 250
asteroseismology 25
asymptotic giant branch (AGB) 2, 11, 15–17, 86, 102, 109
– bump 67, 74
– early (E-AGB) 2, 15, 42, 64
– ejecta 82
– thermally pulsing 16, 42, 64, 118, 120

b
Baade's Window 83, 151
Balmer break 118, 124
baryon to star conversion 252
Big Bang 81, 253
binary population synthesis (BPS) 188, 191, 194
binary systems 19, 31, 35, 64, 105, 128, 133, 140, 176, 184–201, 214
black hole 32, 53, 138, 175, 236, 249
blue cloud 154–157, 231
blue giant 7
blue loops 99
blue stragglers 16, 64, 86, 128, 148
blue supergiant (BSG) 26, 102
bolometric correction 10, 25, 62, 105, 215
BzK criterion 157, 158

c
carbon burning 41, 182, 201
carbon deflagration/detonation 29
carbon ignition 42, 94, 184
carbon star 17, 44, 116, 120
cataclysmic binaries 187
Centaurus A 106, 109
Centaurus group 98
Chabrier IMF 210
Chandra explosion 185, 187, 190
Chandrasekhar limit 29, 179, 184, 185
chemical enrichment 91, 171
chemical evolution 55, 117, 182, 236–258
chemical yields 19, 172, 179–205, 215, 237, 254–257
closed box model 117, 254, 257
clusters of galaxies 61, 67, 110, 146, 208, 213–216, 237–258
cold streams 234, 236
color-magnitude diagram (CMD) 10, 15, 75, 98, 109
Coma cluster 146, 147
common envelope (CE) 31, 129, 186–205
composite stellar populations 11, 91, 93, 97, 113, 116, 121
control time 173
convection 5, 7, 21, 116
convective gradient 21
convective overshooting 24–26
cooling flow 241
cosmic acceleration 171
cosmic variance 165
crowding 61–75, 88

Stellar Populations, First Edition. Laura Greggio and Alvio Renzini.
© 2011 WILEY-VCH Verlag GmbH & Co. KGaA. Published 2011 by WILEY-VCH Verlag GmbH & Co. KGaA.

d

dark matter 212, 219, 245, 252
dark matter halos 219
death rate 176
deflagration 29, 201
delay time 172, 176
density field 163
density parameter 252
detonation 29, 185, 201
disk instabilities 234, 236
double degenerate channel 186, 191–197
dry merging 231
dust emission 133, 136
dust grains 30, 53
dust obscuration 135
dynamical mass 141, 207, 211, 245
dynamical time 163

e

electron capture 175, 184
emission lines 133, 137, 238, 245
empirical spectral libraries 115
energy generation 18, 184
energy transfer 18
envelope ejection 27, 30, 31, 129, 191
envelope inflation/deflation 6
envelope pulsation 31
environment of galaxies 117, 133, 163, 220
environment quenching 163, 227–236, 251, 257
equation of state 18
evolutionary flux 38, 64, 94, 96, 100, 125, 128
evolutionary synthesis of stellar populations 35, 113–131
explosive nucleosynthesis 179
extinction 62, 78, 133, 136–141, 157, 173, 200

f

far-IR SED 136, 161
feedback 224, 234, 251
first blue loop 7
formation redshift 143, 213
fuel consumption 42–45, 114, 128
fuel consumption theorem 39, 48, 114, 128
fundamental plane 251

g

galactic archeology 92
Galactic bulge 61, 83–86, 109, 115, 151, 207, 255
galactic fountains 256
galactic spheroid 109, 257

galaxies
– elliptical 11, 50, 106, 115, 120, 146, 147, 167, 184, 207, 211, 213, 235, 238, 244
– merging 92, 109, 162, 221
– spiral 122, 154, 199, 251, 256
– starburst 163
– submillimeter 162
galaxy group 239
galaxy photometry 56, 141
γ-ray bursts (GRB) 27
gasping star formation 92
globular clusters 9, 16, 23, 28, 43, 64, 83, 86, 150, 207
– ages 77–82
gravitational energy 4, 40
gravitational wave radiation (GWR) 191

h

Hayashi line 1
helium burning 1, 94
helium flash 3, 10, 28, 42
helium ignition 42, 64, 94, 184
helium shell flash 16, 42, 185
helium star 185–188
Hertzsprung–Russel diagram (HRD) 1, 2, 4, 35, 96
hierarchical merging 219
high-mass stars 8, 208
high-mass X-ray binaries (HMXB) 139
horizontal branch (HB) 10, 16, 27, 67, 119, 148
hot-bottom burning (HBB) 16, 29, 109
hydrodynamical simulations 219
hydrogen burning 1
hypernova 27

i

induced collapse (AIC) 184
initial mass–final mass relation 31, 52, 189
initial mass function (IMF) 35, 57, 114, 176, 207–216, 236, 256
– scale factor 38, 51, 96, 114, 209
intergalactic medium (IGM) 55, 117, 171, 252–255
intermediate-mass stars 7, 8, 11, 13, 29, 179, 185
internal energy 253
interstellar medium (ISM) 52, 171
intracluster medium (ICM) 214, 235, 238–258
iron core 8, 175
iron-K line 238, 239, 246
iron-L complex 239, 246

iron mass-to-light ratio 238–258
isochrones 35, 46, 78, 113

l

Large Magellanic Cloud (LMC) 43, 86, 87, 183
Lick/IDS system 147
Local Group 61, 92, 98
long period variables (LPV) 30, 50, 64
lookback time 92, 109, 147, 149, 213, 219, 248
low-mass stars 8, 11, 49, 208, 247
low-mass X-ray binaries (LMXB) 139
luminosity evolution 47, 57, 207
luminosity function 25, 65–67, 69, 89, 98, 117
luminous blue variables (LBV) 27
Lyman-α forest 254, 257
Lyman break 118

m

M/L Ratio 211
M31 61, 92, 109
 – bulge 61, 88, 109
 – giant stream 91
 – halo 61, 90
 – spheroid 88
M33 93
M67 145
macrophysics 20–31
Magellanic Clouds 29, 43, 44, 61, 63, 86, 93, 207
magnetic braking 188
magnetic fields 20
magnetohydrodynamical waves 30
main sequence 64, 114
 – of star-forming galaxies 155–163, 221, 234
 – stellar 4, 97, 189
 – turnoff 9, 16, 35, 64, 78, 81, 83, 88, 99
mass cut 180
mass exchange 128
mass function 164, 166, 167, 222, 225, 229
mass loss 7, 26–31, 52, 175, 183, 256
mass-metallicity relation 238
mass quenching 163, 225–236, 251, 257
mass reduction factor 54
mass return 52–56, 123, 140, 237, 249
mass-specific production 93–109
mass-to-light ratio 56, 93, 198, 200, 207, 245, 252, 255, 257
mass transfer 19, 31, 129

massive stars 7, 25, 42, 133, 175, 176, 179, 245, 256
 – fast rotating 21
merging galaxies 222, 229, 234
meridional circulation 25
metal abundance 8, 37, 244
metal enrichment 84, 152, 179, 207, 252
metal mass-to-light ratio 214, 247, 254, 255
metallicity 10, 73, 143, 238
metallicity gradient 241
metallicity of the universe 252
meteoritic abundances 240
microphysics 19
microwave background 207
Milky Way 73, 80, 85, 92, 204, 207, 250, 254–257
MilkyWay 109
mixing 24–26
mixing length 21
mixing length theory (MLT) 21
model atmosphere 107, 115, 240
model atmosphere libraries 116
molecular bands 10, 118, 121
molecular clouds 135, 162
molecular opacities 19
molecules 19, 30, 116, 137
monolithic collapse 238
Monte Carlo simulations 104, 107, 187
multiple stellar populations 13, 14, 80–82, 116
multiwavelength surveys 220

n

N-body simulations 219
neutrino flux 180
neutrino losses 10, 19, 41
neutron-capture elements 17
neutron star 32, 53, 138, 175, 184, 249
NGC 1399 148
NGC 1705 99
NGC 2808 14, 82
NGC 6528 79, 83, 84
NGC 6553 79, 83, 84
NGC 6752 15, 16
nova 129, 187, 188
nuclear energy generation 19
nuclear statistical equilibrium 201

o

ω Centauri 80, 109
opacity 5, 6, 8, 18, 19, 140
outshine effect 144, 153
overdensity 156, 163, 220, 228, 241, 251
overshooting 13, 43, 140, 180

oxygen mass-to-light ratio 216, 250, 258
oxygen-to-iron mass ratio 247

p

pair instability 175
photometric redshift 141, 158, 166
photospheric abundances 240
planetary nebula 26, 29, 64
Poisson likelihood ratio (PLR) 105
polycyclic aromatic hydrocarbon (PAH) 137
post-AGB 64
post-main sequence 37, 63, 114
preheating 253, 257
pressure scale height 22
primordial perturbation 219
protosolar abundances 240

q

quasar 254
quenched galaxies 145, 147, 149, 153–168, 213, 223–236, 254

r

radiation pressure 30
radiative energy flux 5
radiative gradient 5, 21
radio luminosity 134, 139, 160, 161
radioactive decay 182, 184, 249
ram pressure 236, 250
ram-pressure stripping 235
Rayleigh–Taylor instabilities 255
red clump 10, 27, 73, 101
red giant 5, 7
red giant branch (RGB) 9, 25, 42, 64, 102, 184
 – bump 16, 25, 64–67
 – phase transition 94
 – RGB tip 10, 64–67, 86, 88
red sequence 153, 156, 158, 231
red supergiant (RSG) 8, 102, 140, 175
reddening 79, 105, 118, 134, 141, 158, 200
redshift evolution of the M/L ratio 213
relativistic electrons 133, 139
Roche-lobe 128, 138, 176, 185, 188, 193
rotation of stars 19, 20, 180
RSOphi 187
runaway expansion 6

s

Salpeter-diet IMF 38, 48, 95, 136, 165, 177, 199, 208, 249
satellite quenching 235
Schechter function 165, 166, 220, 226, 229, 231, 235

Schwarzschild criterion 21
second blue loop 7
SED fitting 141, 145, 155
semianalytic models 113, 219
semiconvection 24, 180
SFR relation 155, 157, 159–163, 221
shell burning 4
silicon mass-to-light ratio 216, 247, 250, 258
simple stellar populations (SSP) 35–38, 113, 118, 181
single degenerate channel 186–191
SN 1987A 183
SN 1992am 183
SN 2006X 187
specific evolutionary flux 49, 93
specific production method 102
specific SFR 156
spectral energy distribution (SED) 8, 12, 14, 25, 113, 137
spectral evolution 118–120
spectral libraries 115
spheroids 244
star counts 66, 89, 102, 133, 207
star formation history 1, 56, 92, 109, 116, 237
star formation quenching 121, 126, 142, 220–236
star formation rate (SFR) 1, 25, 54, 75, 113, 133–140, 172, 207
star formation rate density 162, 205, 253
star photometry 61–64, 79, 83, 88
starburst 66, 97, 122, 156, 162, 173, 177, 195, 202, 207, 221, 222, 236
starburst galaxies 234
stellar lifetime 8, 10, 12, 35, 119, 142, 176
stellar mass of galaxies 140–142
stellar mass-to-baryon mass ratio 252
stellar remnant 140
strangulation 235
sub-Chandra explosions 185
subdwarfs 23, 79
subgiant branch (SGB) 9, 42, 64
submillimeter galaxy (SMG) 162, 221, 222
superadiabatic convection 20
supernova 7, 29, 36, 77, 133, 171–205, 234, 238
 – core collapse 7, 115, 139, 171, 175–183, 201, 214, 249
 – distribution of the delay times 172, 187, 201, 205, 258
 – feedback 236
 – productivity 172, 197, 250
 – progenitor 27, 172, 175–177, 188
 – rate 172–205

– thermonuclear   115, 171, 184–205, 246, 250
– Type Ia   29, 115, 129, 171, 174, 184–205, 214, 240, 248
– Type Ib/c   27, 171, 174, 175, 183
– Type II   27, 171, 174–176
– Type IIP   183
– units (SNu)   174, 197
supersoft X-ray sources   129, 187
superwind   29, 64
symbiotic systems   187
synchrotron radiation   133
synthetic CMD   92
synthetic CMD method   104–109

t

thermal energy   4, 20, 193, 205
thermal equilibrium   3, 7, 16, 40
thermal instability   5, 6
thermal runaway   5, 13, 42, 140
thermonuclear runaway   129
TP-AGB   42, 64, 118, 120
47 Tuc   23, 50, 64, 66, 67, 84
turnoff mass   32, 36, 213
turnoff temperature   58

u

ultraviolet continuum   118, 134, 136, 153
UV rising branch   120, 123

v

velocity dispersion   146, 211, 251
Virgo cluster   146, 147
virial radius   243
virial theorem   4

w

warm-hot intergalactic medium (WHIM)   253, 257
white dwarf   7, 36, 64, 79, 128, 179–205
– merging   187
wind
– galactic   117, 171, 224, 238
– radiatively driven   26, 187
– red giant   30
– red supergiant   27
– solar   30
– stellar   7, 26
– thermal   30
Wolf–Rayet (WR) stars   26, 140, 175

x

X-ray luminosity   129, 134, 138, 139, 251
X-ray radiation   133
X-ray spectrum   238

z

zero age main sequence (ZAMS)   1, 40